INVERSE AND ILL POSED PROBLEMS SERIES

Formulas in Inverse and Ill-Posed Problems

Previously published in the Inverse and Ill-Posed Problems Series:

Inverse Logarithmic Potential Problem
V.G. Cherednichenko

Multidimensional Inverse and Ill-Posed Problems for Differential Equations
Yu.E. Anikonov

Ill-Posed Problems with A Priori Information
V.V. Vasin and A.L. Ageev

Integral Geometry of Tensor Fields
V.A. Sharafutdinov

Inverse Problems for Maxwell's Equations
V.G. Romanov and S.I. Kabanikhin

Other related books from VSP:

Inverse Problems of Mathematical Physics
V.G. Romanov

Inverse Sturm-Liouville Problems
B.M. Levitan

Related Journal:

Journal of Inverse and Ill-Posed Problems
Editor-in-Chief: M.M. Lavrent'ev

INVERSE AND ILL-POSED PROBLEMS SERIES

Formulas in Inverse and Ill-Posed Problems

Yu.E. Anikonov

///VSP///

Utrecht, The Netherlands, 1997

VSP BV
P.O. Box 346
3700 AH Zeist
The Netherlands

First published in 1997

ISBN 90-6764-216-9

Printed in The Netherlands by Ridderprint bv, Ridderkerk.

Contents

INTRODUCTION

This monograph is devoted to constructive methods of investigation of the multidimensional, in the main nonlinear, inverse problems for evolution equations. Numerous results on simultaneous determination of the coefficients and solutions of the differential equations and inequalities are cited here. The author discovered various explicit formulas which determine solutions of inverse problems. These formulas contain arbitrary functions and the functions which satisfy some differential equations.

In general such kind of investigations is related to the search for general and particular solutions for multidimensional inverse problems which have applications in the natural sciences and in the social-economical area.

This monograph contains new differential and integro-differential equations and reducing of multidimensional inverse problems to the initial-boundary problems for these equations. The reducing methods, proposed and developed by the author, are the main tools of the proofs of solvability of general and particular multidimensional inverse problems and they lead to various constructive results.

In connection with these methods it should be underlined that general multidimensional inverse problems are, first of all, directly related to mathematical modelling of natural and social processes in the absence of full information on the operators of the substations propagation and interaction, in particular with the determination of the information that emerges from their consequences in control theory and prognoses.

Another wide area of application of multidimensional inverse problems is that of determination of the intrinsic structures and forms of the objects under consideration have in scattering theory problems and tomographical problems with linear and nonlinear variants.

The methods and results of multidimensional inverse problems theory for evolution equations are now and, in the author's opinion, will be in the future an important tool in various kinds of engineering, planning, control and identification situations. This monograph, together with the author's previous monograph, 'Multidimensional Inverse and Ill-posed Problems for Differential Equations', propose a wide collection of investigation methods in this area and the author hopes that they will be useful for the reader who is interested in these and related subjects.

The main results of this book are contained in the author's papers which are in the References section of this book.

This work was supported by the Russian Foundation for Fundamental Research, grant number 96–01–01558.

CHAPTER 1

Equations and relations in inverse and ill-posed problems

1.1 GENERAL APPROACH IN A THEORY OF MULTIDIMENSIONAL INVERSE PROBLEMS

Such a method is suggested by some attributes of solutions of evolution equations: they supplementally should belong to some classes of functions. More strictly and formally the next approach would be: if, in a classical correct initial boundary value problem for evolution differential equations, with coefficients which do not depend on variable t, we know also that the solution has the property that it is periodic for variable t, or it has Fourier transform for t, or a more general solution is found for all t, then we can determine not only solution but one coefficient of an evolution equation, where, in each case, time t is such that $-\infty < t < \infty$.

For example, if we have an initial boundary value problem for a parabolic equation

$$\frac{\partial w}{\partial t} = \Delta w + \lambda(x)w, \quad x \in D, \quad -\infty < t < \infty; \tag{1.1}$$

$$w|_{t=0} = w_0(x) \neq 0, \quad w|_{\partial D} = w_1(s,t), \quad s \in \partial D \tag{1.2}$$

and we know that supplementally there exists

$$\frac{1}{2\pi}\int_{-\infty}^{\infty} w(x,t)\,e^{i\,t\omega}\,dt = \hat{w}(x,\omega), \tag{1.3}$$

then we can define a boundary value problem for function $\hat{w}(x,\omega)$ which does not contain the coefficient $\lambda(x)$ and sometimes it is possible to determine the solution $w(x,t)$ and coefficient $\lambda(x)$. The formal lemma holds.

Lemma 1.1. *If function $\hat{w}(x,\omega)$, $x \in D$, $-\infty < \omega < \infty$ is a solution of the boundary value problem for the nonlinear integro-differential equation*

$$\Delta\hat{w} = -\,i\,\omega\hat{w} + \frac{\int_{-\infty}^{\infty}(-\,i\,\omega)\hat{w}\,d\omega - \Delta w_0}{w_0(x)}\hat{w} \tag{1.4}$$

$$\hat{w}|_{\partial D} = \hat{w}_1(s,t) = \int_{-\infty}^{\infty} w_1(s,t)\, e^{i t\omega}\, dt, \tag{1.5}$$

then the following formulas hold

$$w(x,t) = \int_{-\infty}^{\infty} \hat{w}(x,\omega)\, e^{-i\omega t}\, d\omega,$$

$$\lambda(x) = \frac{\int_{-\infty}^{\infty} (-i\omega)\hat{w}\, d\omega - \Delta w_0}{w_0(x)}.$$

The condition (1.3) allows us to define the boundary value problem (1.4), (1.5). Such problems (with corresponding conditions about data $w_0(x)$ and $w_1(s,t)$) reduce to fixed point theorems and hence we can obtain the existence and uniqueness of the solution for the inverse problem (1.1), (1.2), (1.3).

Now we shall consider a more general evolution equation with initial data only and show that this leads to one equation for function $\hat{w}$ and formula for unknown coefficients. So we have

$$\rho(x)\frac{\partial^m w}{\partial t^m} = \mathbf{A}w + \lambda(x)\mathbf{B}w, \quad x \in D, \quad -\infty < t < \infty, \tag{1.6}$$

$$w|_{t=0} = w_0(x). \tag{1.7}$$

Here, $\mathbf{A}$, $\mathbf{B}$ are linear operators which act for variable x and do not depend on variable t; functions $\rho(x)$ or $\lambda(x)$ will be unknown, $m \geq 1$. We suppose that the supplemental condition

$$\exists \hat{w}(x,\omega) = \frac{1}{2\pi}\int_{-\infty}^{\infty} w(x,t)\, e^{i t\omega}\, dt \tag{1.8}$$

holds and, at first, we shall suppose that $w(x,t)$ and $\lambda(x)$ are unknown functions.

Lemma 1.2. *Let conditions (1.6), (1.7), (1.8) and* $\mathbf{B}w_0 \neq 0$ *hold; then the function* $\hat{w}(x,\omega)$ *is a solution of equation*

$$(-i\omega)^m \rho(x)\hat{w} = \mathbf{A}\hat{w} + \mathbf{B}\hat{w}\frac{\rho(x)\int_{-\infty}^{\infty} (-i\omega)^m \hat{w}\, d\omega - \mathbf{A}w_0}{\mathbf{B}w_0}$$

and

$$\lambda(x) = \frac{\rho(x)\int_{-\infty}^{\infty} (-i\omega)^m \hat{w}\, d\omega - \mathbf{A}w_0}{\mathbf{B}w_0}.$$

Now we shall suppose that function $w(x,t)$ and $\rho(x)$ are unknown.

Lemma 1.3. *Let conditions (1.6), (1.7), (1.8) hold; then the function $\hat{w}(x,\omega)$ is a solution of equation*

$$\frac{(-\mathrm{i}\,\omega)^m \hat{w}(\mathbf{A}w_0 + \lambda \mathbf{B}w_0)}{\int\limits_{-\infty}^{\infty} (-\mathrm{i}\,\omega)^m \hat{w}\, d\omega} = \mathbf{A}\hat{w} + \mathbf{B}\hat{w}$$

and

$$\rho(x) = \frac{\mathbf{A}w_0 + \lambda(x)\mathbf{B}w_0}{\int\limits_{-\infty}^{\infty} (-\mathrm{i}\,\omega)^m \hat{w}\, d\omega}.$$

After that, we want to adopt a general approach, when we have a nonlinear equation which contains some unknown coefficient. We consider equation

$$(\mathbf{A}_x + \mathbf{L}_z w) + \int\limits_{\mathrm{R}^m} F(x,z,w_x,w_z,w)\, dz(\mathbf{B}_x + \mathbf{D}_z)w = 0, \tag{1.9}$$

$x \in D \subset \mathrm{R}^n$, $z \in \mathrm{R}^m$, and D is the domain. Here $\mathbf{A}_x$, $\mathbf{B}_x$ are linear operators which act for variable x and do not depend on variable z and conversely $\mathbf{L}_z$. $\mathbf{D}_z$ are linear operators which act for variable z and do not depend on variable x. F is some function which can contain an unknown element, for example,

$$F = \lambda_0(x) w(x,z)$$

where $\lambda_0(x)$ is the unknown function. We suppose that the following conditions hold

$$w\big|_{z=0} = w_0(x), \tag{1.10}$$

$$\exists \hat{w}(x,\omega) = \frac{1}{(2\pi)^m} \int\limits_{\mathrm{R}^m} w(x,z) e^{\mathrm{i}\,\omega z}\, dz. \tag{1.11}$$

Let $l(\omega)$, $d(\omega)$ be symbols of operators $\mathbf{L}_z$, $\mathbf{D}_z$. Denote by $\lambda(x) = \int\limits_{\mathrm{R}^m} F(x,z,w_x,w_z,w)\, dz$.

Lemma 1.4. *Then the following relations hold*

$$\mathbf{A}_x \hat{w} + l(\omega)\hat{w} - \frac{\mathbf{A}_x w_0 + \int\limits_{\mathrm{R}^m} l(y)\hat{w}(x,y)\, dy}{\mathbf{B}_x w_0 + \int\limits_{\mathrm{R}^m} d(y)\hat{w}(x,y)\, dy} (\mathbf{B}_x \hat{w} + d(\omega)\hat{w}) = 0,$$

$$\lambda(x) = \frac{\mathbf{A}_x w_0 + \int\limits_{\mathrm{R}^m} l(y)\hat{w}(x,y)\, dy}{\mathbf{B}_x w_0 + \int\limits_{\mathrm{R}^m} d(y)\hat{w}(x,y)\, dy}.$$

1.2 CONTROL AND INVERSE PROBLEMS

In this section, we interpret certain control problems as inverse problems for differential equations. The control is by two time-independent coefficients in the evolution equations. Problems of this kind were previously considered in Lions (1973) and Butkovskii (1985). With boundary values given for the controls, the corresponding inverse problems are reducible to boundary-value problems for infinite systems of nonlinear differential equations. This reduction suggests a technique for the analysis of these cases.

The main mathematical difficulties are encountered when we try to establish if these boundary-value problems are well-posed, and especially when we consider the existence of solutions. Once these boundary-value problems have been solved, the control functions are obtained by the formulas in Anikonov (1986) and Bubnov (1986).

Using part of the control boundary values, we will first write out the nonlinear integro-differential equations, and then use the equation, with given boundary values, to state the existence and uniqueness result.

Consider the evolution equation

$$u_t = \mathbf{A}u + f(x,t), \quad x \in D, \quad t \in \mathrm{R}, \tag{1.12}$$

where $\mathbf{A}$ is a linear operator which is independent of t, and $D \subset \mathrm{R}^n$ is a domain with smooth boundary $\Gamma \in C^{2m}$. The operator $\mathbf{A}$ or function $f(x,t)$ may include unknown coefficient, for instance, $c(x)$ and $h(x)$ when $\mathbf{A} = \mathbf{A}_0 + c(x)$, $f = h(x)g(x,t)$, where the operator $\mathbf{A}_0$ and the function $g(x,t)$ are given, and $c(x)$ and $h(x)$ are the unknown control functions. The problems which, in addition to finding the solution $u(x,t)$, also involve determining the coefficients $c(x)$ and $h(x)$ in (1.12), given some information about the solution $u(x,t)$ of equation (1.12), are usually called inverse problems (Lavrent'ev *et al.*, 1986).

In addition to (1.12) the control boundary values $u_0(x)$ and $u_T(x)$ are given:

$$u\big|_{t=0} = u_0(x), \quad u\big|_{t=T} = u_T(x). \tag{1.13}$$

We assume that $u(x,t)$ and $f(x,t)$ have direct and inverse Fourier transformation with respect to the variable t, and differentiation under the sign of the integral is allowed:

$$\begin{aligned} w(x,\lambda) &= \frac{1}{2\pi}\int_{-\infty}^{\infty} u(x,t)\exp(\mathrm{i}\,\lambda t)\,\mathrm{d}t, \\ \hat{f}(x,\lambda) &= \frac{1}{2\pi}\int_{-\infty}^{\infty} f(x,t)\exp(\mathrm{i}\,\lambda t)\,\mathrm{d}t. \end{aligned} \tag{1.14}$$

From (1.12)–(1.14) we obtain new integro-differential equations, which are connected with the control problems by any two coefficients in (1.12) which take $u(x,t)$ from state $u\big|_{t=0} = u_0(x)$ to state $u\big|_{t=T} = u_T(x)$, e.g., the coefficients $c(x)$ and $h(x)$ when $\mathbf{A} = \mathbf{A}_0 + c(x)$, $f = h(x)g(x,t)$. These equations in $w(x,\lambda)$ and any two coefficients have the form

$$\begin{aligned} -\mathrm{i}\lambda w &= \mathbf{A}w + \hat{f}, \\ \int_{-\infty}^{\infty} -\mathrm{i}\lambda w\,\mathrm{d}\lambda &= \mathbf{A}u_0 + f_0, \quad f_0 = f|_{t=0}, \\ \int_{-\infty}^{\infty} -\mathrm{i}\lambda \exp(-\mathrm{i}\lambda T) w\,\mathrm{d}\lambda &= \mathbf{A}u_T + f_T, \quad f_T = f|_{t=T}. \end{aligned} \tag{1.15}$$

Similarly, for the equation

$$u_{tt} = \mathbf{A}u + f$$

with the control boundary values $u|_{t=0} = u_0(x)$, $u|_{t=T} = u_T(x)$, we obtain the system of equations

$$\begin{aligned} -\lambda^2 w &= \mathbf{A}w + \hat{f}, \\ -\int_{-\infty}^{\infty} \lambda^2 w\,\mathrm{d}\lambda &= \mathbf{A}u_0 + f_0, \\ -\int_{-\infty}^{\infty} \lambda^2 w \exp(-\mathrm{i}\lambda T)\,\mathrm{d}\lambda &= \mathbf{A}u_T + f_T. \end{aligned} \tag{1.16}$$

This system, as before, can be considered as a system of equations in $w(x,\lambda)$ with any two coefficients which are included in $\mathbf{A}$ and f and which are responsible for control in taking $u(x,t)$ from state $u_0(x)$ to state $u_T(x)$.

If, instead of condition (1.14), we assume periodic $u(x,t)$ and $f(x,t)$,

$$u = \sum_{k=-\infty}^{\infty} \exp(\mathrm{i}\,kt) u_k(x), \quad f = \sum_{k=-\infty}^{\infty} f_k(x) \exp(\mathrm{i}\,kt),$$

then systems (1.15) and (1.16) reduce to infinite systems of equations in $u_k(x)$ and certain coefficients of $\mathbf{A}$ and f. These systems have the form

$$\begin{aligned} \mathrm{i}\,ku_k &= \mathbf{A}u_k + f_k, \\ \sum_k \mathrm{i}\,ku_k &= \mathbf{A}u_0 + f_0, \\ \sum_k \mathrm{i}\,ku_k \exp(\mathrm{i}\,kT) &= \mathbf{A}u_T + f_T, \end{aligned} \tag{1.17}$$

$$\begin{aligned} -k^2 u_k &= \mathbf{A}u_k + f_k, \\ -\sum_k k^2 u_k &= \mathbf{A}u_0 + f_0, \\ -\sum_k k^2 u_k \exp(\mathrm{i}\,kT) &= \mathbf{A}u_T + f_T, \quad |k| = 0, 1, \dots \end{aligned} \tag{1.18}$$

Sometimes the sought-for coefficients can be eliminated from systems (1.15)–(1.18) which gives equations in $w(x,\lambda)$ or u_k only.

Let $\mathbf{A} = \mathbf{A}_0 + c(x)$, $f = h(x)g(x,t)$ in (1.15), where $h(x)$ and $c(x)$ are the sought-for control functions. $\mathbf{A}_0$ is a linear operator which is independent of t, and g is a given function,

$$\|\mathbf{A}_0 v\|_0 \geq \mu \|v\|_{W_2^m}, \quad v \in \overset{\circ}{W}{}_2^m(D) \cap W_2^{2m}(D),$$

where μ is the constant.

If

$$|M| = |g_T u_0 - u_T g_0| \geq \delta > 0, \quad g_T = g|_{t=T}, \quad g_0 = g|_{t=0},$$

then from (1.15) we obtain

$$h(x) = \frac{1}{M}\left[u_T\left(\int_{-\infty}^{\infty} \mathrm{i}\,\lambda w \,\mathrm{d}\lambda + \mathbf{A}_0 u_0\right) - u_0\left(\int_{-\infty}^{\infty} \mathrm{i}\,\lambda \exp(-\mathrm{i}\,\lambda T) w \,\mathrm{d}\lambda + \mathbf{A}_0 u_T\right)\right]; \tag{1.19}$$

$$c(x) = \frac{1}{M}\left[g_0\left(\int_{-\infty}^{\infty} \mathrm{i}\,\lambda \exp(-\mathrm{i}\,\lambda T) w \,\mathrm{d}\lambda + \mathbf{A}_0 u_T\right) - g_T\left(\int_{-\infty}^{\infty} \mathrm{i}\,\lambda w \,\mathrm{d}\lambda + \mathbf{A}_0 u_0\right)\right]. \tag{1.20}$$

Using (1.19), (1.20) and (1.15), we obtain a nonlinear integro-differential equation in $w(x, \lambda)$:

$$\begin{aligned} &\mathbf{A}_0 w + \mathrm{i}\,\lambda w + \frac{w}{M}\left[g_0\left(\int_{-\infty}^{\infty} \mathrm{i}\,\lambda \exp(-\mathrm{i}\,\lambda T) w \,\mathrm{d}\lambda + \mathbf{A}_0 u_T\right) - g_T\left(\int_{-\infty}^{\infty} \mathrm{i}\,\lambda w \,\mathrm{d}\lambda + \mathbf{A}_0 u_0\right)\right] \\ &+\frac{\hat{g}(x,\lambda)}{M}\left[u_T\left(\int_{-\infty}^{\infty} \mathrm{i}\,\lambda w \,\mathrm{d}\lambda + \mathbf{A}_0 u_0\right) - u_0\left(\int_{-\infty}^{\infty} \mathrm{i}\,\lambda \exp(-\mathrm{i}\,\lambda T) w \,\mathrm{d}\lambda + \mathbf{A}_0 u_T\right)\right] = 0 \end{aligned} \tag{1.21}$$

If we seek a particular solution, then boundary values should be specified for the systems (1.15)–(1.18) and (1.21). Here we will consider only equation (1.21), assuming that the operator $\mathbf{A}_0$ is elliptic of order $2m$ with the coefficients from the class $C^{2m}(\bar{D})$ so that the Dirichlet problem is uniquely solvable.

We can thus analyze (1.21) with the boundary conditions

$$w|_\Gamma = \hat{\varphi}_0, \quad \left.\frac{\partial w}{\partial n}\right|_\Gamma = \hat{\varphi}_1, \ldots, \left.\frac{\partial^{m-1} w}{\partial n^{m-1}}\right|_\Gamma = \hat{\varphi}_{m-1}, \tag{1.22}$$

where

$$\hat{\varphi}_k(s, \lambda) = \frac{1}{2\pi}\int_{-\infty}^{\infty} \left.\frac{\partial^k}{\partial n^k} u\right|_\Gamma \exp(\mathrm{i}\,\lambda t)\,dt.$$

Having found the solution $w(x, \lambda)$ of problem (1.21), (1.22) from the corresponding class, we can find the control functions $h(x)$ and $c(x)$ from (1.19) and (1.20).

Thus, the main task in this case is to establish whether the problem (1.21), (1.22) is well-posed. As an illustration of the potential possibilities, we give the theorem of unique solvability of problem (1.21), (1.22).

Theorem 1.1. *Let g, $\varphi_0, \ldots, \varphi_{m-1}$ be such that*

$$\begin{aligned} E^2(g, \varphi_0, \ldots, \varphi_{m-1}) \;&\equiv\; \int_{-\infty}^{\infty}\int_D (1 + |\lambda|)^{2+2p} |\hat{q}|^2 \,\mathrm{d}\lambda\,\mathrm{d}x \\ &+ \int_{-\infty}^{\infty} (1 + |\lambda|)^{4+2p} \sum_{i=0}^{m-1} \|\hat{\varphi}_i\|^2_{W_2^{2m-i-1/2}(\Gamma)} \,\mathrm{d}\lambda < \infty, \quad 2p > 1, \end{aligned}$$

and let the following conditions hold:

$$n \le 3, \quad u_0, u_T \in C^{2m}(\bar{D}), \quad g_0, g_T \in C(\bar{D}),$$

$$C_0 \frac{E(g, \varphi_0, \ldots, \varphi_{m-1})}{(2p-1)^{1/2}} - \left\| \frac{\Phi_0}{M} \right\|_0 > 0,$$

where C_0 is a constant which depends on u_0, u_T, g_0 and g_T; $\Phi_0 = g_0 \mathbf{A}_0 u_T - g_T \mathbf{A}_0 u_0$. There will then exist a unique solution $w(x,\lambda)$ of the problem (1.21), (1.22) such that

$$\int_{-\infty}^{\infty} (1+\lambda)^{2p+2} \|w\|^2_{L^2(D)} \, d\lambda + \int_{-\infty}^{\infty} (1+\lambda)^{2p} \|w\|^2_{W_2^{2m}(D)} \, d\lambda < \infty.$$

1.3 GENERATING FUNCTIONS, EVOLUTION EQUATIONS, AND INVERSE PROBLEMS

1. Consider a nonlinear system of evolution equations of the Navier-Stokes type:

$$\frac{\partial u_k}{\partial t} = a\Delta u_k + du_k + c\sum_{j=1}^{n} u_j \frac{\partial u_k}{\partial x_j} + \sum_{j=1}^{n} \alpha_j \frac{\partial u_k}{\partial x_j} - \frac{\partial q}{\partial x_k} + b_k(x'')h_k(x',t),$$
$$\sum_{k=1}^{n} \frac{\partial u_k}{\partial x_k} = 0, \tag{1.23}$$

where α_j, a, c, and d are the constants, $x \in \mathrm{R}^n$, R^n is the real unitary space, $x' = (x_1, x_2, \ldots, x_{n-m})$, $x'' = (x_{n-m+1}, \ldots, x_n)$, $m \le n$; q, b_k, h_k are certain functions, and $t \in \mathrm{R}^1$. The systems of equations (1.23) are extensively used in mathematical physics, ecology, biology, and other sciences.

The following inverse problem is proposed: find the functions $u_k(x,t)$, $h_k(x',t)$, $q(x,t)$, $k = 1, 2, \ldots, n$, $x \in \mathrm{R}^n$, $t \in \mathrm{R}^1$ if in (1.23) the constants a, d, c, α_j and the functions $b_k(x'') \ne 0$, $k, j = 1, 2, \ldots, n$ are given, and the following conditions are satisfied:

Boundary data are known

$$f_k(x',t) = u_k|_{x''=0}, \quad x' \in \mathrm{R}^{n-m}, \quad t \in \mathrm{R}^1. \tag{1.24}$$

There are direct and inverse Fourier transforms for the functions u_k, b_k, h_k, and f_k with respect to variables x, x', x'' and t, respectively.

In practice, the inverse problem for the system of evolution equations (1.23) involves finding, together with the substance $u(x,t) = (u_1, u_2, \ldots, u_n)$, the sources J of the form $J = (J_1, J_2, \ldots, J_n)$.

$$J_k = -\frac{\partial q}{\partial x_k} + b_k(x'')h_k(x',t), \quad k = 1, 2, \ldots, n,$$

with *a priori* unknown $q(x,t)$ and $h_k(x'.t)$ for the given control and measurement data:

$$b(x'') = (b_1, b_2, \ldots, b_n), \quad u|_{x''=0} = f(x',t) = (f_1, f_2, \ldots, f_n).$$

For linear equations ($c = 0$) certain inverse problems were discussed in Prilepko and Vasin (1989).

The main result of the present section is the method for obtaining formulas for the required functions q and h_k. It turns out that such formulas are valid within the solution of a system of integral equations of the second kind, which is determined only by the initial data (is independent of q, h_k) and to which conventional methods for investigating the validity of a solution in appropriate classes of functions may be applied. Let $\hat{u}_k(\xi,\tau)$, $\hat{q}(\xi,\tau)$, $\hat{h}_k(\xi',\tau)$, $\hat{b}_k(\xi'')$ be Fourier transforms with respect to (x,t), (x',t), x'' of the functions u_k, q, h_k, b_k and let $R_{lj}(\xi',\tau)$ be well-determined elements of a matrix, inverse to a matrix B, whose elements $B_{lj}(\xi,\tau)$ are determined by the relations

$$B_{lj} = \int\limits_{R^m} \frac{\hat{b}_l(\xi'')\delta_{lj}\xi^2 - \xi_l\hat{b}_j(\xi'')\xi_j}{\xi^2\Big(i\tau + a\xi^2 - d - i(\alpha,\xi)\Big)}\, d\xi'', \tag{1.25}$$

where $\xi^2 = \xi_1^2 + \ldots + \xi_n^2$, $(\alpha,\xi) = \sum\limits_{j=1}^{n} \alpha_j\xi_j$, $l,j = 1,2,\ldots,n$ and δ_{lj} is Kronecker's symbol. In terms of this notation, the inverse problem (1.23)–(1.24) through Fourier transform is initially reduced to finding the functions $\hat{u}_k(\xi,\tau)$, $\hat{q}(\xi,\tau)$, from relations (1.26)–(1.28):

$$\Big[i\tau + a^2\xi^2 - d - i(\alpha,\xi\Big]\hat{u}_k = i\,c\int\limits_{R^n}\int\limits_{-\infty}^{\infty}\sum_{j=1}^{n} \hat{u}_j(\xi-y,\tau-p)y_j\hat{u}_k(y,p)\,dp\,dy + \hat{b}_k(\xi'')\hat{h}_k(\xi',\tau) - i\,\xi_k q, \tag{1.26}$$

$$\sum_{j=1}^{n} i\,\xi_j\hat{u}_j = 0, \tag{1.27}$$

$$\hat{f}_k(\xi',\tau) = \frac{1}{2\pi}\int\limits_{R^m} \hat{u}_k(\xi',\xi'',\tau)\,d\xi''. \tag{1.28}$$

Let

$$F_k(\hat{u}) = \hat{b}_k\hat{h}_k + i\,c\int\limits_{R^n}\int\limits_{-\infty}^{\infty}\sum_{j=1}^{n} \hat{u}_j(\xi - y,\tau - p)y_j\hat{u}_k(y,p)\,dp\,dy.$$

Equations (1.26)–(1.28) lead to

$$\hat{u}_k = \frac{F_k\xi^2 - \xi_k\sum\limits_{j=1}^{n} F_j\xi_j}{\xi^2\Big[i\tau + \xi^2 a - d - i(\alpha,\xi)\Big]}, \tag{1.29}$$

$$q = -\frac{i\sum\limits_{j=1}^{n} F_j\xi_j}{\xi^2}, \tag{1.30}$$

$$\int\limits_{R^m} \frac{F_k\xi^2 - \xi_k\sum\limits_{j=1}^{n} F_j\xi_j}{\xi^2\Big[i\tau + \xi^2 a - d - i(\alpha,\xi)\Big]}\, d\xi'' = f_k(\xi',\tau). \tag{1.31}$$

Excluding $\hat{h}_k$ from (1.28) and (1.30) and combining them with (1.24) we obtain the principal system of nonlinear integral equations of the second kind with respect to the function at $c \neq 0$ and formulas for $\hat{u}_k$ at $c = 0$:

$$\begin{aligned}\hat{u}_k \; = \; & \left\{ \hat{b}_k \sum_{j=1}^{n} R_{kj} \left[(2\pi)^n + c \int_{\mathrm{R}^m} \mathcal{P}_j(\hat{u})\, \mathrm{d}\xi'' \right] \xi^2 \right. \\ & \left. - \xi_k \sum_{l=1}^{n} \sum_{j=1}^{n} \hat{b}_l \xi_l R_{lj} \left[(2\pi)^n \hat{f}_l + c \int_{\mathrm{R}^m} \mathcal{P}_j(\hat{u})\, \mathrm{d}\xi'' \right] \right\} \\ & \times \left\{ \xi^2 \left[\mathrm{i}\,\tau + a\xi^2 - d - \mathrm{i}(\alpha, \xi) \right] \right\},\end{aligned} \tag{1.32}$$

where $\mathcal{P}_k(\hat{u}) = \mathcal{P}_k(\hat{u}_1, \ldots, \hat{u}_n)$ is determined by the formula

$$\begin{aligned}\mathcal{P}_k(\hat{u}) \; = \; & \left\{ -\mathrm{i} \left[\int_{\mathrm{R}^n} \int_{-\infty}^{\infty} \sum_{j=1}^{n} u_j(\xi - y, \tau - p) y_j \hat{u}_k(y, p)\, \mathrm{d}p\, \mathrm{d}y \right. \right. \\ & \left. \left. - \xi_k \int_{\mathrm{R}^n} \int_{-\infty}^{\infty} \sum_{j=1}^{n} \sum_{l=1}^{n} \hat{u}_j(\xi - y, \tau - p) \hat{u}_l(y, p)\, \mathrm{d}p\, \mathrm{d}y \right] \right\} \\ & \times \left\{ \xi^2 \left[\mathrm{i}\,\tau + a\xi^2 - d - \mathrm{i}(\alpha, \xi) \right] \right\}^{-1}\end{aligned}$$

After finding $\hat{u}_k$ from (1.32), we calculate the required functions $\hat{h}_k(\xi', \tau)$, $q(\xi, \tau)$ by

$$\hat{h}_k = \sum_{j=1}^{n} R_{kj} \left[(2\pi)^n \hat{f}_j + c \int_{\mathrm{R}^m} \mathcal{P}_j(\hat{u})\, \mathrm{d}\xi'' \right], \tag{1.33}$$

$$\hat{q} = -\frac{\mathrm{i} \sum_{j=1}^{n} F_j(\hat{u}) \xi_j}{\xi^2}, \tag{1.34}$$

where $F_j(\hat{u})$, $\mathcal{P}_j(\hat{u})$ are already known functions. Relations (1.32)–(1.34) taking into account an inverse Fourier transform, determine the resolvent operator of an inverse problem, which compares the given (f, b) and the solution (u, q, h). For $c = 0$ this operator is linear.

2. A generating function of the sequence $p_n(t)$ is defined, as we know, as the series $\sum_{n=0}^{\infty} p_n y^n$, where p_n and a variable x may have different values and interpretations. For example, in the theory of random processes, p_n is the probability that a random variable takes on integral values of n and t is the time. Generating functions often provide the main tools for investigating many problems of the probability theory combinational analysis, ecology, biology, and sociology (Chin Long Chiang, 1970; Sevostyanov, 1971; Riordan, 1980).

An important analytical property of the generating functions is that they satisfy partial differential equations with the coefficients that are determined by the generating functions themselves. Thus, especially, the group property

$$u(t+\tau,y)=u(t,u(\tau,y)),\quad u(0,y)=y$$

yields the equations of the first and second kind:

$$\frac{\partial u}{\partial t}=c(y)\frac{\partial u}{\partial y},$$

where

$$c(y)=\left.\frac{\partial u}{\partial t}\right|_{t=0},$$

and

$$\frac{\partial^2 u}{\partial t^2}=c_1(y)\frac{\partial^2 u}{\partial y^2}+c_2\frac{\partial u}{\partial y},$$

where

$$c_1(y)=\left(\left.\frac{\partial u}{\partial t}\right|_{t=0}\right)^2,\quad c_2=\left.\frac{\partial^2 u}{\partial t^2}\right|_{t=0}.$$

Here it is clear that the coefficients of the equations for generating functions are generally a *priori* unknown and should also be determined.

The inverse problems for determining coefficients and differential equations related to generating functions are, in our view, of considerable interest in both theory and applications. Initial elements of this direction are given in the present section.

Let us consider differential equations for generating functions that are somewhat more general than certain equations in Chin Long Chiang (1970); Sevostyanov (1971):

$$a(y)\frac{\partial u}{\partial t}=\int\limits_{\mathbf{R}^n}\left[b(x-x',t)u(x',y,t)+c(x-x',t)\frac{\partial u(x',y,t)}{\partial y}\right]dx',\tag{1.35}$$

$$\varphi'(y)\frac{\partial u}{\partial t}=\int\limits_{\mathbf{R}^n}\left[\varphi(y)\mu(x-x',t)+(1-\varphi)\lambda(x-x',t)\frac{\partial u(x',y,t)}{\partial y}\right]dx',\tag{1.36}$$

$$\varphi(y)\psi'(y)\frac{\partial u}{\partial t}=\int\limits_{\mathbf{R}^n}d(x-x',t)\left[\varphi'(y)\psi(y)u(x',y,t)+\varphi\psi\frac{\partial u}{\partial y}\right]dx',\tag{1.37}$$

where a, b, c, φ, ψ, μ. λ, d are certain generally a *priori* unknown, functions, which must be determined together with the function $u(x,y,t)$.

Note that if in (1.36)

$$\varphi=\frac{1}{1-y},\quad \mu=\delta(x-x')\mu_0(t),\quad \lambda=\delta(x-x')\lambda_0,$$

the equation discussed in (Chin Long Chiang, 1970) is obtained.

As regards direct and inverse problems for the differential equations of type (1.35)–(1.36), formulas of general solutions are of interest to us. Assuming that the following operations are correct, we give here such formulas. Below, the circumflex over a letter denotes a Fourier transform of the considered function with respect to x:

$$\hat{a}(\omega) = \int\limits_{\mathrm{R}^n} a(x)\, \mathrm{e}^{\mathrm{i}\,\omega x}\, \mathrm{d}\omega.$$

It turns out that the following formulas satisfy the solutions of (1.35)–(1.37), respectively

$$u = \int\limits_{\mathrm{R}^n} \exp\left[f(A(y) + \hat{\Phi}(\omega, t), t) + \int\limits_{y_0}^{y} g(A(y) - A(\eta) + \hat{\Phi}, \omega)\, \mathrm{d}\eta \right] \mathrm{e}^{\mathrm{i}\,\omega x}\, \mathrm{d}\omega,$$
$$b = -\left(\frac{1}{2\pi}\right)^n \int\limits_{\mathrm{R}^n} g(\hat{\Phi}, \omega) \frac{\partial \hat{\Phi}}{\partial t}\, \mathrm{e}^{\mathrm{i}\,\omega x}\, \mathrm{d}\omega, \quad A' = a, \quad \frac{\partial \hat{\Phi}}{\partial t} = \hat{c}. \tag{1.38}$$

$$u = \int\limits_{\mathrm{R}^n} F\left(\omega, \varphi(y)\, \mathrm{e}^{\gamma(\omega, t)} + \int\limits_{t_0}^{t} \mathrm{e}^{\gamma(\tau, \omega)}\, \lambda(\tau, \omega)\, \mathrm{d}\tau \right) \mathrm{e}^{\mathrm{i}\,\omega x}\, \mathrm{d}\omega,$$
$$\frac{\partial \gamma}{\partial t} = \hat{\mu} - \hat{\lambda}. \tag{1.39}$$

$$u = \int\limits_{\mathrm{R}^n} \varphi(y) F(\omega, \psi(y)\, \mathrm{e}^{D(\omega, t)})\, \mathrm{e}^{\mathrm{i}\,\omega x}\, \mathrm{d}\omega, \quad \frac{\partial D}{\partial t} = \hat{d}(\omega, t), \tag{1.40}$$

where f, g, and F are arbitrary differentiable or, if necessary, analytical functions. Equations (1.38)–(1.40) can be used in the theory and applications of generating functions both in direct and inverse problems.

1.4 ALGEBRAIC QUESTIONS ON THE THEORY OF MULTIDIMENSIONAL INVERSE PROBLEMS FOR NONLINEAR EVOLUTION EQUATIONS

In the present section, we present a method for analyzing inverse problems of nonlinear evolution equations, which are based on algebraic relations and operations.

The main results are formulas found for the solutions of inverse problems, derived up to solutions of scalar equations under the condition that the main boundary information satisfies algebraic relations, for instance, the nilpotency condition in a finite-dimensional algebra.

1. Let R^n and R^m be real Euclidean spaces of variables $x = (x_1, \ldots, x_n)$ and $z = (z_1, \ldots, z_m)$ respectively, $t \in \mathrm{R}$; $\mathbf{A}_x$ and $\mathbf{B}_x$ be linear differential operators, which act only on x and do not depend on z, and $\mathbf{L}_z$ be a linear differential operator that acts only on z and does not depend on x.

For example,

$$\mathbf{A}_x = \sum_{j=1}^{n} \frac{\partial^2}{\partial x_j^2}, \quad \mathbf{L}_z = \sum_{j=1}^{m} \frac{\partial^2}{\partial z_j^2},$$

and $\mathbf{B}_x$ is a unit operator.

We shall consider the nonlinear system of differential equations

$$\frac{\partial u}{\partial t} = \mathbf{A}_x u + \mathbf{L}_z u + u^2 + \lambda(x,t)\mathbf{B}_x u, \quad \frac{\partial \lambda}{\partial z_k} = 0, \quad k = 1, 2, \dots, m, \tag{1.41}$$

where $u(x, z, t)$ and $\lambda(x, t)$ are the desired functional elements belonging to some algebra $\mathcal{U} =< \mathrm{R}^k . \beta >$, i.e. $u(x, z, t)$ and $\lambda(x, t)$ are elements of a vector space (at a fixed x, z, and t), on which is defined the bilinear transformation $\beta : \mathrm{R}^k \times \mathrm{R}^k \to \mathrm{R}^k$, $u^2 = \beta(u, u)$.

Let us assume that the known information is given in the form

$$u|_{z=0} = f(x,t)\varphi_0(t), \tag{1.42}$$

where $f(x, t) \in \mathcal{U}$ and $\varphi_0(t)$ is a real function.

We now state the inverse problem for determining $u(x, z, t)$, and $\lambda(x, t)$ from (1.41) if (1.42) is satisfied.

Theorem 1.2. *Let the element f of the algebra $\mathcal{U}$ in (1.42) be nilpotent or idempotent, i.e., $f^2 = \beta f$, where β is a constant equal to zero or one, and elements, $\mathbf{A}_x f$, $\mathbf{B}_x f$ and $\partial f/\partial t$ exist, and let the real function $\varphi(z, t)$ be a solution to the equation $\partial\varphi/\partial t = \mathbf{L}_z\varphi + \beta^2\varphi^2$, which satisfies the condition $\varphi|_{z=0} = \varphi_0(t)$. Then the elements (u, λ) of the algebra $\mathcal{U}$, which are defined by the relations*

$$u(x,z,t) = f(x,t)\varphi(z,t), \quad \lambda(x,t)\mathbf{B}_x f = \frac{\partial f}{\partial t} - \mathbf{A}_x f,$$

satisfy equation (1.41) and the initial condition (1.42).

By virtue of this theorem, the defining relations for (u, λ) are found up to the solution of the scalar equation $\partial\varphi/\partial t = \mathbf{L}_z\varphi + \beta^2\varphi^2$, which is linear at $\beta = 0$. To fix (u, λ), we need additional information. Here we restrict the discussion to the algebraic aspects. It is well known that, in any finite-dimensional real algebra, there exist either nilpotent or idempotent elements (Röhrl and Wischnewsky, 1976). Let us give some examples of algebras having the product $\beta(u, u)$, which are related to classical models of mathematical physics and control theory and discuss the sets of nilpotent and idempotent elements.

1.4.1 Linear control system with quadratic expenditures

In this model (Levin, 1959), one considers a set of real symmetrical matrices (relative to transposition) having the product $\beta(U, U) = UAU$, where A is a constant symmetric matrix.

We begin by describing a set P of idempotent elements of this algebra. Let the matrix $A = T^{-1}DT$, where D is a diagonal matrix and T is a unitary matrix. Let us consider an arbitrary matrix

$$J_0 = \begin{pmatrix} E & 0 \\ 0 & 0 \end{pmatrix}$$

where E is the unit matrix. Let us choose an invertible matrix

$$C = \begin{pmatrix} C_{11} & C_{12} \\ C_{21} & C_{22} \end{pmatrix}$$

which has the same partition and the condition

$$C_{11}^T D_1 C_{12} + C_{21}^T D_2 C_{22} = 0,$$

where $D = \begin{pmatrix} D_1 & 0 \\ 0 & D_2 \end{pmatrix}$ is partitioned like the matrix C. For instance, this condition is satisfied for any matrix with $C_{12} = 0, C_{21} = 0$.

Then the matrix $P = T^{-1}CJ_0C^{-1}TA$ will satisfy the conditions $PAP = P$ and $P^T = P$. Inversely, any matrix P satisfying these conditions has this form for appropriate C and J_0.

A set of nilpotent elements is described in the following manner. If matrix A is similar to matrix $B = \begin{pmatrix} 0 & B_{12} \\ B_{21} & B_{22} \end{pmatrix}$ under the transformation matrix C, then the nilpotent elements have the form $P = C\begin{pmatrix} M & 0 \\ 0 & 0 \end{pmatrix}$, where M is an arbitrary matrix. If the matrix A is not similar to a matrix of the form $\begin{pmatrix} 0 & B_{12} \\ B_{21} & B_{22} \end{pmatrix}$, then there are no nilpotent elements.

1.4.2 The Lorenz model for heat interaction

In the Lorenz model, there is the product of the form

$$\beta(u,u) = \begin{pmatrix} 0 \\ -u_1u_3 \\ u_1u_2 \end{pmatrix} \quad \text{for} \quad u = \begin{pmatrix} u_1 \\ u_2 \\ u_3 \end{pmatrix}$$

If e_1, e_2, e_3 is the standard basis, then the multiplication table has the form

	e_1	e_2	e_3
e_1	0	$1/2e_3$	$-1/2e_2$
e_2	$1/2e_3$	0	0
e_3	$-1/2e_2$	0	0

From the multiplication table it follows that $A^2 = \mathrm{R}\, e_1 \oplus \mathrm{R}\, e_2$ and $A^2A^2 = 0$. Therefore, the algebra A is solvable and has no idempotents. It is easy to see that a set of nilpotent elements has the form

$$\{\alpha e_1/\alpha \in \mathrm{R}\} \cup \{\alpha e_2 + \beta e_3/\alpha, \beta \in \mathrm{R}\}$$

1.4.3

The Euler equations of motion for a rotating rigid body in the absence of external forces define a product of the type

$$\beta(u,u) = \begin{pmatrix} \dfrac{\gamma-\beta}{\alpha}u_2u_3 \\[2ex] \dfrac{\alpha-\gamma}{\beta}u_1u_3 \\[2ex] \dfrac{\beta-\alpha}{\gamma}u_1u_2 \end{pmatrix} \quad \text{for} \quad u = \begin{pmatrix} u_1 \\ u_2 \\ u_3 \end{pmatrix}$$

If $u = (u_1, u_2, u_3)$ are the coordinates of the idempotent e, then u is a solution of the system

$$\frac{\gamma-\beta}{\alpha}u_2u_3 = u_1,$$
$$\frac{\alpha-\gamma}{\beta}u_1u_3 = u_2,$$
$$\frac{\beta-\alpha}{\gamma}u_1u_2 = u_3,$$

where α, β, and γ are constants, which is the same as the system

$$u_1^2 = \frac{\gamma\beta}{(\alpha-\gamma)(\beta-\alpha)},$$
$$u_2^2 = \frac{\alpha\gamma}{(\gamma-\beta)(\beta-\alpha)},$$
$$u_3^2 = \frac{\alpha\beta}{(\gamma-\beta)(\alpha-\gamma)}.$$

It appears that there are no $\alpha, \beta, \gamma \in \mathbf{R}$ such that right-hand sides of the latter system are simultaneously positive. Therefore, this algebra has no idempotents.

A set of nilpotent elements of this algebra are made from αe_2, γe_3; α, β, $\gamma \in \mathbf{R}$.

2. Let us proceed to other nonlinearities. In the complex case, instead of (1.41) we consider the nonlinear evolution equation

$$\frac{\partial u}{\partial t} = \mathbf{A}_x u + \mathbf{L}_z u + g(|u|^2)u + \lambda(x,t)\mathbf{B}_x u, \tag{1.43}$$

where $\mathbf{A}_x$, $\mathbf{L}_z$, and $\mathbf{B}_x$ are linear operators with the same properties as in (1.41), $g(p)$, $p \geq 0$, is a function, and $u(x,z,t)$ and $\lambda(x,t)$ are the desired functions. Equations of type (1.43) are widely used in theory and applications. In particular, for different choices of $g(p)$ and the operators $\mathbf{A}_x$, $\mathbf{B}_x$, and $\mathbf{L}_z$, from (1.43) one obtains the nonlinear Schrödinger equation, the Kuramoto-Suzuki equation, and the Fitzhugh-Nagumo equation.

Theorem 1.3. *Let us assume that* $|f| = 1$ *in (1.42), that* $\mathbf{A}_x f$, $\mathbf{B}_x f$, $(\mathbf{B}_x f)^{-1}$, *and* $\partial f/\partial t$ *exist, and that the real function* $\varphi(z,t)$ *is a solution of the equation* $\dfrac{\partial\varphi}{\partial t} = \mathbf{L}_z\varphi + g(\varphi^2)\varphi$ *and that it satisfies the condition* $\varphi|_{z=0} = \varphi_0(t)$. *The functions* (u, λ), *defined by the equations*

$$u(x,z,t) = f(x,t)\varphi(z,t), \quad \lambda(x,t) = \left(\frac{\partial f}{\partial t} - \mathbf{A}_x f\right)(\mathbf{B}_x f)^{-1},$$

will then satisfy (1.43) and (1.42).

For another example which applies the symmetry of the general boundary condition, we consider, again in the complex case, an inverse problem for determining the two functions $u(x,t)$, $\lambda(x,z)$, $x \in \mathbf{R}^n$, $t \in \mathbf{R}$, $z \in \mathbf{R}$, $\dfrac{\partial\lambda(x,z)}{\partial t} = 0$, when the following conditions are valid:

$$\frac{\partial u}{\partial t} = \Delta u + \lambda(x,u), \quad \Delta = \sum_{j=1}^{n} \frac{\partial^2}{\partial x_j^2}, \tag{1.44}$$

$$u|_{x_n=0} = f(x',t), \quad x' = (x_1, x_2, \dots, x_{n-1}), \tag{1.45}$$

$$\lambda(x,0) = \lambda(x,k) = 0, \tag{1.46}$$

where k is some number.

The nonlinear equations of type (1.44) have applications in chemistry, biology, and ecology, while relations of type (1.46) are also frequently used (Murray, 1977; Davydov, 1986; Svirizhev, 1987). The main condition on the boundary information is that the given function $f(x',t)$ has the representation

$$f(x',t) = R_0(x')F(v_0(x') + \varphi(t)), \tag{1.47}$$

where $R_0(x')$, $F(z)$, $v_0(x')$, and $\varphi(t)$ are some functions, $x' \in \mathrm{R}^{n-1}$, $t \in \mathrm{R}$, and $z \in \mathrm{R}$. For specific functions in (1.47), the given representation may characterize a trace of spiral waves in R^n of the hyperplane $x_n = 0$.

Theorem 1.4. *Let all the functions in (1.47) be twice differentiable and $F(0) = 0$, $F'(0) = \alpha \neq 0$. The following equations will then be valid:*

$$u(x,t) = R(x)F(v(x) + \varphi(t)),$$

$$\begin{aligned}\lambda(x,z) &= F'\left(F^{-1}\left(\frac{z}{R(x)}\right)\right)\left[R(x)b\left(F^{-1}\left(\frac{z}{R(x)}\right) - v\right) - R\Delta v - 2(\operatorname{grad} R, \operatorname{grad} v)\right] \\ &\quad - F''\left(F^{-1}\left(\frac{z}{R(x)}\right)\right) R\,|\operatorname{grad} v|^2 - \frac{z}{R(x)}\Delta R,\end{aligned}$$

where the function $b(za)$ is defined by the equality $\varphi'(t) = b(\varphi)$, and the functions $R(x)$ and $v(x)$, where $x \in \mathrm{R}^n$, are a solution of the system of equations

$$\Delta v = \Phi_1(v, R, v_x, R_x),$$

$$\Delta R = \Phi_2(v, R, v_x, R_x),$$

$$\Phi_1 = -\frac{1}{R}\left[2(\operatorname{grad} v, \operatorname{grad} R) + F''(0)(\operatorname{grad} v)^2 - \alpha b(-v)\right]$$

$$\begin{aligned}\Phi_2 &= \left\{F'\left(F^{-1}\left(\frac{k}{R}\right)\right)\left[b\left(F^{-1}\left(\frac{k}{R}\right) - v\right) + \Phi_1 - 2(\operatorname{grad} R, \operatorname{grad} v)\right]\right. \\ &\quad \left. - F''\left(F^{-1}\left(\frac{k}{R}\right)\right)|\operatorname{grad} v|^2\right\}\end{aligned}$$

These functions also satisfy the conditions

$$v|_{x_n=0} = v_0(x'), \quad R|_{x_n=0} = R_0(x').$$

REMARK 1.1. Other representations for $f(x',t)$, for example, in the form $f(x',t) = R(x')f(v(x)\varphi(t))$, result in different equations for $R(x)$ and $v(x)$.

Various representations of solutions and coefficients for evolution equations pertaining to inverse problems are given in (Anikonov, 1990b, 1991a, b, 1992c, d, e, 1994a, 1995e).

Here, we present one generalization which is associated with the second-order evolution equations.

Let

$$\mathbf{L} = \sum_{i,j=1}^{n} a_{ij}(x,t)\frac{\partial^2}{\partial x_i \partial x_j} + \sum_{i=1}^{n} a_i(x,t)\frac{\partial}{\partial x_i} + b(x,t)$$

be an arbitrary second-order operator, and the function $F(z,t)$, $z \in \mathrm{R}$, $t \in \mathrm{R}$, be a solution of the parabolic equation $\dfrac{\partial F}{\partial t} = \dfrac{\partial^2 F}{\partial z^2}$. We denote

$$|\mathrm{grad}\, v|^2 = \sum_{i,j=1}^{n} a_{ij}\frac{\partial v}{\partial x_i}\frac{\partial v}{\partial x_j}.$$

Theorem 1.5. *Let the functions $\varphi(x,t) \neq 0$ and $R(x,t)$ be a solution of the system of equations*

$$\left|\mathrm{grad}\,\frac{R}{\varphi}\right|^2 \frac{\partial R}{\partial t} = \mathbf{L}R, \quad \left|\mathrm{grad}\,\frac{R}{\varphi}\right|^2 \frac{\partial \varphi}{\partial t} = \mathbf{L}\varphi. \tag{1.48}$$

Then the function $u(x,t) = \varphi(x,t)f(t, R/\varphi)$ and the coefficient $\rho = |\mathrm{grad}\, R/\varphi|^2$ satisfy the equation $\rho\partial u/\partial t = \mathbf{L}u$.

COROLLARY 1.1. *Let the operator $\mathbf{L}$ be independent of t. Then the functions $R(x)$ and $\varphi(x)$, $x \in \mathrm{R}^n$, which satisfy the equation $\mathbf{L}v = 0$, are a solution of the system (1.48).*

From Theorem 1.5 and Corollary 1.1 it follows:

Theorem 1.6. *If the function $f(x',t) = u|_{x_n=0}$ has the form*

$$f(x',t) = \varphi_0(x')F\left(\frac{R_0(x')}{\varphi_0(x')}, t\right), \qquad \text{then } \varphi_0(x') \neq 0,$$

where the function $F(z,t)$ is a solution of the parabolic equation $\dfrac{\partial F}{\partial t} = \dfrac{\partial^2 F}{\partial z^2}$. In the inverse problem of finding $u(x,t)$ and $\rho(x)$:

$$\rho\frac{\partial u}{\partial t} = \mathbf{L}u, \quad u|_{x_n=0} = f(x',t),$$

we will then have

$$u(x,t) = \varphi(x)F\left(\frac{R(x)}{\varphi(x)}, t\right), \quad \rho(x) = \left|\mathrm{grad}\,\frac{R}{\varphi}\right|^2;$$

here, the functions $\varphi(x)$ and $R(x)$ satisfy the equation $\mathbf{L}\varphi = \mathbf{L}R = 0$ and obey the conditions $\varphi|_{x_n=0} = \varphi_0(x')$, $R|_{x_n=0} = R_0(x')$.

More common results and proofs one can see in Chapter 6.

1.5 INVERSE PROBLEM FOR THE MAXWELL EQUATION

In this section, we consider the system of Maxwell equations:

$$\begin{cases} \operatorname{div}(\varepsilon\mathbf{E}) = \rho, & \operatorname{div}(\mu\mathbf{H}) = 0 \\ \operatorname{curl}\mathbf{E} = -\mu\dfrac{\partial\mathbf{H}}{\partial t}, & \operatorname{curl}\mathbf{H} = \varepsilon\dfrac{\partial\mathbf{E}}{\partial t} + \mathbf{J}. \end{cases} \tag{1.49}$$

Here $\mathbf{E}(x,y,z,t) = (E_1, E_2, E_3)$, $\mathbf{H} = (H_1, H_2, H_3)$ are electric and magnetic fields, ρ is the density of the medium, $\varepsilon(x,y,z,t)$, $\mu(x,y,z,t)$ are functions which characterized the medium. $\mathbf{J}(x,y,z,t)$ is source.

The inverse problem is to find functions $\mathbf{E}(x,y,z,t)$, $\mathbf{H}(x,y,z,t)$, $\varepsilon(x,y,z,t)$, $\mu(x,y,z,t)$ in the neighborhood of zero if we know information

$$\begin{aligned} \mathbf{E}\big|_{z=0} &= \mathbf{a}(x,y,t), \qquad \mathbf{H}\big|_{z=0} = \mathbf{b}(x,y,t) \\ E_3\big|_{t=0} &= E_0(x,y,z) \quad H_3\big|_{t=0} = H_0(x,y,z) \end{aligned} \tag{1.50}$$

and functions $\mathbf{a} = (a_1, a_2, a_3)$, $\mathbf{b} = (b_1, b_2, b_3)$, E_0, H_0, $\mathbf{J}$, ρ are known.

Theorem 1.7. *Let vector $(E_1, E_2, H_1, H_2, A, B)$ be a solution of the Cauchy problem for variable z.*

$$\frac{\partial E_1}{\partial z} = \frac{\partial}{\partial x}\left(E_0 \exp\int_0^t \frac{d\xi}{A}\right) - \frac{\left(\dfrac{\partial E_1}{\partial y} - \dfrac{\partial E_2}{\partial x}\right) B}{H_0 \exp\displaystyle\int_0^t \frac{d\xi}{B}} \frac{\partial H_2}{\partial t},$$

$$\frac{\partial E_2}{\partial z} = -\frac{\partial}{\partial y}\left(E_0 \exp\int_0^t \frac{d\xi}{A}\right) + \frac{\left(\dfrac{\partial E_1}{\partial y} - \dfrac{\partial E_2}{\partial x}\right) B}{H_0 \exp\displaystyle\int_0^t \frac{d\xi}{B}} \frac{\partial H_1}{\partial t},$$

$$\frac{\partial H_1}{\partial z} = \frac{\partial}{\partial x}\left(H_0 \exp\int_0^t \frac{d\xi}{B}\right) + \frac{\left(\dfrac{\partial H_2}{\partial x} - \dfrac{\partial H_1}{\partial y} - J_3\right) A}{E_0 \exp\displaystyle\int_0^t \frac{d\xi}{A}} \frac{\partial E_2}{\partial t} + J_2,$$

$$\frac{\partial H_2}{\partial z} = -\frac{\partial}{\partial y}\left(H_0 \exp\int_0^t \frac{d\xi}{B}\right) + \frac{\left(\dfrac{\partial H_2}{\partial x} - \dfrac{\partial H_1}{\partial y} - J_3\right) A}{E_0 \exp\displaystyle\int_0^t \frac{d\xi}{A}} \frac{\partial E_1}{\partial t} - J_1,$$

$$\frac{\partial A}{\partial z} = \frac{1}{\dfrac{\partial H_2}{\partial x} - \dfrac{\partial H_1}{\partial y} - J_3} \left\{ \frac{\partial}{\partial x} \left[\frac{A\left(\dfrac{\partial H_2}{\partial x} - \dfrac{\partial H_1}{\partial y} - J_3\right)}{E_0 \exp \displaystyle\int_0^t \frac{d\xi}{A}} \left(\frac{\partial E_1}{\partial t} - E_1\right) \right] \right.$$

$$+ \frac{\partial}{\partial y} \left[\frac{A\left(\dfrac{\partial H_2}{\partial x} - \dfrac{\partial H_1}{\partial y} - J_3\right)}{E_0 \exp \displaystyle\int_0^t \frac{d\xi}{A}} \left(\frac{\partial E_2}{\partial t} - E_2\right) \right]$$

$$\left. + \frac{\partial J_1}{\partial x} + \frac{\partial J_2}{\partial y} + \rho + A\frac{\partial J_3}{\partial z} \right\}$$

$$\frac{\partial B}{\partial z} = \frac{1}{\dfrac{\partial E_1}{\partial y} - \dfrac{\partial E_2}{\partial x}} \left\{ \frac{\partial}{\partial x} \left[\frac{B\left(\dfrac{\partial E_1}{\partial y} - \dfrac{\partial E_2}{\partial x}\right)}{H_0 \exp \displaystyle\int_0^t \frac{d\xi}{B}} \left(\frac{\partial H_1}{\partial t} - H_1\right) \right] \right.$$

$$\left. + \frac{\partial}{\partial y} \left[\frac{B\left(\dfrac{\partial E_1}{\partial y} - \dfrac{\partial E_2}{\partial x}\right)}{H_0 \exp \displaystyle\int_0^t \frac{d\xi}{B}} \left(\frac{\partial H_2}{\partial t} - H_2\right) \right] \right\}$$

$$E_1|_{z=0} = a_1, \quad E_2|_{z=0} = a_2, \quad H_1|_{z=0} = b_1. \quad H_2|_{z=0} = b_2.$$

$$A|_{z=0} = \frac{a_3}{\partial a_3/\partial t}, \quad B|_{z=0} = \frac{b_3}{\partial b_3/\partial t}$$

Then the following formulas hold

$$E_3 = E_0 \exp \int_0^t \frac{d\xi}{A}, \quad H_3 = H_0 \exp \int_0^t \frac{d\xi}{B},$$

$$\varepsilon = \frac{\left(\dfrac{\partial H_2}{\partial x} - \dfrac{\partial H_1}{\partial y} - J_3\right) A}{E_0 \exp \displaystyle\int_0^t \frac{d\xi}{A}},$$

$$\mu = \frac{\left(\dfrac{\partial E_1}{\partial y} - \dfrac{\partial E_2}{\partial x}\right) B}{H_0 \exp \displaystyle\int_0^t \frac{d\xi}{B}}$$

Thus the theorem is proved and we have found functions $\varepsilon(x, y, z, t)$, $\mu(x, y, z, t)$ from (1.49) by using $A = \dfrac{E_3}{\partial E_3/\partial t}$, $B = \dfrac{H_3}{\partial H_3/\partial t}$ and the data in (1.50).

COROLLARY 1.2. *In the analytical case, if* $H_0 \neq 0$, $E_0 \neq 0$, $A|_{z=0} \neq 0$, $B|_{z=0} \neq 0$ *in the neighborhood of point* $x = 0$, $y = 0$, $z = 0$, $t = 0$ *the Cauchy problem exists and has no more than unique analytical solution.*

CHAPTER 2

Formulas in inverse problems for evolution equations of the general type

2.1 INVERSE PROBLEM. FORMAL RESULTS

Here, we consider the inverse problem for a multidimensional general evolution equation: to find complex-value functions $w(x,t)$, $\lambda(x)$, $\frac{\partial \lambda}{\partial t} = 0$, $x \in D \subset \mathrm{R}^n$, $a \le t \le b$ such that

$$\frac{\partial w}{\partial t} = \mathbf{A}(t)w + \lambda(x)f(t), \tag{2.1}$$

$$w\big|_{t=a} = w_a(x), \quad w\big|_{t=b} = w_b(x). \tag{2.2}$$

Here, $\mathbf{A}(t)$ is a linear operator, which acts in variables $x = (x_1, \ldots, x_n)$, $x \in D \subset \mathrm{R}^n$ and D is a domain in real Euclidean space R^n, $n \ge 1$. The functions $f(t) \not\equiv 0$, $w_a(x)$, $w_b(x)$ and constants a, b, $a < b$ are supposed to be known.

If domain D has a boundary, then this boundary is suggested to be smooth and functions $f(t)$, w_a, w_b are supposed to be smooth enough.

The inverse problem (2.1), (2.2) can be interpreted as a control problem of translation of w_a to w_b. The control element is the function $\lambda(x)$, $x \in D$.

REMARK 2.1. Here, we do not suppose any boundary conditions in the inverse problem (2.1), (2.2). Later when it is necessary we shall impose some conditions.

The principal aim of this chapter is to derive some of the formulas in the multidimensional inverse problem (2.1), (2.2).

At first, we suppose that operator $\mathbf{A}$ does not depend on the variable x and has symbol $\tilde{A}(\xi,t)$, $\xi \in \mathrm{R}^n$, $a \le t \le b$:

$$\tilde{A}(\xi,t) \in C^\infty(\mathrm{R}^n \times [a,b]), \quad \mathbf{A}(t)\,\mathrm{e}^{\mathrm{i}\,x\xi} = \mathrm{e}^{\mathrm{i}\,x\xi}\,\tilde{A}(\xi,t).$$

Let $B(\xi,t) \in C^\infty(\mathrm{R}^n \times [a,b])$ be the function such that $\frac{\partial B}{\partial t} = \tilde{A}(\xi,t)$.

For example, if $\tilde{A} = A_0(\xi)$, then we can take $B(t,\xi) = A_0(\xi)t$.

Theorem 2.1. *Let the information $w_a(x)$, $w_b(x)$ be represented in the form:*

$$w_a(x) = \int\limits_{\mathrm{R}^n} \hat{w}_a(\xi)\,\mathrm{e}^{\mathrm{i}\,x\xi}\,\mathrm{d}\xi, \quad w_b(x) = \int\limits_{\mathrm{R}^n} \hat{w}_b(\xi)\,\mathrm{e}^{\mathrm{i}\,x\xi}\,\mathrm{d}\xi,$$

where functions $\hat{w}_a(\xi)$, $\hat{w}_b(\xi)$ *are some complex-valued functions,* $\int_a^b e^{-B(\xi,p)} f(p)\,dp \neq 0$, $\xi \in R^n$.

Then the functions $w(x,t)$, $\lambda(x)$:

$$w(x,t) = \int_{R^n} \left[\frac{\hat{w}_a(\xi)\, e^{B(\xi,t)-B(\xi,a)} \int_t^b e^{-B(\xi,p)} f(p)\,dp}{\int_a^b e^{-B(\xi,p)} f(p)\,dp} + \frac{\hat{w}_b(\xi)\, e^{B(\xi,t)-B(\xi,b)} \int_a^t e^{-B(\xi,p)} f(p)\,dp}{\int_a^b e^{-B(\xi,p)} f(p)\,dp} \right] e^{ix\xi}\,d\xi, \tag{2.3}$$

$$\lambda(x) = \int_{R^n} \frac{\hat{w}_b(\xi)\, e^{-B(\xi,b)} - \hat{w}_a(\xi)\, e^{-B(\xi,a)}}{\int_a^b e^{-B(\xi,p)} f(p)\,dp} e^{ix\xi}\,d\xi \tag{2.4}$$

formally satisfy (2.1), (2.2).

We emphasize that the formulas (2.3), (2.4) give the formal solution of the inverse problem (2.1), (2.2).

Now, we shall suppose that the operator **A** does not depend on variable t, but the operator **A** can depend on variable $x \in D \subset R^n$: for example, $\mathbf{L} = \sum_{i,j=1}^{n} a_{ij} \frac{\partial^2}{\partial x_i \partial x_j} + \sum_{j=1}^{n} a_j(x) \frac{\partial}{\partial x_j} + a(x)$ and coefficients $a_{ij}(x)$, $a_j(x)$, $a(x)$ are smooth.

We suppose also that for some boundary operator Γ, $\Gamma w|_{\partial D} = 0$, for example, $w|_{\partial D} = 0$.

Denote by $T_k(x)$ eigenfunctions and by λ_k eigenvalues:

$$\mathbf{A} T_k = \lambda_k T_k, \quad x \in D, \quad \Gamma T_k = 0, \quad x \in \partial D, \quad k = 1, 2, \ldots$$

In this case, we can also write formal formulas for the inverse problem (2.1), (2.2).

Theorem 2.2. *Let the information* $w_a(x)$, $w_b(x)$ *be represented by formulas* $w_a(x) = \sum_{k=1}^{\infty} a_k T_k(x)$, $w_b(x) = \sum_{k=1}^{\infty} b_k T_k(x)$, $x \in D$, *where* a_k, b_k *are constants, and* $\int_a^b e^{\lambda_k p} f(p)\,dp \neq 0$, *for all* k.

Then the functions $w(x,t)$, $\lambda(x,t)$:

$$w(x,t)=\sum_{k=1}^{\infty}\left[\frac{a_k e^{-\lambda_k b}\int\limits_t^b e^{\lambda_k(t-p)} f(p)\,dp + b_k e^{-\lambda_k a}\int\limits_a^t e^{\lambda_k(t-p)} f(p)\,dp}{\int\limits_a^b e^{-\lambda_k p} f(p)\,dp}\right] T_k(x). \quad (2.5)$$

$$\lambda(x)=\sum_{k=1}^{\infty}\left[\frac{b_k e^{-\lambda_k b} - a_k e^{-\lambda_k a}}{\int\limits_a^b e^{-\lambda_k p} f(p)\,dp}\right] T_k(x) \quad (2.6)$$

are the formal solution of the inverse problem (2.1), (2.2).

2.2 SUFFICIENT CONDITIONS FOR CORRECTNESS OF THE FORMULAS (2.3), (2.4)

Here we formulate and prove the following existence result:

Theorem 2.3. *Let the functions $w_a(x)$, $w_b(x)$, $x\in D\subset \mathrm{R}^n$ belong to the class of entire functions $\{\varphi(z),\quad z=x+\mathrm{i}\,y,\quad x\in\mathrm{R}^n,\quad y\in\mathrm{R}^n,\quad \mathrm{i}^2=-1\}$ such that*

$$|\varphi(x+\mathrm{i}\,y)|\le C_\varepsilon e^{(N+\varepsilon)|y|}(1+|x|)^m,\quad C_\varepsilon>0,\quad N>0,\quad m\ge 0,\quad \varepsilon>0.$$

and let the functions $B(\xi,t)$, $f(t)$ be infinitely differentiable and such that

$$\int\limits_a^b e^{-B(\xi,p)} f(p)\,dp\ne 0,\quad \text{for all } \xi: \text{such that}\quad |\xi|<N+\varepsilon_0,\quad \varepsilon_0>0.$$

Then the formulas (2.3), (2.4) uniquely determine solution $w(x,t)$, $\lambda(x)$ of the inverse problem (2.1), (2.2), where $w(x,t)$, $\lambda(x)$ are entire functions with respect to variable x.

PROOF. The functions $w_a(x)$, $w_b(x)$, $x\in D\subset\mathrm{R}^n$ are entire. Hence we can consider any $x\in\mathrm{R}^n$ as the argument of these functions. By the Paley-Wiener-Schwarz theorem, the functions $\hat{w}_a(\xi)$, $\hat{w}_b(\xi)$ are uniquely determined by formulas

$$\hat{w}_a(x)=\frac{1}{(2\pi)^n}\int\limits_{\mathrm{R}^n} w_a(\xi)e^{-\mathrm{i}x\xi}\,d\xi,\quad \hat{w}_b(x)=\frac{1}{(2\pi)^n}\int\limits_{\mathrm{R}^n} w_b(\xi)e^{-\mathrm{i}x\xi}\,d\xi$$

and these functions are distributions with compact supports which are contained in the set $\{\,\xi\in\mathrm{R}^n\ :\ |\xi_j|\le N,\ \ j=1,2,\dots,n\}$. Recalling that $\int\limits_a^b e^{-B(\xi,p)} f(p)\,dp\ne 0$,

$|\xi| < N + \varepsilon_0$, we can see that functions $w(x,t)$, $\lambda(x)$ in the formulas (2.3), (2.4) are correct, determined and entire for variable x.

We have

$$w|_{t=a} = \int_{\mathrm{R}^n} \hat{w}_a(\xi)\,\mathrm{e}^{\mathrm{i}x\xi}\,\mathrm{d}\xi = w_a(x),$$

$$w|_{t=b} = \int_{\mathrm{R}^n} \hat{w}_b(\xi)\,\mathrm{e}^{\mathrm{i}x\xi}\,\mathrm{d}\xi = w_b(x).$$

Now we should prove that functions $w(x,t)$, $\lambda(x)$ satisfy the equation

$$\frac{\partial w}{\partial t} = \mathbf{A}w + \lambda(x) f(t).$$

It follows from (2.3) that

$$\begin{aligned} w(x,t) \;=\; & \int_{\mathrm{R}^n} \left[\frac{\tilde{A}(\xi,t)\hat{w}_a(\xi)\,\mathrm{e}^{B(\xi,t)-B(\xi,a)} \int_t^b \mathrm{e}^{-B(\xi,p)} f(p)\,\mathrm{d}p}{\int_a^b \mathrm{e}^{-B(\xi,p)} f(p)\,\mathrm{d}p} \right. \\ & \left. + \frac{\hat{w}_b(\xi)\,\mathrm{e}^{B(\xi,t)-B(\xi,b)} \int_a^t \mathrm{e}^{-B(\xi,p)} f(p)\,\mathrm{d}p}{\int_a^b \mathrm{e}^{-B(\xi,p)} f(p)\,\mathrm{d}p} \right] \mathrm{e}^{\mathrm{i}x\xi}\,\mathrm{d}\xi \\ & + f(t) \int_{\mathrm{R}^n} \left[\frac{\hat{w}_b(\xi)\,\mathrm{e}^{-B(\xi,b)} - \hat{w}_a(\xi)\,\mathrm{e}^{-B(\xi,a)}}{\int_a^b \mathrm{e}^{-B(\xi,p)} f(p)\,\mathrm{d}p} \right] \mathrm{e}^{\mathrm{i}x\xi}\,\mathrm{d}\xi. \end{aligned}$$

It is clear that the right part of this relation is equal to: $\mathbf{A}(t)w + \lambda(x)f(t)$. Thus, Theorem 2.3 is proved.

Theorem 2.4. *Let $B(\xi,t) = A_0(\xi)t$ and $A_0(\xi)t$ be real infinitely differentiable function such that $|A_0(\xi)| \le (A_1|\xi|^m + B_1)$, $A_1 > 0$, $B_1 > 0$, $m > 0$, $D = \mathrm{R}^n$ and the functions $w_a(x)$, $w_b(x)$ in the inverse problem (2.1), (2.2) be fast decreasing functions and belong to C^∞. Let the function $f(t)$ be continuous and strictly positive for $t \in [a,b]$.*

Then the formulas (2.3), (2.4) are correct and they give the solution of the inverse problem (2.1), (2.2).

Proof. Let $f_0 = \min_{t\in[a,b]} f(t)$, $f_1 = \max_{t\in[a,b]} f(t)$. We have inequalities $0 < f_0 \le f(t) \le f_1$.

At first, if points $\xi \in \mathrm{R}^n$ satisfy $A_0(\xi) = 0$, we have such points

$$\int_a^b \mathrm{e}^{-A_0(\xi)} f(p)\,\mathrm{d}p = \int_a^b f(p)\,\mathrm{d}p \neq 0,$$

and

$$\left| \frac{\hat{w}_a(\xi)\,\mathrm{e}^{A_0(\xi)(t-a)} \int_t^b \mathrm{e}^{-A_0(\xi)p} f(p)\,\mathrm{d}p + \hat{w}_b(\xi)\,\mathrm{e}^{A_0(\xi)(t-b)} \int_a^t \mathrm{e}^{-A_0(\xi)p} f(p)\,\mathrm{d}p}{\int_a^b \mathrm{e}^{-A_0(\xi)} f(p)\,\mathrm{d}p} \right|$$

$$= \left| \frac{\hat{w}_a(\xi) \int_t^b f(p)\,\mathrm{d}p + \hat{w}_b(\xi) \int_a^t f(p)\,\mathrm{d}p}{\int_a^b f(p)\,\mathrm{d}p} \right| \leq |\hat{w}_a(\xi)| + |\hat{w}_b(\xi)|.$$

We have also

$$\left| \frac{\hat{w}_a(\xi)\,\mathrm{e}^{-A_0(\xi)b} - \hat{w}_b(\xi)\,\mathrm{e}^{-A_0(\xi)a}}{\int_a^b \mathrm{e}^{-A_0(\xi)} f(p)\,\mathrm{d}p} \right| = \left| \frac{\hat{w}_a(\xi) - \hat{w}_b(\xi)}{\int_a^b f(p)\,\mathrm{d}p} \right| \leq C\left(|\hat{w}_a(\xi)| + |\hat{w}_b(\xi)|\right),$$

$$C = \left[\int_a^b f(p)\,\mathrm{d}p \right]^{-1}.$$

After that we must estimate

$$I_1 = \frac{\hat{w}_a(\xi)\,\mathrm{e}^{-A_0(\xi)a} \int_t^b \mathrm{e}^{A_0(\xi)(t-p)} f(p)\,\mathrm{d}p + \hat{w}_b(\xi)\,\mathrm{e}^{-A_0(\xi)b} \int_a^t \mathrm{e}^{A_0(\xi)(t-p)} f(p)\,\mathrm{d}p}{\int_a^b \mathrm{e}^{-A_0(\xi)p} f(p)\,\mathrm{d}p}$$

for $\xi \in \mathrm{R}^n$, $A_0(\xi) \neq 0$.

At first, we note that

$$\int_a^b \mathrm{e}^{-A_0(\xi)p} f(p)\,\mathrm{d}p \geq f_0 \frac{\mathrm{e}^{-A_0(\xi)b} - \mathrm{e}^{-A_0(\xi)a}}{A_0(\xi)}.$$

Using this inequality we have

$$|I_1| \le \frac{f_1}{f_0}\left\{\left|\frac{|\hat{w}_a|\,\mathrm{e}^{-A_0(\xi)a}\,\mathrm{e}^{A_0(\xi)t}\left(\mathrm{e}^{-A_0(\xi)b}-\mathrm{e}^{-A_0(\xi)t}\right)}{\mathrm{e}^{-A_0(\xi)b}-\mathrm{e}^{-A_0(\xi)a}}\right| + \left|\frac{|\hat{w}_b|\,\mathrm{e}^{-A_0(\xi)b}\,\mathrm{e}^{A_0(\xi)t}\left(\mathrm{e}^{-A_0(\xi)t}-\mathrm{e}^{-A_0(\xi)a}\right)}{\mathrm{e}^{-A_0(\xi)b}-\mathrm{e}^{-A_0(\xi)a}}\right|\right\}$$

Hence

$$\begin{aligned}|I_1| &\le \frac{f_1}{f_0}|\hat{w}_a|\left|\frac{\mathrm{e}^{A_0(\xi)(t-a)}-\mathrm{e}^{-A_0(\xi)(b-a)}}{1-\mathrm{e}^{A_0(\xi)(b-a)}}\right| + \frac{f_1}{f_0}|\hat{w}_b|\left|\frac{\mathrm{e}^{A_0(\xi)(a-b)}-\mathrm{e}^{A_0(\xi)(t-b)}}{1-\mathrm{e}^{A_0(\xi)(b-a)}}\right| \\ &\le C(|\hat{w}_a(\xi)|+|\hat{w}_b(\xi)|),\end{aligned}$$

where C is some constant.

Now we shall estimate

$$I_2(\xi) = \frac{\hat{w}_b\,\mathrm{e}^{-A_0(\xi)b} - \hat{w}_a\,\mathrm{e}^{-A_0(\xi)a}}{\int_a^b \mathrm{e}^{-A_0(\xi)p} f(p)\,\mathrm{d}p} \qquad \text{for } A_0(\xi) \ne 0.$$

We have

$$\begin{aligned}|I_2(\xi)| &\le \frac{|A_0(\xi)|\left(|\hat{w}_a|\,\mathrm{e}^{-A_0(\xi)a} + |\hat{w}_b|\,\mathrm{e}^{-A_0(\xi)b}\right)}{f_0(\mathrm{e}^{-A_0(\xi)b}-\mathrm{e}^{-A_0(\xi)a})} \\ &= \frac{|A_0(\xi)|\,|\hat{w}_a|}{|(1-\mathrm{e}^{-A_0(\xi)(b-a)})|} + \frac{|A_0(\xi)|\,|\hat{w}_b|}{|(1-\mathrm{e}^{-A_0(\xi)(b-a)})|} \\ &\le C_1(|A_0(\xi)|+\tilde{A}_1)\,|\hat{w}_a| + C_2(|A_0(\xi)|+\tilde{A}_2)\,|\hat{w}_b|\end{aligned}$$

where C_1, $\tilde{A}_1$, $\tilde{A}_2$ are some constants. According to the conditions of Theorem 2.4, the functions $w_a(x)$, $w_b(x)$ are infinitely differentiable and fast decreasing. Together with $w_a(x)$ and $w_b(x)$, the functions $\hat{w}_a(\xi)$, $\hat{w}_b(\xi)$ are fast decreasing too. Furthermore $|A_0(\xi)| \le (A_1|\xi|^m + B_1)$. Hence, the functions $I_1(\xi)$, $I_2(\xi)$, $\xi \in \mathrm{R}^n$ are fast decreasing and the formulas (2.3), (2.4) are correct. Theorem 2.4 is proved.

2.3 THE FORMULAS FOR SOLUTION OF THE INVERSE PROBLEM IN BOUNDED DOMAIN

Here we consider the inverse problem (2.1), (2.2) in a bounded domain $D \subset \mathrm{R}^n$ with smooth boundary ∂D and we shall prove well-posedness of the formulas (2.5), (2.6).

We consider the inverse problem: to find two functions $w(x,t)$, $\lambda(x)$, $x \in D$, $a \le t \le b$ if the following conditions hold

$$\frac{\partial w}{\partial t} = \mathbf{A}w + \lambda(x)f(t), \quad \Gamma w|_{\partial D} = 0, \quad w|_{t=a} = w_a(x), \quad w|_{t=b} = w_b(x). \tag{2.7}$$

Here, $\mathbf{A}$ is a linear operator of the second order with continuous coefficients such that

$$\mathbf{A} = \sum_{i,j=1}^{n} \frac{\partial}{\partial x_i}\left(a_{ij}(x)\frac{\partial}{\partial x_j}\right) + a(x),$$

$$\mu^{-1}|\xi|^2 \le \sum_{i,j=1}^{n} a_{ij}(x)\xi_i\xi_j \le \mu|\xi|^2, \quad \mu > 0, \quad \xi \in \mathrm{R}^n, \quad x \in D.$$

The boundary operator Γ has the form

$$\Gamma = \alpha_1\left(\sum_{i,j=1}^{n} a_{ij}(x)\frac{\partial}{\partial x_j}\cos(\mathbf{n}, x_i) + \sigma(x)\right) + \alpha_0,$$

$\alpha_1 \ge 0$, $\alpha_1^2 + \alpha_0^2 > 0$, $\mathbf{n}$ is normal to ∂D. It is well known that the spectrum of problem

$$\mathbf{A}T = \lambda T, \quad x \in D, \quad \Gamma T = 0, \quad x \in \partial D \tag{2.8}$$

is discrete (Ladyzhenskaya *et al.*, 1968).

Denote by $\lambda_1 > \lambda_2 \ge \lambda_3 \ge \ldots$ the eigenvalues of the spectral problem (2.8). Let $T_1(x), T_2(x), \ldots, T_k(x), \ldots$ be a system of eigenfunctions of the problem (2.8). It is well known (Ladyzhenskaya *et al.*, 1968) that if $\psi(x) \in W_2^m(D)$, then

$$\psi(x) = \sum_{k=1}^{\infty} a_k T_k(x), \quad a_k = \int_D \psi(x)T_k(x)\,\mathrm{d}x$$

and the following estimates hold

$$\begin{gathered}\left\|\sum_{k=1}^{\infty} a_k T_k(x)\right\|_{W_2^m} \le C_m \sum_{k=1}^{\infty} a_k^2(1 + |\lambda_k|^m)^{\frac{1}{2}}, \\ |a_k| \le C_m'\alpha_k|\lambda_k|^{-\frac{m}{2}}, \quad \left(\sum_{k=1}^{\infty} a_k^2\right)^{\frac{1}{2}} \le \|\psi\|_{W_2^m} \le C_m'',\end{gathered} \tag{2.9}$$

for some constants C_m, C_m', C_m''.

Theorem 2.5. *Suppose that* $\lambda_1 < 0$ *and* $f(t)$ *is continuous,* $t \in [a, b]$ *and* $w_a(x) \in W_2^m(D)$, $w_a(x) = \sum_{k=1}^{\infty} a_k T_k(x)$, $w_b(x) \in W_2^m(D)$, $w_b(x) = \sum_{k=1}^{\infty} b_k T_k(x)$, $m \ge 2$. *Then the functions* $w(x,t)$, $\lambda(x)$

$$w(x,t) = \sum_{k=1}^{\infty} T_k(x)\left(\frac{a_k\,\mathrm{e}^{-\lambda_k a}\int_t^b \mathrm{e}^{\lambda_k(t-\tau)} f(\tau)\,\mathrm{d}\tau + b_k\,\mathrm{e}^{-\lambda_k b}\int_a^t \mathrm{e}^{\lambda_k(t-\tau)} f(\tau)\,\mathrm{d}\tau}{\int_a^b \mathrm{e}^{-\lambda_k\tau} f(\tau)\,\mathrm{d}\tau}\right), \tag{2.10}$$

$$\lambda(x) = \sum_{k=1}^{\infty} T_k(x)\left[\frac{b_k\,\mathrm{e}^{-\lambda_k b} - a_k\,\mathrm{e}^{-\lambda_k a}}{\int_a^b \mathrm{e}^{-\lambda_k\tau} f(\tau)\,\mathrm{d}\tau}\right] \tag{2.11}$$

are solution of the inverse problem (2.7).

PROOF. We consider a term

$$w_k(x,t) = T_k(x)\frac{a_k\,\mathrm{e}^{-\lambda_k a}\int\limits_t^b \mathrm{e}^{\lambda_k(t-\tau)} f(\tau)\,\mathrm{d}\tau + b_k\,\mathrm{e}^{-\lambda_k b}\int\limits_a^t \mathrm{e}^{\lambda_k(t-\tau)} f(\tau)\,\mathrm{d}\tau}{\int\limits_a^b \mathrm{e}^{-\lambda_k\tau} f(\tau)\,\mathrm{d}\tau}.$$

If we put $w(x,t)$ in (2.7), we have the relations

$$\frac{\partial w_k}{\partial t} = \mathbf{A}w_k + \frac{b_k\,\mathrm{e}^{-\lambda_k b} - a_k\,\mathrm{e}^{-\lambda_k a}}{\int\limits_a^b \mathrm{e}^{-\lambda_k\tau} f(\tau)\,\mathrm{d}\tau} f(t), \quad x \in D, \quad a \le t \le b,$$

$$\Gamma w_k = 0, \quad x \in \partial D, \quad w_k\big|_{t=a} = a_k T_k, \quad w_k\big|_{t=b} = b_k T_k(x).$$

Denote by $\lambda_k(x)$ function

$$\lambda_k(x) = \frac{b_k\,\mathrm{e}^{-\lambda_k b} - a_k\,\mathrm{e}^{-\lambda_k a}}{\int\limits_a^b \mathrm{e}^{-\lambda_k\tau} f(\tau)\,\mathrm{d}\tau} T_k(x).$$

The formal solution of (2.7) is

$$w = \sum_{k=1}^{\infty} w_k(x), \quad \lambda = \sum_{k=1}^{\infty} \lambda_k(x), \tag{2.12}$$

and we must prove that series (2.12) are convergent.

At first, we consider $\sum\limits_{k=1}^{\infty} w_k(x)$. The following inequality holds:

$$\frac{\mathrm{e}^{-\lambda_k a}\int\limits_t^b \mathrm{e}^{\lambda_k(t-\tau)} f(\tau)\,\mathrm{d}\tau}{\int\limits_a^b \mathrm{e}^{-\lambda_k\tau} f(\tau)\,\mathrm{d}\tau} = \frac{\mathrm{e}^{\lambda_k(t-a)}\int\limits_t^b \mathrm{e}^{-\lambda_k\tau} f(\tau)\,\mathrm{d}\tau}{\int\limits_a^b \mathrm{e}^{-\lambda_k\tau} f(\tau)\,\mathrm{d}\tau} \le 1$$

because $\lambda_k < 0$. On the other hand we can write

$$\frac{\mathrm{e}^{-\lambda_k b}\int\limits_a^t \mathrm{e}^{\lambda_k(t-\tau)} f(\tau)\,\mathrm{d}\tau}{\int\limits_a^b \mathrm{e}^{-\lambda_k\tau} f(\tau)\,\mathrm{d}\tau} = \frac{\mathrm{e}^{\lambda_k(t-b)}\int\limits_a^t \mathrm{e}^{-\lambda_k\tau} f(\tau)\,\mathrm{d}\tau}{\int\limits_a^b \mathrm{e}^{-\lambda_k\tau} f(\tau)\,\mathrm{d}\tau} \le \mathrm{e}^{\lambda_k(t-b)}\frac{f_1(\mathrm{e}^{\lambda_k t} - \mathrm{e}^{\lambda_k a})\dfrac{1}{\lambda_k}}{f_0(\mathrm{e}^{\lambda_k b} - \mathrm{e}^{\lambda_k a})\dfrac{1}{\lambda_k}}$$

$$\le \frac{f_1}{f_0}\frac{1}{1 - \mathrm{e}^{\lambda_k(b-a)}} \le \frac{f_1}{f_0}\frac{1}{1 - \mathrm{e}^{\lambda_1(b-a)}} = \hat{k}$$

Hence, $|w_k(x,t)| \le |a_k| + \hat{k}\,|b_k|$, $\hat{k} > 0$. From inequality (2.9) it follows

$$\begin{aligned}\left\|\sum_{k=1}^{\infty} w_k(x,t)\right\|_{W_2^m(D)} &\le C_m \left(\sum_{k=1}^{\infty} a_k^2\,(|\lambda_k|^m + 1)\right)^{\frac{1}{2}} + \hat{k}\left(\sum_{k=1}^{\infty} b_k^2\,(|\lambda_k|^m + 1)\right)^{\frac{1}{2}} \\ &\le \tilde{C}\left(\|w_a\|_{W_2^m(D)} + \|w_b\|_{W_2^m(D)}\right).\end{aligned}$$

Finally, we should show that series $\sum_{k=1}^{\infty} \lambda_k T_k(x)$ is convergent too. We have

$$\begin{aligned}\left|\frac{b_k\,\mathrm{e}^{-\lambda_k b} - a_k\,\mathrm{e}^{-\lambda_k a}}{\int_a^b \mathrm{e}^{-\lambda_k \tau} f(\tau)\,\mathrm{d}\tau}\right| &\le \frac{|b_k|\,\mathrm{e}^{-\lambda_k b} + |a_k|\,\mathrm{e}^{-\lambda_k a}}{-\frac{1}{\lambda_k}(\mathrm{e}^{-\lambda_k b} - \mathrm{e}^{-\lambda_k a}) f_0} \\ &\le \frac{|\lambda_k|\,|b_k| + |\lambda_k|\,|a_k|\,\mathrm{e}^{\lambda_k(b-a)}}{(1 - \mathrm{e}^{\lambda_k(b-a)}) f_0} \le \hat{C}\,|\lambda_k|\,(|b_k| + |a_k|).\end{aligned}$$

Hence,

$$\begin{aligned}\|\lambda(x)\|_{W_2^{m-2}(D)} &\le C_m \hat{C}\left(\left(\sum_{k=1}^{\infty} a_k^2(|\lambda_k|^m + 1)\right)^{\frac{1}{2}} + \left(\sum_{k=1}^{\infty} b_k^2(|\lambda_k|^m + 1)\right)^{\frac{1}{2}}\right) \\ &\le \tilde{\tilde{C}}\left(\|w_a\|_{W_2^m(D)} + \|w_b\|_{W_2^m(D)}\right).\end{aligned}$$

Theorem 2.5 is proved.

2.4 FORMULAS IN THE INVERSE PROBLEM FOR A GENERAL EVOLUTION EQUATION OF THE SECOND ORDER

We consider the following multidimensional inverse problem for an evolution equation: to find complex-valued functions

$$w(x,t), \quad \lambda(x), \quad \mu(x), \quad \frac{\partial \lambda}{\partial t} = 0, \quad \frac{\partial \mu}{\partial t} = 0, \quad x \in \mathrm{R}^n, \quad a \le t \le b$$

such that

$$\frac{\partial^2 w}{\partial t^2} + \mathbf{A}w = \lambda(x) f(t) + \mu(x)\varphi(t), \tag{2.13}$$

$$w|_{t=a} = w_a(x), \quad \left.\frac{\partial w}{\partial t}\right|_{t=a} = w_a'(x). \quad w|_{t=b} = w_b(x), \quad \left.\frac{\partial w}{\partial t}\right|_{t=b} = w_b'(x). \tag{2.14}$$

Here, $\mathbf{A}$ is linear operator acting in the variable $x = (x_1, x_2, \ldots, x_n)$ of real Euclidean space R^n, whose symbol is $\tilde{A}(\xi)$, $\xi \in \mathrm{R}^n$:

$$\tilde{A}(\xi) \in C^\infty(\mathrm{R}^n). \quad \mathbf{A}\,\mathrm{e}^{\mathrm{i}x\xi} = \mathrm{e}^{\mathrm{i}x\xi}\,\tilde{A}(\xi).$$

Differentiable functions $f(t)$, $\varphi(t)$, $w_a(x)$, $w'_a(x)$, $w_b(x)$, $w'_b(x)$, $x \in \mathrm{R}^n$ and constants a, b are assumed to be known. The inverse problem (2.13), (2.14) can be interpreted as a problem of control when transferring the state $(w_a(x), w'_a(x))$ into the state $(w_b(x), w'_b(x))$; here the functions $\lambda(x)$, $\mu(x)$, $x \in \mathrm{R}^n$ are the control.

Let us introduce the following notations

$$\Delta_r(\xi) = \begin{vmatrix} \int\limits_a^b \frac{1}{\sqrt{\tilde{A}(\xi)}} \sin(\sqrt{\tilde{A}(\xi)}(r-p)) f(p)\,\mathrm{d}p & \int\limits_a^b \cos(\sqrt{\tilde{A}(\xi)}(r-p)) f(p)\,\mathrm{d}p \\ \int\limits_a^b \frac{1}{\sqrt{\tilde{A}(\xi)}} \sin(\sqrt{\tilde{A}(\xi)}(r-p)) \varphi(p)\,\mathrm{d}p & \int\limits_a^b \cos(\sqrt{\tilde{A}(\xi)}(r-p)) \varphi(p)\,\mathrm{d}p \end{vmatrix} \tag{2.15}$$

$$\begin{aligned} P_r^\alpha(\xi) &= \hat{w}_r(\xi) \int\limits_a^b \cos(\sqrt{\tilde{A}(\xi)}(r-p))\alpha(p)\,\mathrm{d}p \\ &\quad - \hat{w}'_r(\xi) \int\limits_a^b \frac{\sin(\sqrt{\tilde{A}(\xi)}(r-p))}{\sqrt{\tilde{A}(\xi)}} \alpha(p)\,\mathrm{d}p. \end{aligned} \tag{2.16}$$

Here $|\cdot|$ denotes the determinant. The number r takes only two values, either a or b; the function $\alpha(p)$ can be equal to either $f(p)$ or $\varphi(p)$, $a \le p \le b$; the functions $\hat{w}_r(\xi)$, $\hat{w}'_r(\xi)$ are the Fourier images of $w_a(x)$, $w'_a(x)$, $w_b(x)$, $w'_b(x)$ at $r = a$ and $r = b$ respectively:

$$w_r(x) = \int\limits_{\mathrm{R}^n} \hat{w}_r(\xi)\, \mathrm{e}^{\mathrm{i}x\xi}\,\mathrm{d}\xi, \quad w'_r(x) = \int\limits_{\mathrm{R}^n} \hat{w}'_r(\xi)\, \mathrm{e}^{\mathrm{i}x\xi}\,\mathrm{d}\xi.$$

Obviously, we assume that the functions appearing in (2.15), (2.16) are such that $\Delta_r(\xi)$ and $P_r^\alpha(\xi)$ are defined correctly.

Analogous problems have been considered in Anikonov(1993a, 1994a); see also references therein.

Theorem 2.6. *Let the functions $w_a(x)$, $w'_r(x)$, $w_b(x)$, $w'_r(x)$ in (2.14) belong to the class of entire functions $w(x)$, satisfying the estimate*

$$|w(x+iy)| \le c_\varepsilon\, \mathrm{e}^{(N+\varepsilon)|y|}(1+|x|^m), \quad c_\varepsilon > 0, \quad N > 0, \quad m \ge 0, \quad \forall \varepsilon > 0, \quad \mathrm{i}^2 = -1$$

anf let the following relations hold

$$\Delta(\xi) = \Delta_a(\xi) = \Delta_b(\xi) \ne 0, \quad \forall \xi \in \mathrm{R}^n; \quad |\xi| \le N + \varepsilon_0, \quad \varepsilon_0 > 0.$$

Then there exists the solution $w(x,t)$, $\lambda(x)$, $\mu(x)$ of the inverse problem (2.13), (2.14), which is entire on x and is defined by the formulas

$$w(x,t) = \int\limits_{\mathrm{R}^n} \frac{1}{\Delta(\xi)} \left\{ \begin{vmatrix} \int\limits_t^b \frac{\sin(\sqrt{\tilde{A}(\xi)}(t-p))}{\sqrt{\tilde{A}(\xi)}} f(p)\mathrm{d}p & \int\limits_t^b \frac{\sin(\sqrt{\tilde{A}(\xi)}(t-p))}{\sqrt{\tilde{A}(\xi)}} \varphi(p)\mathrm{d}p \\ P_a^f(\xi) & P_a^\varphi(\xi) \end{vmatrix} \right.$$

$$+\left|\begin{array}{cc} \int\limits_a^t \frac{\sin(\sqrt{\tilde{A}(\xi)}(t-p))}{\sqrt{\tilde{A}(\xi)}} f(p)\mathrm{d}p & \int\limits_a^t \frac{\sin(\sqrt{\tilde{A}(\xi)}(t-p))}{\sqrt{\tilde{A}(\xi)}} \varphi(p)\mathrm{d}p \\ P_b^f(\xi) & P_b^\varphi(\xi) \end{array}\right|\Bigg\} e^{ix\xi}\mathrm{d}\xi,$$

$$\lambda(x) = \int\limits_{\mathrm{R}^n} \left[\frac{P_b^\varphi(\xi) - P_a^\varphi(\xi)}{\Delta(\xi)}\right] e^{ix\xi}\,\mathrm{d}\xi,$$

$$\mu(x) = \int\limits_{\mathrm{R}^n} \left[\frac{P_a^f(\xi) - P_b^f(\xi)}{\Delta(\xi)}\right] e^{ix\xi}\,\mathrm{d}\xi.$$

Theorem 2.6 is proved by direct verification with the help of the well-known Paley-Wiener-Schwarz theorem on analytical functions, the Fourier transform of which has a finite support.

REMARK 2.2. Theorem 2.6 guarantees only sufficient conditions for correct applicability of the above formulas which were obtained for the solution to the inverse problem (2.13), (2.14). Obviously, these formulas can give the solution in a more general situation as well.

Suppose now that $x \in D \subset \mathrm{R}^n$, where D is a domain with a smooth boundary, and that the operator $\mathbf{A}$ in (2.13) can depend on x, for example

$$\mathbf{A} = \sum_{i,j=1}^{n} a_{ij}(x)\frac{\partial^2}{\partial x_i \partial x_j} + \sum_{j=1}^{n} b_j(x)\frac{\partial}{\partial x_j}.$$

Let $T_k(x)$ and λ_k be the eigenfunctions and the eigenvalues of the operator $\mathbf{A}$, which are subject to some boundary condition, for example,

$$\mathbf{A}T_k(x) + \lambda_k T_k(x) = 0, \quad x \in D, \quad T_k(x)|_{\partial D} = 0.$$

It turns out that, in this case, when the operator $\mathbf{A}$ depends on $x \in D \subset \mathrm{R}^n$, there is an analog of the formulas of Theorem 2.6 for the inverse problem (2.13), (2.14).

Let us suppose that the data (2.14) of the inverse problem (2.13), (2.14) are represented as

$$w_a(x) = \sum_{k=1}^{\infty} a_k T_k(x), \quad w_a'(x) = \sum_{k=1}^{\infty} a_k' T_k(x),$$

$$w_b(x) = \sum_{k=1}^{\infty} b_k T_k(x), \quad w_b'(x) = \sum_{k=1}^{\infty} b_k' T_k(x),$$

where a_k, a_k', b_k, b_k' are constants.

Let us introduce the sequences Δ_{rk} and P_{rk}^{α}, substituting in (2.15), (2.16) the eigenvalues λ_k of the operator $\mathbf{A}$ for the symbol $\tilde{A}(\xi)$, and substituting constants a_k, a_k' when $r = a$ and constants b_k, b_k' when $r = b$ for the functions $\hat{w}_r(\xi)$, $\hat{w}_r'(\xi)$.

Thus we arrive at the relations

$$\Delta_{rk} = \begin{vmatrix} \int\limits_a^b \frac{\sin(\sqrt{\lambda_k}(r-p))}{\sqrt{\lambda_k}} f(p)\,dp & \int\limits_a^b \cos(\sqrt{\lambda_k}(r-p)) f(p)\,dp \\ \int\limits_a^b \frac{\sin(\sqrt{\lambda_k}(r-p))}{\sqrt{\lambda_k}} \varphi(p)\,dp & \int\limits_a^b \cos(\sqrt{\lambda_k}(r-p)) \varphi(p)\,dp \end{vmatrix},$$

$$P_{rk}^{\alpha} = r_k \int\limits_a^b \cos(\sqrt{\lambda_k}(r-p))\alpha(p)\,dp - r_k' \int\limits_a^b \frac{\sin(\sqrt{\lambda_k}(r-p))\alpha(p)}{\sqrt{\lambda_k}}\,dp, \quad r = a, b.$$

Theorem 2.7. *Let $\Delta_k = \Delta_{ak} = \Delta_{bk} \neq 0$, $k = 1, 2, \dots$ Then the following formulas hold for the formal solution $w(x,t)$, $\lambda(x)$, $\mu(x)$ to the inverse problem (2.13), (2.14):*

$$w(x,t) = \sum_{k=1}^{\infty} \left\{ \begin{vmatrix} \int\limits_t^b \frac{\sin(\sqrt{\lambda_k}(r-p))}{\sqrt{\lambda_k}} f(p)\,dp & \int\limits_t^b \frac{\sin(\sqrt{\lambda_k}(r-p))}{\sqrt{\lambda_k}} \varphi(p)\,dp \\ P_{ak}^f & P_{ak}^{\varphi} \end{vmatrix} + \begin{vmatrix} \int\limits_a^t \frac{\sin(\sqrt{\lambda_k}(r-p))}{\sqrt{\lambda_k}} f(p)\,dp & \int\limits_a^t \frac{\sin(\sqrt{\lambda_k}(r-p))}{\sqrt{\lambda_k}} \varphi(p)\,dp \\ P_{bk}^f & P_{bk}^{\varphi} \end{vmatrix} \right\} \frac{T_k(x)}{\Delta_k}$$

$$\lambda(x) = \sum_{k=1}^{\infty} \left\{ \frac{P_{bk}^{\varphi} - P_{ak}^{\varphi}}{\Delta_k} \right\} T_k(x), \quad \mu(x) = \sum_{k=1}^{\infty} \left\{ \frac{P_{ak}^{f} - P_{bk}^{f}}{\Delta_k} \right\} T_k(x).$$

REMARK 2.3. If $f(t) = \delta(t-t_0)$, $\varphi(t) = \delta(t-t_1)$, where $\delta(t)$ is the Dirac delta-function, then

$$\Delta_k = \Delta_{ak} = \Delta_{bk} = \frac{1}{\sqrt{\lambda_k}} \sin\left[\sqrt{\lambda_k}(t_1 - t_0)\right].$$

Therefore, in this case, the inequality $\sqrt{\lambda_k}(t_1 - t_0) \neq m\pi$, where m is an integer, is the condition of controllability. In particular, when $\sqrt{\lambda_k} = k$, the number $\frac{1}{\pi}(t_1 - t_0)$ should not be rational.

2.5 REDUCTION OF MULTIDIMENSIONAL INVERSE PROBLEMS TO INITIAL-BOUNDARY VALUE PROBLEMS IN HILBERT SPACES

These are the problems of simultaneous determination of solutions and coefficients of the equations under consideration from some additional information about these solutions that are usually called inverse problems for the differential or other equations. The main mathematical problems of this subject relate to the questions about well-posedness of the problems, in particular, to the questions about solvability.

We consider a nonlinear multidimensional inverse problem for an evolution equation of a general type. The thrust of the problem is as follows: find complex-valued functions $w(x,z,t)$, $x \in \mathrm{R}^n$, $z \in \mathrm{R}^m$, $t \geq 0$, and $\lambda(x,t)$, $x \in \mathrm{R}^n$, $t \geq 0$, with $\dfrac{\partial \lambda}{\partial z_j} = 0$, $j = 1,\dots,m$, such that

$$\frac{\partial w}{\partial t} = \mathbf{A}(x,t)w + \mathbf{L}(z,t)w + \lambda(x,t)\mathbf{B}(x,t)w, \quad w|_{z=0} = Q(x,t); \quad w|_{t=0} = w_0(x,z),$$

where $\mathbf{A}(x,t)$ and $\mathbf{B}(x,t)$ are operators linear in x and not depending on z and $\mathbf{L}(z,t)$ is an operator linear in z and not depending on x.

Statement and investigation of such inverse problems for parabolic equations can be found in Tikhonov and Arsenin (1977), Bubnov (1985, 1986), Anikonov (1990a), Isakov (1991), Nakagiry and Yamamoto (1991), Prilepko and Kostin (1992), Vishnevskii (1992) and the bibliography therein.

Here, we will prove a theorem on reduction of the considered inverse problem to two problems with initial data for evolution equations in a Hilbert space, one of the two being nonlinear. In other words, we shall prove that a solution of the inverse problem can be constructed if we know a solution of two evolution problems, one of which is linear and the other nonlinear. Moreover, the initial data for the evolution problems are uniquely determined by the information that is given in the inverse problem. This fact constitutes an important constructive element in the new approach proposed here to the study of multidimensional inverse problems for evolution equations.

It turns out in particular that if the dimension of the Hilbert space under consideration equals 1, which corresponds to factorization of the initial state $w_0(x,z)$ (in other words, $w_0(x,z) = c(x)b(z)$, where $c(x)$ and $b(z)$ are some functions; moreover, $c(x)$ depends only on x, and $b(z)$ only on z), then we can obtain explicit formulas for a solution of the inverse problem. If the Hilbert space in which we consider the evolution equation is separable, then a solution of the inverse problem can be written as a convergent function series. In this case, the reduction method proposed can be treated as a generalization of the method of separation of variables and so it can be applied to solving nonlinear multidimensional problems for general evolution equations.

We point out that the idea of reducing an inverse problem to an evolution problem in a Hilbert space with a specially-chosen basis was used earlier in the articles by Bubnov (1985, 1986).

In 2.5.1 we state the problem together with some examples of the inverse problems of the considered type. We prove a theorem on reduction of the inverse problem to two Cauchy problems for evolution equations in a Hilbert space. Moreover, as was already mentioned, one of the problems is linear and the other is nonlinear. At the end of the section, we provide examples in which an explicit formula for a solution of the inverse problem can be written.

In 2.5.2 we consider the case in which $\mathbf{A}(x,t)$ and $\mathbf{L}(z,t)$ are elliptic operators in x and z, and $\mathbf{B}(x,t)$ is a first-order operator in x. In this case, solvability of the linear Cauchy problem is well known. Here, we also demonstrate that the nonlinear evolution Cauchy problem has a unique local solution. It should be noted that the character of nonlinearity in the nonlinear evolution problem is such that global solvability (without additional assumptions on the character of nonlinearity) can fail.

In 2.5.3 we prove a theorem on solvability of an abstract nonlinear evolution problem in a separable Hilbert space.

2.5.1 Statement of the Problem. A Reduction Theorem

Let R^n and R^m denote the real Euclidean spaces of variables $x = (x_1, \dots, x_n) \in R^n$ and $z = (z_1, \dots, z_m) \in R^m$, $n \geq 1$, $m \geq 1$. Consider the evolution equation

$$\frac{\partial w}{\partial t} = \mathbf{A}(x,t)w + \mathbf{L}(z,t)w + \lambda(x,t)\mathbf{B}(x,t)w, \quad x \in R^n, \quad z \in R^m, \quad t \geq 0. \tag{2.17}$$

Here, $\mathbf{A}(x,t)$ and $\mathbf{B}(x,t)$ are linear differential operators in the variables x and not depending on z, and $\mathbf{L}(z,t)$ is a linear differential operator in the variable z and not depending on x. We suppose also that the complex-valued function $w(x,z,t)$ satisfies the following initial-boundary conditions:

$$w(x,z,0) = w_0(x,z), \quad x \in R^n, \quad z \in R^m, \tag{2.18}$$

$$w(x,0,t) = Q(x,t), \quad x \in R^n, \quad t \geq 0. \tag{2.19}$$

THE INVERSE PROBLEM: to find a pair of functions $w(x,z,t)$ and $\lambda(x,t)$ $(\partial\lambda/\partial z_j = 0$, $j = 1, \dots, m)$ satisfying equation (2.17) and initial-boundary conditions (2.18), (2.19).

Suppose that the functions $w_0(x,z)$ and $Q(x,t)$ satisfy the necessary agreement conditions

$$w_0(x,0) = Q(x,0). \tag{2.20}$$

Observe that for this inverse problem we can take condition (2.19) and the function $Q(x,t)$ as the additional information that allows us to determine the function $\lambda(x,t)$. We give some examples of the inverse evolution problems of the indicated type.

Example 2.1. Let

$$\mathbf{A}(x,t) = \sum_{i=1}^{n} \frac{\partial^2}{\partial x_i^2}, \quad \mathbf{L}(z,t) = \sum_{j=1}^{m} \frac{\partial^2}{\partial z_j^2}, \quad \mathbf{B}(x,t) = \mathbf{E},$$

where $\mathbf{E}$ is the identity operator. Then equation (2.17) takes the form

$$\frac{\partial w}{\partial t} = \Delta w + \lambda(x,t)w, \quad \Delta = \sum_{i=1}^{n} \frac{\partial^2}{\partial x_i^2} + \sum_{j=1}^{m} \frac{\partial^2}{\partial z_j^2}.$$

Example 2.2. Let

$$\mathbf{A}(x,t) = \mathrm{i}\sum_{i=1}^{n} \frac{\partial^2}{\partial x_i^2}, \quad \mathbf{L}(z,t) = \mathrm{i}\sum_{j=1}^{m} \frac{\partial}{\partial z_j^2}, \quad \mathbf{B}(x,t) = \mathbf{E} \cdot \mathrm{i},$$

with i the imaginary unity. In this case, equation (2.17) is written as

$$-\mathrm{i}\frac{\partial w}{\partial t} = \Delta w + \lambda(x,t)w.$$

Let H denote a Hilbert space over the field of complex numbers with a scalar product $(\cdot,\cdot)$. We shall also assume that $w(x,z,t)$ and $\lambda(x,t)$ are complex-valued functions.

Alongside inverse problem (2.17)–(2.19), consider the Cauchy problem

$$\frac{\partial P}{\partial t} = \mathbf{A}(x,t)P + p(x,t)\mathbf{B}(x,t)P, \quad x \in R^n, \quad t \geq 0,$$
$$P(x,0) = P_0(x). \tag{2.21}$$

Let $p(x,t)$ be a continuous complex-valued function and let the initial data $P_0(x)$ belong to the domain of definition of the operator $\mathbf{A}(x,0)$. Suppose that the Cauchy problem (2.21) has a unique solution for every continuous complex-valued function $p(x,t)$ and every function $P_0(x)$.

Theorem 2.8. *Assume $w_0(x,z) = \big(c(x), b(z)\big)$, where $c(x)$ and $b(z)$ are elements of a Hilbert space H, $x \in \mathrm{R}^n$, $z \in \mathrm{R}^m$, $t \geq 0$; and let $\mathbf{B}(x,t)Q(x,t) > 0$ for $x \in \mathrm{R}^n$ and $t \geq 0$. Suppose that in the Hilbert space H there exist solutions $\varphi(z,t)$ and $f(x,t)$ of the following Cauchy problems:*

$$\frac{\partial\varphi}{\partial t} = \mathbf{L}(z,t)\varphi, \quad z \in \mathrm{R}^m, \quad t \geq 0; \quad \varphi(z,0) = b(z), \quad z \in \mathrm{R}^m, \tag{2.22}$$

$$\begin{gathered}\frac{\partial f}{\partial t} = \mathbf{A}(x,t)f + \frac{Q_t - \mathbf{A}(x,t)Q - \big(f, \varphi_t(0,t)\big)}{\mathbf{B}(x,t)Q}\mathbf{B}(x,t)f, \\ x \in \mathrm{R}^n, \quad t \geq 0, \quad f(x,0) = c(x); \quad \varphi_t(0,t) = \left.\frac{\partial\varphi}{\partial t}\right|_{z=0}.\end{gathered} \tag{2.23}$$

Then the functions $w(x,z,t)$ and $\lambda(x,t)$ defined by the formulas

$$w(x,z,t) = \big(f(x,t), \varphi(z,t)\big), \tag{2.24}$$

$$\lambda(x,t) = \frac{Q_t - \mathbf{A}(x,t)Q - \big(f(x,t), \varphi_t(0,t)\big)}{\mathbf{B}(x,t)Q} \tag{2.25}$$

constitute a solution of inverse problem (2.17)–(2.19).

PROOF. Suppose that all conditions of Theorem 2.8 hold and let $\varphi(z,t)$ and $f(x,t)$ be solutions to problems (2.22), (2.23). We show that the functions $w(x,z,t)$ and $\lambda(x,t)$ defined by formulas (2.24), (2.25) satisfy equation (2.17) and conditions (2.18), (2.19).

Differentiating equality (2.24) with respect to the time variable, we obtain

$$\frac{\partial w}{\partial t}(x,z,t) = \left(\frac{\partial f}{\partial t}(x,t), \varphi(z,t)\right) + \left(f(x,t), \frac{\partial\varphi}{\partial t}(z,t)\right).$$

Apply the operator $\mathbf{A}(x,t)$ to both sides of equality (2.24):

$$\mathbf{A}(x,t)w(x,z,t) = \big(\mathbf{A}(x,t)f(x,t), \varphi(z,t)\big).$$

In the same way, we obtain

$$\mathbf{L}(z,t)w(x,z,t) = \big(f(x,t), \mathbf{L}(z,t)\varphi(z,t)\big),$$

$$\lambda(x,t)\mathbf{B}(x,t)w(x,z,t) = \big(\lambda(x,t)\mathbf{B}(x,t)f(x,t), \varphi(z,t)\big).$$

Hence

$$\begin{aligned}&\frac{\partial w}{\partial t} - \mathbf{A}(x,t)w - \mathbf{L}(z,t)w - \lambda(x,t)\mathbf{B}(x,t)w \\ &= \left(\frac{\partial f}{\partial t} - \mathbf{A}(x,t)f - \lambda(x,t)\mathbf{B}(x,t)f, \varphi\right) + \left(f, \frac{\partial\varphi}{\partial t} - \mathbf{L}(z,t)\varphi\right).\end{aligned}$$

Inserting the $\lambda(x,t)$ that is given by (2.25) into the previous equality and taking into account that $f(x,t)$ is a solution of problem (2.23) and $\varphi(z,t)$ is a solution of problem (2.22), we obtain

$$\frac{\partial w}{\partial t} - \mathbf{A}(x,t)w - \mathbf{L}(z,t)w - \lambda(x,t)\mathbf{B}(x,t)w = 0;$$

i. e., the functions $w(x,z,t)$ and $\lambda(x,t)$ satisfy equation (2.17). Obviously, the function $w(x,z,t)$ satisfies the initial condition (2.18):

$$w(x,z,0) = \big(f(x,0),\varphi(z,0)\big) = \big(c(x),b(z)\big) = w_0(x,z).$$

Now we prove that $w(x,z,t)$ satisfies relation (2.19). Towards the end of the proof, denote by $\tilde{Q}(x,t)$ the function $w(x,0,t) = \big(f(x,t),\varphi(0,t)\big)$. We have to prove that $Q(x,t) = \tilde{Q}(x,t)$, $x \in \mathrm{R}^n$, $t \geq 0$. From the agreement conditions (2.20) we have

$$\tilde{Q}(x,0) = w(x,0,0) = w_0(x,0) = Q(x,0).$$

Now we prove that the difference $\tilde{Q}(x,t) - Q(x,t)$ satisfies the evolution equation (2.21) with some function $p(x,t)$ and zero initial data. Then the uniqueness of a solution of the Cauchy problem (2.21) would imply that $\tilde{Q}(x,t) = Q(x,t)$ for $x \in \mathrm{R}^n$ and $t \geq 0$. Multiplying the equation (2.23) by function $\varphi(0,t)$ scalarly we have

$$\left(\frac{\partial f}{\partial f} - \mathbf{A}(x,t)f,\varphi(0,t)\right) = \big(\lambda(x,t)\mathbf{B}(x,t)f,\varphi(0,t)\big).$$

Insert the expression for the function $\lambda(x,t)$ given by (2.25) into this formula. Observe that

$$\frac{\partial}{\partial t}(f,\varphi(0,t)) = \left(\frac{\partial f}{\partial t},\varphi(0,t)\right) + (f,\varphi_t(0,t)),$$

$$\big(\mathbf{A}(x,t)f,\varphi(0,t)\big) = \mathbf{A}(x,t)\big(f,\varphi(0,t)\big),$$

$$\big(\lambda(x,t)\mathbf{B}(x,t)f,\varphi(0,t)\big) = \lambda(x,t)\mathbf{B}(x,t)\big(f,\varphi(0,t)\big).$$

By definition, $\tilde{Q}(x,t) = \big(f,\varphi(0,t)\big)$. Hence

$$\frac{\partial\tilde{Q}}{\partial t} = \mathbf{A}(x,t)\tilde{Q} - \big(f,\varphi_t(0,t)\big) + \lambda(x,t)\mathbf{B}(x,t)\tilde{Q}. \tag{2.26}$$

Insert into (2.26) the value of $\lambda(x,t)$ that is given by (2.25):

$$\frac{\partial\tilde{Q}}{\partial t} = \mathbf{A}(x,t)\tilde{Q} - \big(f,\varphi_t(0,t)\big) - \frac{\dfrac{\partial Q}{\partial t} - \mathbf{A}(x,t)Q - \big(f,\varphi_t(0,t)\big)}{\mathbf{B}(x,t)Q}\mathbf{B}(x,t)\tilde{Q}.$$

Rewrite the last equality as

$$\frac{\partial(\tilde{Q}-Q)}{\partial t} = \mathbf{A}(x,t)(\tilde{Q}-Q) - \frac{Q_t - \mathbf{A}(x,t)Q - \big(f,\varphi_t(0,t)\big)}{\mathbf{B}(x,t)Q}\mathbf{B}(x,t)(\tilde{Q}-Q).$$

Denote the function $\Big(Q_t - \mathbf{A}(x,t)Q - \big(f,\varphi_t(0,t)\big)\Big)/\mathbf{B}(x,t)Q$ by $p(x,t)$. We infer that $\tilde{Q} - Q$ is a solution of the Cauchy problem (2.21) with the initial data $(\tilde{Q}-Q)(x,0) = \tilde{Q}(x,0) - Q(x,0) = 0$. Therefore, $\tilde{Q}(x,t) \equiv Q(x,t)$. The theorem is proved.

Theorem 2.9. *Assume that the dimension of the space H equals 1. Then, under the conditions of Theorem 2.8, the following formulas hold:*

$$w(x,z,t) = f(x,t)\varphi(z,t), \quad \lambda(x,t) = \frac{Q_t - \mathbf{A}(x,t)Q - f(x,t)\varphi_t(0,t)}{\mathbf{B}(x,t)Q}. \tag{2.27}$$

PROOF. Since the dimension of the Hilbert space H equals unity, the scalar product of functions transforms into the conventional product of functions, and formula (2.27) ensues immediately from (2.24) and (2.25). Theorem 2.9 is proved.

We now consider examples in which the formulas derived in Theorem 2.9 allow us to find a solution of the inverse problem in explicit form.

Example 2.3. Let

$$\mathbf{L}(z,t) = \sum_{j=1}^{m} \frac{\partial^2}{\partial z_j^2}, \quad z \in \mathrm{R}^m, \quad t \geq 0,$$

let $H = 1$ and let function $b(z)$ be bounded, positive, and continuous. Suppose $Q(x,t)$ to be a sufficiently smooth function; moreover, $\mathbf{B}(x,t)Q(x,t) \neq 0$ for $x \in \mathrm{R}^n$ and $t \geq 0$. From condition (2.20) we infer that $c(x) = Q(x,0)/b(0)$. In this case, we obtain the following formulas for a solution of the inverse problem:

$$\lambda(x,t) = \frac{(Q_t - \mathbf{A}(x,t)Q)\varphi(0,t) - Q(x,t)\varphi_t(0,t)}{\varphi(0,t)\mathbf{B}(x,t)Q},$$

$$w(x,z,t) = \frac{Q(x,t)\displaystyle\int_{\mathrm{R}^m} b(\xi)\exp\left(-\frac{|\xi - z|^2}{4t}\right) d\xi}{\displaystyle\int_{\mathrm{R}^m} b(\xi)\exp\left(-\frac{|\xi|^2}{4t}\right) d\xi},$$

$$\varphi(0,t) = \frac{1}{(2\sqrt{\pi t})^m}\int_{\mathrm{R}^m} b(\xi)\exp\left(-\frac{|\xi|^2}{4t}\right) d\xi.$$

Example 2.4. Let $u(z)$ and $v(z)$ be harmonic functions, $z \in \mathrm{R}^m$. Also, suppose that $v(z) \neq 0$.

Consider a function $F(t,y)$, $t \geq 0$, $y \in \mathrm{R}$, satisfying the heat equation $F_t = F_{xx}$. Then the function $\varphi(t,z) = v(z)F(t, u(z)/v(z))$ satisfies the equation

$$\frac{\partial \varphi}{\partial t} = \frac{1}{|\operatorname{grad} u(z)/v(z)|^2}\Delta\varphi = \mathbf{L}(z,t)\varphi.$$

Let the operator $\mathbf{L}(z,t)$ of equation (2.17) have the following form

$$\mathbf{L} = \frac{1}{|\operatorname{grad} u(z)/v(z)|^2}\Delta_z, \quad \text{where} \quad \Delta_z = \sum_{j=1}^{m} \frac{\partial^2}{\partial z_j}.$$

Suppose that the dimension of the space H equals 1. In this case, the following formulas are valid for a solution of the inverse problem:

$$w(x,z,t) = \frac{Q(x,t)v(z)F(t,u(z)/v(z))}{\varphi_0(t)}, \quad \varphi_0(t) = F\left(t, \frac{u(0)}{v(0)}\right),$$

$$\lambda(x,t) = \frac{(Q_t - \mathbf{A}(x,t)Q)\varphi_0(t) - Q(x,t)\varphi_0'(t)}{\varphi_0(t)\mathbf{B}(x,t)Q},$$

$$w\big|_{z=0} = Q(x,t), \quad w\big|_{t=0} = w_0(x,z) = \frac{Q(x,0)F\left(0, \dfrac{u(z)}{v(z)}\right)}{\varphi_0(0)}.$$

Example 2.5. Let $\mathbf{L}(z,t)$ be a pseudodifferential operator with symbol $l(\omega)$, $\omega \in \mathrm{R}^n$. The function

$$\varphi(z,t) = \int\limits_{\mathrm{R}^m} \exp\big(\mathrm{i}\,\omega z + l(\omega)t\big) g(\omega)\,\mathrm{d}\omega$$

is a solution of the Cauchy problem

$$\frac{\partial \varphi}{\partial t} = \mathbf{L}(z,t)\varphi, \quad \varphi(z,0) = b(z) = \int\limits_{\mathrm{R}^m} g(\omega)\exp(\mathrm{i}\,\omega z)\,\mathrm{d}\omega.$$

Suppose that the operators $\mathbf{A}(x,t)$ and $\mathbf{B}(x,t)$ and the function $Q(x,t)$ satisfy the same assumptions. For the solution $\big(w(x,z,t), \lambda(x,t)\big)$ to inverse problem (2.17)–(2.19), the following formulas hold:

$$\varphi_0(t) = \int\limits_{\mathrm{R}^m} g(\omega)\exp\big(l(\omega)t\big)\,\mathrm{d}\omega,$$

$$w(x,z,t) = \frac{Q(x,t)\displaystyle\int\limits_{\mathrm{R}^m} g(\omega)\exp\big(\mathrm{i}\,\omega z + c(\omega)t\big)\,\mathrm{d}\omega}{\varphi_0(t)},$$

$$\lambda(x,t) = \frac{(Q_t - \mathbf{A}(x,t)Q)\varphi_0(t) - Q\varphi_0'(t)}{\varphi_0(t)\mathbf{B}(x,t)Q}.$$

Evolution problems of the form (2.17) in which the sought unknown coefficient is a functional (possibly nonlinear) in the solution and its derivatives have a great scientific and practical importance. For instance,

$$\lambda(x,t) = \int\limits_{\mathrm{R}^m} F\big(x,z,t,w(x,z,t), D^\alpha w(x,z,t)\big)\,\mathrm{d}z.$$

Here F is some, in general nonlinear, sufficiently smooth function and D^α is the differentiation with respect to the variables (x,z,t). In particular, the evolution equations of the form

$$\frac{\partial w}{\partial t} = -\frac{\partial w}{\partial z} + q(z,t)w + \mathbf{A}(x,t)w + \mu(t, P(x,t))w \tag{2.28}$$

are under consideration in ecology and biology; here $P(x,t) = \displaystyle\int\limits_0^\infty w(x,z,t)\,\mathrm{d}z$ and $\mathbf{A}(x,z)$ is the propagation operator. We refer to Ianelli *et al.* (1992) and Prüss (1981) for more details about the equations of the form (2.28).

In particular, the function $w(x,z,t)$ and the coefficients $q(z,t)$ and $\mu\big(t, P(x,t)\big)$ can be interpreted as follows: $w(x,z,t)$ is the number of specimens of some population at the age of z located at the point x at the moment of time t, and $q(z,t)$ and $\mu(t, P(x,t))$ are the coefficients of the deathrate and the birthrate of the specimens.

Equation (2.28) takes the form (2.17) if we put in (2.17) $\mathbf{L}(z,t) = -\dfrac{\partial}{\partial z} + q(z,t)$, $\mathbf{B}(x,t) = \mathbf{E}$, $\mathbf{E}$ is the identity operator, and $\mu(t, P(x,t)) = \lambda(x,t)$.

Observe that in this case the function $Q(x,t) = w(x,0,t)$ equals the number of the newborn specimens at the point x at the moment of time t. It will be shown below that, by making use of Theorem 2.8, we can solve the corresponding inverse problem and find the functions $w(x,z,t)$ and $\lambda(x,t)$; afterwards one can raise the question of finding the function $\mu(\xi,\eta)$ from the relation

$$\lambda(x,t) = \mu\left(t, \int_0^\infty w(x,z,t)\,\mathrm{d}z\right).$$

On the other hand, in ecology the function $Q(x,t)$ is represented as

$$Q(x,t) = \int_0^\infty \beta(z, P(x,t))w(x,z,t)\,\mathrm{d}z,$$

where $\beta(z, P(x,t))$ is the birthrate coefficient. Therefore, after finding the solution $(w(x,z,t), \lambda(x,t))$ of the inverse problem, we obtain the following equation of the first kind for the birthrate coefficient:

$$Q(x,t) = \int_0^\infty \beta(z, P(x,t))w(x,z,t)\,\mathrm{d}z,$$

which is of great importance.

2.5.2 Some Applications and Examples Connected with a Second-Order Parabolic Equation

In this subsection, we consider several examples of second-order parabolic equations for which the application of the scheme of studying the inverse problem of 2.5.1 succeeds completely.

Consider the parabolic equation

$$w_t = \mathbf{A}(x,t)w + \mathbf{L}(z,t)w + \lambda(x,t)\mathbf{B}(x,t)w, \quad x \in \mathrm{R}^n, \quad z \in \mathrm{R}^m, \quad t \geq 0, \tag{2.29}$$

with the operators of the form

$$\mathbf{A}(x,t) = \sum_{i,j=1}^n a_{ij}(x,t)\frac{\partial^2}{\partial x_i \partial x_j} + \sum_{i=1}^n a_i(x,t)\frac{\partial}{\partial x_i} + a(x,t),$$

$$\mathbf{B}(x,t) = \sum_{i=1}^n b_i(x,t)\frac{\partial}{\partial x_i} + b(x,t),$$

$$\mathbf{L}(z,t) = \sum_{i,j=1}^m c_{ij}(z,t)\frac{\partial^2}{\partial z_i \partial z_j} + \sum_{i=1}^m c_i(z,t)\frac{\partial}{\partial z_i} + c(z,t).$$

Denote by Q_T^1 the strip $\mathrm{R}^n \times (0,T)$; by Q_T^2, the strip $\mathrm{R}^m \times (0,T)$; and by Q_T, the strip $\mathrm{R}^n \times \mathrm{R}^m \times (0,T)$. Assume that the following conditions hold:

(A1) the coefficients of the operators $\mathbf{A}(x,t)$, $\mathbf{B}(x,t)$, and $\mathbf{L}(z,t)$ belong to $C^\beta(Q_T^1)$ or $C^\beta(Q_T^2)$, where $0 < \alpha < \beta < 1$;

(A2) the following relations hold:

$$\mu^{-1}|\xi|^2 \le \sum_{i,j=1}^{n} a_{ij}(x,t)\xi_i\xi_j \le \mu|\xi|^2, \quad x \in \mathrm{R}^n, \quad \xi \in \mathrm{R}^n, \quad t \ge 0, \quad \mu > 0,$$

$$\mu^{-1}|\xi|^2 \le \sum_{i,j=1}^{m} c_{ij}(z,t)\xi_i\xi_j \le \mu|\xi|^2, \quad z \in \mathrm{R}^n, \quad \xi \in \mathrm{R}^m, \quad t \ge 0, \quad \mu > 0.$$

As earlier, we assume that at $t = 0$ we have the initial data

$$w(x,z,0) = w_0(x,z), \quad x \in \mathrm{R}^n, \quad z \in \mathrm{R}^m, \tag{2.30}$$

and at $z = 0$, the additional information

$$w(x,0,t) = Q(x,t), \quad x \in \mathrm{R}^n, \quad t \ge 0. \tag{2.31}$$

We should solve the following inverse problem: to find a pair $\big(w(x,z,t),\ \lambda(x,t)\big)$ satisfying equation (2.29) and conditions (2.30), (2.31).

Suppose that the initial data and additional information satisfy the necessary agreement conditions:

(A3) $Q(x,0) = w_0(x,0)$.

We shall seek a solution $\big(w(x,z,t),\lambda(x,t)\big)$ to the inverse problem in the classes $C^{2+\alpha}(\overline{Q}_T)$ and $C^\alpha(\overline{Q}_T)$, $0 < \alpha < \beta < 1$.

Suppose that

(A4) $w_0(x,z) = \sum_{i=1}^{\infty} f_i(x)g_i(z)$ and the series $\sum_{i=1}^{\infty} \|f_i\|_\beta^{\mathrm{R}^n} \|g_i\|_\beta^{\mathrm{R}^m}$ is convergent

and

(A5) $Q(x,t) \in C^{2+\beta}(Q_T^1)$ and $\mathbf{B}(x,t)Q(x,t) \ge \mu_0 > 0$ for $(x,t) \in Q_T^1$.

Denote by $\varphi_i(z,t)$ ($i \ge 1$ is an integer) a solution of the linear Cauchy problem

$$\varphi_{it} = \mathbf{L}(z,t)\varphi_i, \quad (z,t) \in Q_T^2; \quad \varphi_i(z,0) = g_i(z), \quad z \in \mathrm{R}^m. \tag{2.32}$$

Denote by $u_i(x,t)$ ($i \ge 1$ is an integer) the functions that are solutions of the following infinite-dimensional system of parabolic equations:

$$u_{it} = \mathbf{A}(x,t)u_i + \frac{Q_t - \mathbf{A}(x,t)Q - \sum_{j=1}^{\infty} u_j(x,t)\varphi_{jt}(0,t)}{\mathbf{B}(x,t)Q}\mathbf{B}(x,t)u_i, \quad (x,t) \in Q_T^1; \tag{2.33}$$

$$u_i(x,0) = f_i(x), \quad x \in \mathrm{R}^n \quad (i \ge 1 \text{ is integral}).$$

We assume that the series $\sum_{j=1}^{\infty} u_j(x,t)\varphi_{jt}(0,t)$ of formula (2.33) converges in $C^\alpha(\overline{Q}_T^1)$.

Theorem 2.10. *Suppose that all conditions* (A1)–(A5) *hold. Consider the functions* w *and* λ:

$$w(x,z,t)=\sum_{i=1}^{\infty}u_i(x,t)\varphi_i(z,t); \tag{2.34}$$

moreover, the series $\sum_{i=1}^{\infty}\|u_i(x,t)\|_{2+\alpha}^{Q_T^1}\|\varphi_i(z,t)\|_{2+\alpha}^{Q_T^2}$ *is convergent, and*

$$\lambda(x,t)=\frac{Q_t-\mathbf{A}(x,t)Q-\sum_{j=1}^{\infty}u_j(x,t)\varphi_j'(0,t)}{\mathbf{B}(x,t)Q}. \tag{2.35}$$

The functions $\varphi_i(z,t)$ *and* $u_i(x,t)$ *satisfy relations (2.32) and (2.33). Then* $\big(w(x,z,t),\lambda(x,t)\big)$ *is a solution of the above-stated inverse problem for (2.29)–(2.31).*

PROOF. The proof of Theorem 2.10 mostly repeats that of Theorem 2.8 of 2.5.1. First, we observe that the convergence of the series $\sum_{i=1}^{\infty}\|u_i(x,t)\|_{2+\alpha}^{Q_T^1}\|\varphi_i(z,t)\|_{2+\alpha}^{Q_T^2}$ implies that the function $w(x,z,t)$ belongs to $C^{2+\alpha}(\overline{Q}_T)$. Prove that the pair $(w(x,z,t),\ \lambda(x,t))$ satisfies equation (2.29). To this end, multiply each of the equations (2.32) by $u_i(x,t)$, multiply each of the equalities (2.33) by $\varphi_i(z,t)$, and sum the results. We obtain

$$\begin{aligned}\sum_{i=1}^{\infty}(u_{it}\varphi_i+\varphi_{it}u_i) &= \sum_{i=1}^{\infty}\Bigg(\mathbf{A}(x,t)u_i\varphi_i+\mathbf{L}(z,t)u_i\varphi_i\\ &\quad+\frac{Q_t-\mathbf{A}(x,t)Q-\sum_{j=1}^{\infty}u_j(x,t)\varphi_{jt}(0,t)}{\mathbf{B}(x,t)Q}\mathbf{B}(x,t)u_i\varphi_i\Bigg).\end{aligned}$$

Considering that

$$w_t=\sum_{i=1}^{\infty}(u_{it}\varphi_i+u_i\varphi_{it}),\quad \mathbf{A}(x,t)w=\sum_{i=1}^{\infty}\mathbf{A}(x,t)u_i\cdot\varphi_i,\quad \mathbf{L}(z,t)w=\sum_{i=1}^{\infty}\mathbf{L}(z,t)u_i\varphi_i;$$

$$\lambda(x,t)\mathbf{B}(x,t)w=\lambda(x,t)\sum_{i=1}^{\infty}\mathbf{B}(x,t)u_i\varphi_i=\sum_{i=1}^{\infty}\frac{Q_t-\mathbf{A}(x,t)Q-\sum_{j=1}^{\infty}u_j\varphi_{jt}(0,t)}{\mathbf{B}(x,t)Q}\mathbf{B}(x,t)u_i\varphi_i,$$

we infer that the pair $\big(w(x,z,t),\ \lambda(x,t)\big)$ satisfies equation (2.29). Observe that

$$w(x,z,0)=\sum_{i=1}^{\infty}u_i(x,0)\varphi_i(z,0)=\sum_{i=1}^{\infty}f_i(x)g_i(z)=w_0(x,z).$$

Now we prove validity of equality (2.31). We have to demonstrate that $Q(x,t)=\check{Q}(x,t)$ for $(x,t)\in Q_T^1$, where $\check{Q}(x,t)$ denotes the function $w(x,0,t)$. Multiply each of the equations of system (2.33) by $\varphi_i(0,t)$ and add and subtract the product

$u_i\varphi_{it}(0,t)$. Summing up the obtained results, we arrive at

$$\sum_{i=1}^{\infty}\left(u_{it}+\varphi_i(0,t)+u_i\varphi_{it}(0,t)\right)-\mathbf{A}(x,t)u_i\varphi_i(0,t)$$

$$=\sum_{i=1}^{\infty}u_i\varphi_i(0,t)+\sum_{i=1}^{\infty}\frac{Q_t-\mathbf{A}(x,t)Q-\sum\limits_{j=1}^{\infty}u_j\varphi_{jt}(0,t)}{\mathbf{B}(x,t)Q}\mathbf{B}(x,t)u_i\varphi_i(0,t).$$

Let $w(x,0,t)=\sum\limits_{i=1}^{\infty}u_i\varphi_i(0,t)=\tilde{Q}(x,t)$. Then

$$\begin{aligned}
w_t(x,0,t) &= \sum_{i=1}^{\infty}\left(u_{it}\varphi_i(0,t)+u_i\varphi_{it}(0,t)\right)=\tilde{Q}_t(x,t),\\
\mathbf{A}(x,t)w(x,0,t) &= \sum_{i=1}^{\infty}\mathbf{A}(x,t)u_i\varphi_i(0,t)=\mathbf{A}(x,t)\tilde{Q}_t(x,t),\\
\mathbf{B}(x,t)w(x,0,t) &= \sum_{i=1}^{\infty}\mathbf{B}(x,t)u_i\varphi_i(0,t)=\mathbf{B}(x,t)\tilde{Q}.
\end{aligned}$$

Considering these relations, we obtain

$$\tilde{Q}_t-\mathbf{A}(x,t)\tilde{Q}=\sum_{j=1}^{\infty}u_j\varphi_{jt}(0,t)+\frac{\mathbf{B}(x,t)\tilde{Q}}{\mathbf{B}(x,t)Q}\left(Q_t-\mathbf{A}(x,t)Q-\sum_{j=1}^{\infty}u_j\varphi_{jt}(0,t)\right).$$

Denote the ratio

$$\frac{Q_t-\mathbf{A}(x,t)Q-\sum\limits_{j=1}^{\infty}u_j\varphi_{jt}(0,t)}{\mathbf{B}(x,t)Q}$$

by $r(x,t)$. We obtain

$$(\tilde{Q}-Q)_t=\mathbf{A}(x,t)(\tilde{Q}-Q)+r(x,t)\mathbf{B}(x,t)(\tilde{Q}-Q). \tag{2.36}$$

Moreover, by virtue of condition (A3), we have

$$(\tilde{Q}-Q)(x,0)=w(x,0,0)-Q(x,0)=w_0(x,0)-Q(x,0)=0.$$

Hence, the difference $(\tilde{Q}-Q)(x,t)$ is a solution of the homogeneous Cauchy problem with zero initial data. Such a solution must be identical to zero; therefore, $(\tilde{Q}-Q)(x,t)\equiv 0$, $(x,t)\in Q_T^1$. Theorem 2.10 is proved.

In the proof of Theorem 2.10, we supposed that systems (2.32) and (2.33) are solvable in Q_T^2 and Q_T^1 for some $T>0$. We now demonstrate that if assumptions (A1)–(A5) hold, then the supposition on solvability will always be valid for some T.

Theorem 2.11. *Let assumptions* (A1)–(A5) *hold. Then there exists a positive number* T *such that system (2.32) has a unique solution in* Q_T^2, *system (2.33) has a unique solution in* Q_T^1, *and the series* $\sum\limits_{i=1}^{\infty}\|u_i(x,t)\|_{2+\alpha}^{Q_T^1}\,\|\varphi_i(z,t)\|_{2+\alpha}^{Q_T^2}$ *converges.*

PROOF. We employ the fixed point method. First of all, observe that system (2.32) splits into independent equations. Every equation in (2.32) yields a solution of the Cauchy problem which exists, is unique, and satisfies the estimate (Ladyzhenskaya *et al.*, 1968).

$$\|\varphi_i(z,t)\|_{2+\beta}^{Q_T^2} \le C\,\|g_i(z)\|_{2+\beta}^{\mathrm{R}^m}, \quad 0<\alpha<\beta<1. \tag{2.37}$$

The constant C in estimate (2.37), as is shown in Ladyzhenskaya *et al.* (1968), is independent of the index i.

Consider the space X consisting of the sequences of functions $\xi = (\xi_1,\ldots,\xi_n,\ldots)$, where $\xi_i(x,t)$ belongs to $C^{2+\alpha}(\overline{Q}_T^1)$ and the series $\sum_{i=1}^{\infty} \|\xi_i(x,t)\|_{2+\alpha}^{Q_T^1} \|\varphi_i(z,t)\|_{2+\beta}^{Q_T^2}$ converges. Introduce the distance in the space X by the formula

$$d(\xi,\eta) = \sum_{i=1}^{\infty} \|\xi_i(x,t)-\eta_i(x,t)\|_{2+\alpha}^{Q_T^1} \|\varphi_i(z,t)\|_{2+\beta}^{Q_T^2}.$$

Prove that the space X is complete. Assume ξ^n to be a Cauchy sequence in X. Then, for each i, the sequence ξ_i^n is a Cauchy sequence and therefore converges. Assign $\xi_i = \lim_{n\to\infty} \xi_i^n$ in $C^{2+\alpha}(\overline{Q}_T^1)$. Denote the sequence $(\xi_1,\ldots,\xi_n,\ldots)$ by ξ. We have to demonstrate that

$$\sum_{i=1}^{\infty} \|\xi_i(x,t)\|_{2+\alpha}^{Q_T^1} \|\varphi_i(z,t)\|_{2+\beta}^{Q_T^2} < \infty;$$

i. e., $\xi \in X$, $\lim_{n\to\infty} d(\xi^n,\xi) = 0$.

Indeed, for an arbitrarily-fixed natural k and all $\varepsilon > 0$ we have

$$\sum_{i=1}^{k} \|\xi_i^m - \xi_i^n\|_{2+\alpha}^{Q_T^1} \|\varphi_i(z,t)\|_{2+\beta}^{Q_T^2} < \varepsilon \quad \text{for} \quad m,n > N(\varepsilon).$$

In this sum, the number of summands is finite; so we can fix n and pass to the limit as $m \to \infty$. We obtain

$$\sum_{i=1}^{k} \|\xi_i^n - \xi_i\|_{2+\alpha}^{Q_T^1} \|\varphi_i\|_{2+\beta}^{Q_T^2} \le \varepsilon \quad \text{for} \quad n > N(\varepsilon).$$

The inequality is valid for every k. Let us restore the infinite series. Taking the limit as $k \to \infty$, we obtain

$$\sum_{i=1}^{\infty} \|\xi_i^n - \xi_i\|_{2+\alpha}^{Q_T^1} \|\varphi_i\|_{2+\beta}^{Q_T^2} \le \varepsilon \quad \text{for} \quad n > N(\varepsilon).$$

The first claim ensues from the inequality $\|\xi_i\|_{2+\alpha}^{Q_T^1} \le \|\xi_i^n - \xi_i\|_{2+\alpha}^{Q_T^1} + \|\xi_i^n\|_{2+\alpha}^{Q_T^1}$ and convergence of the series $\sum_{i=1}^{\infty} \|\xi_i^n\|_{2+\alpha}^{Q_T^1} \|\varphi_i\|_{2+\beta}^{Q_T^2}$. The second ensues from the inequality $d(\xi^n,\xi) \le \varepsilon$ for $n > N(\varepsilon)$, since ε is arbitrarily small.

Thus, X is a complete metric space. Denote by $B(K)$ the set of elements ξ of X such that $d(\xi,0) \le K$. Define a mapping G on the set $B(K)$ as follows: If a point

$\xi = (\xi_1, \dots, \xi_n, \dots)$ belongs to $B(K)$, then

$$\lambda(\xi) = \frac{Q_t - \mathbf{A}(x,t)Q - \sum_{j=1}^{\infty} \xi_j(x,t)\varphi_{jt}(0,t)}{\mathbf{B}(x,t)Q}, \tag{2.38}$$

$$\eta_{it} = \mathbf{A}(x,t)\eta_i + \frac{Q_t - \mathbf{A}(x,t)Q - \sum_{j=1}^{\infty} \xi_j(x,t)\varphi_{jt}(0,t)}{\mathbf{B}(x,t)Q}\mathbf{B}(x,t)\eta_i, \tag{2.39}$$

$$\eta_i(x,0) = f_i(x), \quad i \geq 1, \quad i \text{ is natural.}$$

Observe that $\lambda(\xi)$ belongs to $C^\beta(\overline{Q}_T^1)$; moreover,

$$\|\lambda(\xi)\|_\beta^{Q_T^1} \leq C_1^1 + C_2^1 K = C_1.$$

Denote the quantity $\sum_{i=1}^{\infty} \|f_i\|_{2+\beta}^{\mathrm{R}^n} \|g_i\|_{2+\beta}^{\mathrm{R}^m}$ by M and put $K = 2M$. It was demonstrated in Ladyzhenskaya *et al.* (1968) that

$$\|\eta_i(x,t) - f_i(x)\|_{2+\alpha}^{Q_T^1} \leq C_2 T^{(\beta-\alpha)/2} \|f_i\|_{2+\beta}^{\mathrm{R}^n}. \tag{2.40}$$

The constants C_1 and C_2 remain bounded as $T \to 0$. Therefore, (2.40) implies that

$$\sum_{i=1}^{\infty} \|\eta_i\|_{2+\alpha}^{Q_T^1} \|\varphi_i\|_{2+\beta}^{Q_T^2} \leq M + C_2 C_3 T^{(\beta-\alpha)/2} M. \tag{2.41}$$

Choose $T > 0$ so as to have $C_3 C_2 T^{(\beta-\alpha)/2} < 1$. This is always possible since $\beta - \alpha > 0$. From (2.41) we infer

$$\sum_{i=1}^{\infty} \|\eta_i\|_{2+\alpha}^{Q_T^1} \|\varphi_i\|_{2+\beta}^{Q_T^2} \leq 2M = K.$$

This inequality shows that the mapping G transforms the ball $B(K)$ into itself. We shall prove that G is a contraction on $B(K)$. Let $\tilde{\eta} = G(\tilde{\xi})$. Denote the difference $\eta_i(x,t) - \tilde{\eta}_i(x,t)$ by $\zeta_i(x,t)$. Then we obtain the following system of parabolic equations for $\zeta_i(x,t)$:

$$\begin{aligned} \zeta_{it} &= \mathbf{A}(x,t)\zeta_i + \frac{Q_t - \mathbf{A}(x,t)Q - \sum_{j=1}^{\infty} \xi_j(x,t)\varphi_{jt}(0,t)}{\mathbf{B}(x,t)Q}\mathbf{B}(x,t)\zeta_i \\ &\quad + \frac{\sum_{j=1}^{\infty} \theta_j(x,t)\varphi_{jt}(0,t)}{\mathbf{B}(x,t)Q}, \quad (x,t) \in Q_T^1; \end{aligned} \tag{2.42}$$

$$\theta_j(x,t) = \xi_j(x,t) - \tilde{\xi}_j(x,t), \quad \zeta_j(x,0) = 0, \quad x \in \mathrm{R}^n; \quad i = 1, 2, \dots.$$

The previous estimates and those for a solution of the Cauchy problem in the Hölder classes (see Ladyzhenskaya *et al.*, 1968) imply

$$\|\zeta_i(x,t)\|_{2+\alpha}^{Q_T^1} \leq C_4 T^{(\beta-\alpha)/2} \left(\sum_{j=1}^{\infty} \|\theta_j\|_{2+\alpha}^{Q_T^1} \|\varphi_j\|_{2+\beta}^{Q_T^2} \right) \|\eta_i\|_{2+\alpha}^{Q_T^1};$$

therefore,

$$\sum_{i=1}^{\infty} \|\zeta_i\|_{2+\alpha}^{Q_T^1} \|\varphi_i\|_{2+\beta}^{Q_T^2} \leq C_4 T^{(\beta-\alpha)/2} \left(\sum_{j=1}^{\infty} \|\theta_j\|_{2+\alpha}^{Q_T^1} \|\varphi_j\|_{2+\beta}^{Q_T^2}\right) \left(\sum_{i=1}^{\infty} \|\eta_i\|_{2+\alpha}^{Q_T^1} \|\varphi_i\|_{2+\beta}^{Q_T^2}\right)$$
$$\leq C_4 T^{(\beta-\alpha)/2} K \sum_{j=1}^{\infty} \|\theta_j\|_{2+\alpha}^{Q_T^1} \|\varphi_j\|_{2+\beta}^{Q_T^2} = C_4 T^{(\beta-\alpha)/2} K d(\xi, \tilde{\xi}).$$

It means that if $C_4 T^{(\beta-\alpha)/2} K < 1$ then the mapping G is a contraction on $B(K)$. Finally, we choose T such that

$$T^{(\beta-\alpha)/2} < \min(1/C_2C_3, 1/C_4K).$$

For this choice of T, the mapping G carries the set $B(K)$ into itself and is a contraction on it. The contraction mapping principle implies that G has a unique fixed point u in $B(K)$: $u = G(u)$. Then

$$u_{it} = \mathbf{A}(x,t)u_i + \frac{Q_t - \mathbf{A}(x,t)Q - \sum_{j=1}^{\infty} u_j(x,t)\varphi_{it}(0,t)}{\mathbf{B}(x,t)Q} \mathbf{B}(x,t)u_i, \quad (x,t) \in Q_T^1; \tag{2.43}$$

$$u_i(x,0) = f_i(x), \quad x \in \mathrm{R}^n, \quad i = 1,2\ldots.$$

Put

$$\lambda(x,t) = \frac{Q_t - \mathbf{A}(x,t)Q - \sum_{j=1}^{\infty} u_j(x,t)\varphi_{it}(0,t)}{\mathbf{B}(x,t)Q}. \tag{2.44}$$

It follows from (2.43) that $u = (u_1, \ldots, u_n, \ldots)$ is a solution of system (2.33). Furthermore,

$$\sum_{i=1}^{\infty} \|u_i\|_{2+\alpha}^{Q_T^1} \|\varphi_i\|_{2+\beta}^{Q_T^2} \leq K = 2M, \tag{2.45}$$

since u does not belong to $B(K)$. Theorem 2.11 is proved.

Theorems 2.10 and 2.11 imply that the inverse problem (2.29)–(2.31) is solvable in Q_T provided conditions (A1)–(A5) hold. Indeed, (2.32) is always solvable in Q_T^2 for every $T > 0$ and (2.33) is solvable in Q_T by Theorem 2.11. The validity of estimate (2.45) was established above. Hence, applying Theorem 2.10, we conclude that the function $w(x,z,t)$ defined by the formula $w(x,z,t) = \sum_{i=1}^{\infty} u_i(x,t)\varphi_i(z,t)$ and the function $\lambda(x,t)$ given by formula (2.44) determine the pair $\big(w(x,z,t), \lambda(x,t)\big)$ which is a solution of the inverse problem (2.29)–(2.31).

Note that in the proofs of Theorems 2.10 and 2.11 we used estimates for solutions to a linear Cauchy problem in Hölder function classes; these estimates can be found, for instance, in Ladyzhenskaya *et al.* (1968). Analogous results can be obtained in the case when we use estimates for u_i and φ_i in the spaces $W_{p,\mathrm{loc}}^{2,1}(Q_T^1)$ and $W_{p,\mathrm{loc}}^{2,1}(Q_T^2)$ rather than in Hölder spaces and use estimates for f_i and g_i in the spaces $W_{p-2/p,\mathrm{loc}}^{2}(\mathrm{R}^n)$ and $W_{p-2/p,\mathrm{loc}}^{2}(\mathrm{R}^m)$.

It is natural that, for inferring such estimates, we can impose some weaker conditions on the coefficients of the operators $\mathbf{A}(x,t)$, $\mathbf{B}(x,t)$, and $\mathbf{L}(z,t)$ as compared with (A1)–(A5).

Now we consider the boundary value problems for equation (2.29) instead of the Cauchy problem. We will demonstrate that, in this case, some theorems analogous to Theorems 2.10 and 2.11 are valid. Let Ω_1 be a bounded domain in R^n with boundary $\partial\Omega_1$ of class C^3 (see Ladyzhenskaya *et al.*, 1968) and let Ω_2 be a bounded domain in R^m with boundary $\partial\Omega_2$. Denote by $\tilde{Q}_T^1$ the cylinder $\Omega_1 \times (0,T)$; by $\tilde{Q}_T^2$, the cylinder $\Omega_2 \times (0,T)$; and by $\tilde{Q}_T$, the cylinder $\Omega_1 \times \Omega_2 \times (0,T)$. Also, suppose that the point 0 belongs to the domain Ω_2. On the boundary of the domain $\tilde{Q}_T^1$, we stipulate boundary conditions of the form

$$\mathbf{D}^1(x,t)w(x,z,t) = 0, \quad (x,t) \in \Gamma_T^1 = \partial\Omega_1 \times (0,T), \ z \in \Omega_2, \tag{2.46}$$

where $\mathbf{D}^1(x,t)$ is an operator of the form

$$\beta_1 \left(\sum_{i=1}^{n} d_i^1(x,t) \frac{\partial}{\partial x_i} + d_0^1(x,t) \right) + \beta_0.$$

The functions $d_i^1(x,t)$ belong to the space $C^{2+\beta}(\tilde{\Gamma}_T^1)$; β_1 and β_0 are constants, $\beta_1^2 + \beta_0^2 > 0$; moreover, $\sum_{i=1}^{n} d_i^1(x,t)\cos(\mathbf{n}^1, x_i) \geq \mu_2 > 0$, where $\mathbf{n}_1$ is a normal to $\partial\Omega_1$ at the point $(x,t) \in \tilde{\Gamma}_T^1$. We pose analogous boundary conditions on the boundary $\Gamma_T^2 = \partial\Omega_2 \times (0,T)$ of the domain $\tilde{Q}_T^2$:

$$\begin{gathered} \mathbf{D}^2(z,t)w(x,z,t) = 0, \quad (z,t) \in \tilde{\Gamma}_T^2, \quad x \in \Omega_1, \\ \mathbf{D}^2(z,t) = \gamma_1 \left(\sum_{i=1}^{m} d_i^2(z,t) \frac{\partial}{\partial z_i} + d_0^2(z,t) \right) + \gamma_0, \\ \sum_{i=1}^{m} d_i^2(z,t)\cos(\mathbf{n}^2, z_i) \geq \mu_2, \quad (z,t) \in \Gamma_T^2, \quad d_i^2 \in C^{2+\beta}(\Gamma_T^2). \end{gathered} \tag{2.47}$$

Suppose the following agreement conditions to be held:

(A6) $\mathbf{D}^1(x,t)w_0(x,z) = 0$, $x \in \partial\Omega_1$, $z \in \Omega_2$; $\mathbf{D}^2(z,t)w_0(x,z) = 0$, $x \in \Omega_1$, $z \in \partial\Omega_2$.

Consider the inverse problem that consists in finding the pair $(w(x,z,t), \lambda(x,t))$. Here we shall assume that the function $w(x,z,t)$ belongs to the class $C^{2+\alpha}(\tilde{Q}_T)$, the function $\lambda(x,t)$ belongs to $C^\alpha(\tilde{Q}_T^1)$, and the functions $w(x,z,t)$ and $\lambda(x,t)$ satisfy (2.29)–(2.41), (2.46), (2.47).

Henceforth, while speaking about validity of conditions (A1)–(A5), we shall consider the domain Ω_1 instead of R^n, the domain Ω_2 instead of R^m, and the cylinders $\tilde{Q}_T^1$ and $\tilde{Q}_T^2$ instead of the cylinders Q_T^1 and Q_T^2.

The following theorem is valid:

Theorem 2.12. *Let conditions* (A1)–(A6) *hold. The functions* $\varphi_i(z,t)$, $(z,t) \in \tilde{Q}_T^2$, $i = 1,2,\ldots$, *are solutions of the boundary value problems*

$$\begin{gathered} \varphi_{it} = \mathbf{L}(z,t)\varphi_i, \quad (z,t) \in \tilde{Q}_T^2; \\ \mathbf{D}^2(z,t)\varphi_i(z,t) = 0, \quad (z,t) \in \Gamma_T^2, \quad \varphi_i(z,0) = g_i(z), \quad z \in \Omega_2, \quad i = 1,2,\ldots. \end{gathered} \tag{2.48}$$

The functions $u_i(x,t)$ are solutions of the infinite-dimensional evolution problem

$$u_{it} = \mathbf{A}(x,t)u_i + \frac{Q_t - \mathbf{A}(x,t)Q - \sum\limits_{j=1}^{\infty} u_j(x,t)\varphi_{jt}(0,t)}{\mathbf{B}(x,t)Q}\mathbf{B}(x,t)u_i, \quad (x,t) \in \tilde{Q}_T^2; \quad (2.49)$$

$$\mathbf{D}^1(x,t)u_i = 0, \quad (x,t) \in \Gamma_T^2; \quad u_i(x,0) = f_i(x), \quad x \in \Omega_1, \quad i = 1,2,\ldots.$$

Suppose that the series $\sum\limits_{i=1}^{\infty} \|u_i\|_{2+\alpha}^{\tilde{Q}_T^1} \|\varphi_i\|_{2+\beta}^{\tilde{Q}_T^2}$ *converges. Then the functions*

$$w(x,z,t) = \sum_{i=1}^{\infty} u_i(x,t)\varphi_i(z,t), \quad \lambda(x,t) = \frac{Q_t - \mathbf{A}(x,t)Q - \sum\limits_{j=1}^{\infty} u_j(x,t)\varphi_{jt}(0,t)}{\mathbf{B}(x,t)Q} \quad (2.50)$$

constitute a solution of the inverse problems in $\tilde{Q}_T$.

The proof of Theorem 2.12 is analogous to that of Theorem 2.10 and is therefore omitted.

Theorem 2.13. *Let conditions* (A1)–(A6) *hold. There exists a positive number* T *such that systems (2.48) and (2.49) are solvable in* $\tilde{Q}_T^2$ *and* $\tilde{Q}_T^1$; *moreover, the series* $\sum\limits_{i=1}^{\infty} \|u_i\|_{2+\alpha}^{\tilde{Q}_T^1} \|\varphi_i\|_{2+\alpha}^{\tilde{Q}_T^2}$ *converges.*

PROOF. The proof of Theorem 2.13 is analogous to that of Theorem 2.11, the only difference being in the use of the contraction mapping method.

For any i the problem (2.48) is a linear boundary value problem. Therefore, the functions $\varphi_i(z,t)$ are uniquely determined by the initial data $g_i(z)$ and satisfy the estimate (see Ladyzhenskaya *et al.*, 1968)

$$\|\varphi_i\|_{2+\beta}^{\tilde{Q}_T^2} \leq C\,\|g_i\|_{2+\beta}^{\Omega_2}. \quad (2.51)$$

The constant C in (2.51) is independent of i and remains bounded as $T \to 0$.

As in Theorem 2.11, we consider the space $\tilde{X}$ consisting of the function sequences $\xi = (\xi_1,\ldots;\xi_n,\ldots)$ such that $\xi_i(x,t)$ belongs to $C^{2+\alpha}(\tilde{Q}_T^1)$ for every i and the series $\sum\limits_{i=1}^{\infty} \|\xi_i(x,t)\|_{2+\alpha}^{\tilde{Q}_T^1} \|\varphi_i(z,t)\|_{2+\beta}^{\tilde{Q}_T^2}$ converges. Define the distance between elements ξ and η of the space $\tilde{X}$ by the formula

$$d(\xi,\eta) = \sum_{i=1}^{\infty} \|\xi_i - \eta_i\|_{2+\alpha}^{\tilde{Q}_T^1} \|\varphi_i\|_{2+\beta}^{\tilde{Q}_T^2}. \quad (2.52)$$

The space $\tilde{X}$ endowed with the distance-function (2.52) is complete. This fact can be proved in the same way as completeness of the space X in Theorem 2.11; for this reason we omit the corresponding arguments.

Put $B(K) = \{\xi \in \tilde{X}, d(\xi,0) \leq K\}$; $B(K)$ is a bounded closed set. Define a mapping G on the set $B(K)$ as follows: If a point $\xi = (\xi_1,\ldots,\xi_n,\ldots)$ belongs to $B(K)$, then we define

$$\lambda(\xi) = \frac{Q_t - \mathbf{A}(x,t)Q - \sum\limits_{j=1}^{\infty} \xi_j(x,t)\varphi_{jt}(0,t)}{\mathbf{B}(x,t)Q}; \quad (2.53)$$

$$\eta_{it} = \mathbf{A}(x,t)\eta_i + \frac{Q_t - \mathbf{A}(x,t)Q - \sum_{j=1}^{\infty} \xi_j(x,t)\varphi_{jt}(0,t)}{\mathbf{B}(x,t)Q}\mathbf{B}(x,t)\eta_i, \quad (x,t) \in \tilde{Q}_T^1; \tag{2.54}$$

$$\mathbf{D}^1(x,t)\eta_i = 0, \quad (x,t) \in \Gamma_T^1, \quad \eta_i(x,0) = f_i(x).$$

In this case, $\lambda(\xi)$ belongs to $C^\beta(\tilde{Q}_T^1)$ and the following estimate holds:

$$\|\lambda(\xi)\|_\beta^{\tilde{Q}_T^1} \leq C_1. \tag{2.55}$$

Let M denote the quantity $\sum_{i=1}^{\infty} \|f_i\|_{2+\beta}^{\Omega_1} \|g_i\|_{2+\beta}^{\Omega_2}$ and put $K = 2M$.

The functions $\eta_i(x,t)$ are solutions of the linear boundary value problem (2.54) given initial data $f_i(x)$ in $C^{2+\beta}(\overline{\Omega}_1)$. Therefore, $\eta_i(x,t)$ belongs to $C^{2+\beta}(\tilde{Q}_T^1)$ and, as proved in Ladyzhenskaya *et al.* (1968),

$$\|\eta_i(x,t) - f_i(x)\|_{2+\alpha}^{\tilde{Q}_T^1} \leq C_2 T^{(\beta-\alpha)/2} \|f_i\|_{2+\beta}^{\Omega_1}. \tag{2.56}$$

Observe that

$$\sum_{i=1}^{\infty} \|\eta_i\|_{2+\alpha}^{\tilde{Q}_T^1} \|\varphi_i\|_{2+\beta}^{\tilde{Q}_T^2} \leq \sum_{i=1}^{\infty} (\|f_i\|_{2+\alpha}^{\Omega_1} + \|\eta_i - f_i\|_{2+\alpha}^{\tilde{Q}_T^1}) \|\varphi_i\|_{2+\beta}^{\tilde{Q}_T^2} \leq C_2 T^{(\beta-\alpha)/2} M + M.$$

If $C^2 T^{(\beta-\alpha)/2} < 1$ then the mapping G transforms the set $B(K)$ into itself.

In the same manner as in Theorem 2.11, we make sure that if $C_4 K T^{(\beta-\alpha)/2} < 1$ then G is a contraction on $B(K)$. It means that if $T^{(\beta-\alpha)/2} < \min(1/C_2, 1/C_4 K)$ then the mapping G carries the bounded closed set $B(K)$ of a complete space into itself and is a contraction on this set. Hence G has a unique fixed point $u = (u_1, \ldots, u_n, \ldots)$ on the set $B(K)$. Furthermore,

$$u_i = \mathbf{A}(x,t)u_i + \frac{Q_t - \mathbf{A}(x,t)Q - \sum_{j=1}^{\infty} u_j(x,t)\varphi_{jt}(0,t)}{\mathbf{B}(x,t)Q}\mathbf{B}(x,t)u_i, \quad (x,t) \in \tilde{Q}_T^1;$$

$$\mathbf{D}^1(x,t)u_i = 0, \quad (x,t) \in \Gamma_T^1; \quad u_i(x,0) = f_i(x), \quad x \in \Omega_1; \quad i = 1,2,\ldots,n,\ldots.$$

The point u belongs to $B(K)$; therefore,

$$\sum_{i=1}^{\infty} \|u_i\|_{2+\alpha}^{\tilde{Q}_T^1} \|\varphi_i\|_{2+\beta}^{Q_T^2} \leq K = 2M.$$

Theorem 2.13 is proved.

2.5.3 Nonlinear Evolution Problems in Hilbert Space

We begin with some preliminaries. Let V and H be Hilbert spaces. Let V' and H' denote the dual spaces of V and H; moreover, the space H is identified with H'. We shall assume that $V \subset H \subset V'$. Also, suppose that a family of bilinear continuous forms is given on the space V. Regarding this family, we assume the following:

(B1) for arbitrary $\varphi, \psi \in V$, the function $t \to a(t,\varphi,\psi)$ is measurable on $(0,T)$ and $|a(t,\varphi,\psi)| \le C\,\|\varphi\|\,\|\psi\|$;

(B2) there is a number λ such that $a(t,\varphi,\varphi)+\lambda\,|\varphi|^2 \ge \alpha\,\|\varphi\|^2$, $\alpha > 0\ \forall\varphi \in V$, $t \in (0,T)$.

In conditions (B1) and (B2), we denote by $|\cdot|$ the norm in the space H. For every $t \in (0,T)$, we can write

$$a(t,\varphi,\psi) = \big(\mathbf{A}(t)\varphi,\psi\big), \tag{2.57}$$

where $\mathbf{A}(t)\varphi \in V'$ and the parentheses stand for the scalar product of elements in V and V'.

Consider the space $L^2((0,T);V)$ (Lions, 1968; Lions and Magenes, 1972) constituted by the functions $t \to f(t)$ that take the interval $(0,T)$ into the space V; moreover, the functions $f(t)$ are measurable and $\left(\int\limits_0^T \|f(t)\|^2\,\mathrm{d}t\right)^{1/2} < \infty$. Define the space $L^2((0,T);V')$ analogously.

Introduce the space

$$W(0,T) = \left\{f \mid f \in L^2((0,T);V);\ \frac{\mathrm{d}f}{\mathrm{d}t} \in L^2((0,T);V')\right\}.$$

The norm $\|f\|_{W(0,T)} = \left(\int\limits_0^T \|f(t)\|_V^2\,\mathrm{d}t + \int\limits_0^T \left\|\frac{\mathrm{d}f}{\mathrm{d}t}\right\|_{V'}^2 \mathrm{d}t\right)^{1/2}$, determines a Hilbert space structure in it. It is demonstrated in Lions and Magenes (1972, Chapter 1) that every function f in $W(0,T)$, appropriately modified (if necessary) on some set of measure zero, becomes a continuous function $(0,T) \to H$; i.e.,

$$W(0,T) \subset C^0([0,T];H). \tag{2.58}$$

Consider the following evolution problem: find a function y in $W(0,T)$ satisfying the equation

$$\mathbf{A}(t)y + \frac{\mathrm{d}y}{\mathrm{d}t} = f, \tag{2.59}$$

where f is a given function of the space $L^2((0,T);V')$, and meeting the initial condition

$$y(0) = y_0, \tag{2.60}$$

where y_0 is a given element of the space H.

It was demonstrated in Lions (1968) and Lions and Magenes (1972) that if conditions (B1) and (B2) hold, then the problem (2.59), (2.60) has a unique solution in $W(0,T)$. The solution satisfies the following inequality

$$\|y(t)\|_{W(0,T)} \le C(|y_0| + \|f\|_{L_2((0,T);V')}). \tag{2.61}$$

In 2.5.1 we considered some evolution equation in a Hilbert space; now denote this space by E. In Theorem 2.8, solution of the inverse problem was reduced to the solution of two evolution problems in the Hilbert space E:

$$\varphi_t - \mathbf{L}(z,t)\varphi = 0, \quad (z,t) \in Q_T^2; \quad \varphi(z,0) = \varphi_0(z), \quad z \in \mathrm{R}^m; \tag{2.62}$$

$$u_t = \mathbf{A}(x,t)u + \frac{Q_t - \mathbf{A}(x,t)Q - \big(u(x,t), \varphi_t(0,t)\big)}{Q}u, \tag{2.63}$$

$$(x,t) \in Q_T^1; \quad u(x,0) = u_0(x), \quad x \in \mathrm{R}^n.$$

The operators $\mathbf{L}(z,t)$ and $\mathbf{A}(x,t)$ act from Q_T^2 and Q_T^1 into E. In problem (2.63), we assume that the operator $\mathbf{B}(x,t)$ equals the identity operator and we denote by $(\cdot,\cdot)$ the scalar product of two elements in the Hilbert space E.

Suppose that the operators $\mathbf{A}(x,t)$ and $\mathbf{L}(z,t)$ have the following form:

$$\mathbf{A}(x,t) = \sum_{i,j=1}^{n} \frac{\partial}{\partial x_i}\left(a_{ij}(x,t)\frac{\partial u}{\partial x_j}\right), \quad \mathbf{L}(z,t) = \sum_{i,j=1}^{m} \frac{\partial}{\partial z_i}\left(c_{ij}(z,t)\frac{\partial}{\partial z_j}\right),$$

where the functions $a_{ij}(x,t)$ and $c_{ij}(z,t)$ are continuous for $x \in \mathrm{R}^n$, $z \in \mathrm{R}^m$, and $t \geq 0$.

Denote by H_i, $i = 1,2$, the space $L^2(\mathrm{R}^n, E)$ or $L^2(\mathrm{R}^m, E)$ that consists of the functions carrying R^n or R^m into E and such that $\left(\int_{\mathrm{R}^n} \|u(x)\|_E^2 \,\mathrm{d}x\right)^{1/2} < \infty$ or $\left(\int_{\mathrm{R}^m} \|v(z)\|_E^2 \,\mathrm{d}z\right)^{1/2} < \infty$ (see, for instance, (Lions, 1973)). Denote by V_i, $i = 1,2$, the space of functions from R^n or R^m into E such that the function itself and all its first-order derivatives belong to H_i, $i = 1,2$. If functions φ and ψ belong to V_1 and $\tilde{\varphi}$ and $\tilde{\psi}$ belong to V_2, then we can define two bilinear forms:

$$a(t,\varphi,\psi) = \sum_{i,j=1}^{n} \int_{\mathrm{R}^n} a_{ij}(x,t)\left(\frac{\partial \varphi}{\partial x_i}, \frac{\partial \psi}{\partial x_j}\right) \mathrm{d}x, \tag{2.64}$$

$$c(t,\tilde{\varphi},\tilde{\psi}) = \sum_{i,j=1}^{m} \int_{\mathrm{R}^m} c_{ij}(z,t)\left(\frac{\partial \tilde{\varphi}}{\partial z_i}, \frac{\partial \tilde{\psi}}{\partial z_j}\right) \mathrm{d}z. \tag{2.65}$$

We suppose the following condition:

(C1)

$$\mu^{-1}|\xi|^2 \leq \sum_{i,j=1}^{n} a_{ij}(x,t)\xi_i\xi_j \leq \mu|\xi|^2, \quad (x,t) \in Q_T^1; \quad \xi \in \mathrm{R}^n;$$

$$\mu^{-1}|\eta|^2 \leq \sum_{i,j=1}^{m} c_{ij}(z,t)\eta_i\eta_j \leq \mu|\eta|^2, \quad (z,t) \in Q_T^2; \quad \eta \in \mathrm{R}^m; \quad \mu > 0;$$

$$Q(x,t) \geq \mu_0 > 0, \quad (x,t) \in Q_T^1.$$

It is easy to prove that if condition (C1) is valid, then conditions (B1) and (B2) hold for the bilinear forms. In a natural way, we assume

$$|\varphi|_{H_1} = \left(\int_{\mathrm{R}^n} \|\varphi\|_E^2\right)^{1/2}, \quad |\varphi|_{V_1} = \left(\int_{\mathrm{R}^n} \|\varphi\|_E^2 + \sum_{i=1}^{n} \int_{\mathrm{R}^n} \left\|\frac{\partial \varphi}{\partial x_i}\right\|_E^2 \mathrm{d}x\right)^{1/2}.$$

We define the norms in H_2 and V_2 analogously.

Denote by $W_i(0,T)$ the function space

$$\left\{ f \in L_2((0,T);V_i); \frac{\mathrm{d}f}{\mathrm{d}t} \in L_2((0,T);V_i') \right\}$$

with the norm

$$\|f\|_{W_i(0,T)} = \left(\int_0^T \|f(t)\|_{V_i}^2 \,\mathrm{d}t + \int_0^T \left\| \frac{\mathrm{d}f}{\mathrm{d}t} \right\|_{V_i'}^2 \mathrm{d}t \right)^{1/2},$$

where $i = 1, 2$.

(C2) Let $\varphi_0(z)$ and $\dfrac{\partial \varphi_0}{\partial z_i}(z)$ belong to V_2.

Problem (2.62) has a solution in $W_2(0,T)$.

In addition, we suppose that

(C3) the functions $c_{ij}(z,t)$ have continuous and bounded first-order derivatives with respect to the time for $(z,t) \in \mathrm{R}^m \times \mathrm{R}_+$.

Consider the problem

$$\frac{\mathrm{d}\psi}{\mathrm{d}t} = \mathbf{L}(z,t)\psi + \mathbf{L}_t(z,t)\varphi; \quad \psi(z,0) = \mathbf{L}(z,0)\varphi_0(z).$$

A solution $\psi(z,t)$ of the problem exists and belongs to $W_2(0,T)$ as well, since $\mathbf{L}_t(z,t)\varphi$ lies in $L_2((0,T);V_2')$. Furthermore, $\psi(z,t) = \dfrac{\mathrm{d}\varphi}{\mathrm{d}t}(z,t)$. It is demonstrated in Lions (1973) and Lions and Magenes (1972) that $W_2(0,t) \subset C^0([0,T],H_2)$; therefore, $\varphi_t(0,t) = \psi(0,t)$ belongs to H_2 for all $t \in [0,T]$ and hence $\varphi_t(0,T)$ are elements of the Hilbert space E for all $t \in [0,T]$.

Consider the nonlinear evolution problem (2.63). The following theorem is valid:

Theorem 2.14. *Let conditions* (C1)–(C3) *hold,* u_0 *belong to* H_1, *and* $|u_0| < C_1$. *There exists a positive number* T *such that problem (2.63) has a unique solution in* $W_1(0,T)$.

PROOF. *Uniqueness*: Suppose that problem (2.63) has two solutions: $u_1(x,t)$ and $u_2(x,t)$. Denote the difference $u_1(x,t) - u_2(x,t)$ by $v(x,t)$. Then

$$v_t = \mathbf{A}(x,t)v - \frac{Q_t - \mathbf{A}(x,t)Q - (u_1(x,t),\varphi_t(0,t))}{Q}v + u_2(x,t)\frac{(v(x,t),\varphi_t(0,t))}{Q},$$

$$(x,t) \in Q_T^1; \quad v(x,0) = 0.$$

Make the substitution $v(x,t) = w(x,t)\exp(-kt)$; the value $k > 0$ will be specified later. We obtain the following problem for $w(x,t)$:

$$w_t = \mathbf{A}(x,t)w - kw - \frac{Q_t - \mathbf{A}(x,t)Q - (u_1(x,t),\varphi_t(0,t))}{Q}w + u_2(x,t)\frac{(w(x,t),\varphi_t(0,t))}{Q},$$

$$(x,t) \in Q_T^1; \quad w(x,0) = 0.$$

Multiply the preceding equality by $w(x,t)$ scalarly to obtain

$$\left(\frac{\mathrm{d}w}{\mathrm{d}t}, w\right) + a(t,w,w) + k(w,w) = \frac{Q_t - \mathbf{A}(x,t)Q - (u_1(x,t),\varphi_t(0,t))}{Q}(w,w) + \frac{(w(x,t),\varphi_t(0,t))(w(x,t),u_2(x,t))}{Q}.$$

Since $w(x,0)=0$, in the same manner as in (Lions, 1968, Chapter 3) we obtain

$$\int_0^T \frac{\mathrm{d}}{\mathrm{d}t}(w(x,t),w(x,t))\,\mathrm{d}t = \frac{1}{2}|w(x,T)|^2.$$

Also, observe that $|(w(x,t),\varphi_t(0,t))| \le C_1|w|$ and $|(w(x,t),u_2(x,t))| \le C_2|w|$, $(w,w)=|w|^2$. Therefore, if

$$k > \frac{1}{\mu_0}\Big(C_1C_2 + \max_{t\in[0,T]}\big|Q_t - \mathbf{A}(x,t)Q - (u_1(x,t),\varphi_t(0,t))\big|\Big),$$

then

$$\alpha\int_0^T \|w(x,t)\|^2\,\mathrm{d}t + \frac{1}{2}|w(x,T)|^2 \le 0,$$

which is possible only if $w(x,t)\equiv 0$.

Existence: We shall prove this part of the theorem by the method of successive approximation. Let M denote the bounded closed set of functions in $W_1(0,T)$ which is bounded from above in the norm of $W_1(0,T)$ by some constant K that will be indicated later. If $w(x,t)$ belongs to M, then by the embedding $W_1(0,T)\subset C^0([0,T],H_1)$ (cf. Lions, 1968; Lions and Magenes, 1972) we have $|w(x,t)|<K$ for all $t\in[0,T]$.

Consider the mapping G transforming a function $w(x,t)$ in $W_1(0,T)$ into the function $v(x,t)$ that is a solution of the problem

$$v_t - \mathbf{A}(x,t)v = \frac{Q - \mathbf{A}(x,t)Q - (w(x,t),\varphi_t(0,t))}{Q}v; \quad (x,t)\in Q_T^1; \quad v(x,0)=u_0(x).$$

It is easy to see that

$$\max_{t\in[0,T]}\left|\frac{Q_t - \mathbf{A}(x,t)Q - (w(x,t),\varphi_t(0,t))}{Q}\right| \le C_3 + C_4K;$$

therefore, in virtue of (2.61) we obtain

$$\|v(x,t)\|_{W_1(0,T)} \le C_5 e^{C_4KT}|u_0|. \tag{2.66}$$

Choose $K = 2C_5|u_0|$ and $T>0$ so as to have $\exp(C_42C_5|u_0|T)\le 2$. Then $G(w)$ belongs to M. Now, choose $T>0$ so that the mapping G become a contraction on M. Let $v_1 = G(w_1)$, $v_2 = g(w_2)$, $w = w_1 - w_2$, and $v = v_1 - v_2$. Then

$$v_t - \mathbf{A}(x,t)v = \frac{Q_t - \mathbf{A}(x,t)Q - (w_1(x,t),\varphi_t(0.t))}{Q}v + \frac{(w(x,t),\varphi_t(0,t))}{Q}v_1(x,t),$$

$$(x,t)\in Q_T^1, \quad w(x,0)=0. \tag{2.67}$$

By applying estimate (2.61) to problem (2.67), we obtain

$$\|v\|_{W_1(0,T)} \leq C_6 \left(\int_0^T |w(x,t)|^2 \, dt \right)^{1/2} \leq C_6 T \, \|w\|_{W_1(0,T)} .$$

It means that if $C_6 T < 1$ then the mapping G is a contraction on M.

Thus, if $T < \min(\ln 2/2C_4C_5\,|u_0|, 1/C_6)$, then the mapping G takes the set M into itself and is a contraction on it. Therefore, the mapping G has a unique fixed point on M which gives a solution of the inverse problem (2.63). Theorem 2.14 is proved.

2.6 REDUCTION OF MORE COMMON INVERSE PROBLEMS TO INITIAL-BOUNDARY VALUE PROBLEMS

It is usual to call *identification problems* those problems (for instance of differential type) in which we have to find out, beside the so-called solution (which usually describes the *state of the physical system*), also some *unknown terms* by means of some additional information concerning the solution itself.

Many interesting theoretical and practical problems are connected with recovering unknown coefficients in differential, integral and integrodifferential equations. As a rule, such problems are multidimensional, nonlinear and *ill-posed* according to Hadamard's classical definition. The ill-posedness causes, as is well-known, serious difficulties in the mathematical treatment of identification problems.

A special interest is addressed nowadays to the problem of identifying coefficients depending on a vector-space variable (briefly *multidimensional coefficients*) entering some differential equation. For this purpose, we stress that the question of the solvability strictly depends on a *suitable* (and physically meaningful) choice of the additional information.

In the present section, we consider a multidimensional nonlinear identification problem related to a general (second-order) linear parabolic equation with separated vector-variables, containing two *unknown* coefficients, each of which depends on one vector-variable, only.

The basic idea in the present section is somewhat different: it consists in reducing our identification problem to the analysis of a *infinitely dimensional* nonlinear Cauchy (*direct*) problem, also *containing* the additional information. This procedure allows us to prove the *existence, uniqueness* and *stability* (with respect to the data) of the solution of our identification problem.

We state, now, our problem: determine three functions $w : [0,T] \times \Omega_1 \times \Omega_2 \to \mathrm{R}$, $\lambda : [0,T] \times \Omega_1 \to \mathrm{R}$, $\mu : [0,T] \times \Omega_2 \to \mathrm{R}$ such that

$$\begin{aligned} D_t w(t,x,y) \; = \; & \mathbf{A}(x,D_x)w(t,x,y) + \lambda(t,x)\mathbf{B}(x,D_x)w(t,x,y) + \mathbf{C}(y,D_y)w(t,x,y) \\ & + \mu(t,y)\mathbf{D}(y,D_y)w(t,x,y), \quad (t,x,y) \in [0,T] \times \Omega_1 \times \Omega_2, \end{aligned} \tag{2.68}$$

$$w(0,x,y) = \Psi\big(u_0(x), v_0(y)\big), \quad (x,y) \in \Omega_1 \times \Omega_2, \tag{2.69}$$

$$w(t,x,y) = 0, \quad (t,x,y) \in ([0,T] \times \partial\Omega_1 \times \Omega_2) \cup ([0,T] \times \Omega_1 \times \partial\Omega_2), \tag{2.70}$$

$$\tilde{\Phi}_2\big[w(t,x,\cdot)\big] = p(t,x), \quad (t,x) \in [0,T] \times \Omega_1, \tag{2.71}$$

$$\tilde{\Phi}_1\big[w(t,\cdot,y)\big] = q(t,y), \quad (t,y) \in [0,T] \times \Omega_2. \tag{2.72}$$

We have used the following notation: $\Omega_1 \subset \mathrm{R}^m$ and $\Omega_2 \subset \mathrm{R}^n$ $(m, n \in \mathbf{N})$ denote two open bounded domains of class $C^{2+\varepsilon}$ $(\varepsilon \in (0,1))$.

$\mathbf{A}(x, D_x)$ and $\mathbf{C}(y, D_y)$ denote second-order uniformly elliptic linear operators with continuous coefficients acting on the variables $x \in \Omega_1$ and $y \in \Omega_2$:

$$\mathbf{A}(x, D_x) = \sum_{i,j=1}^{m} a_{ij}(x) D_{x_i} D_{x_j}, \quad \mathbf{C}(y, D_y) = \sum_{i,j=1}^{n} c_{ij}(y) D_{y_i} D_{y_j}.$$

$$a_{ij}(x) = a_{ji}(x) \quad \text{for all } x \in \overline{\Omega}_1,\ i,j = 1, \dots m,$$

$$c_{ij}(x) = c_{ji}(x) \quad \text{for all } y \in \overline{\Omega}_2,\ i,j = 2, \dots n,$$

$$\sum_{i,j=1}^{m} a_{ij}(x) \xi_i \xi_j \geq \mu_1 |\xi|^2 \quad \text{for all } (x, \xi) \in \overline{\Omega}_1 \times \mathrm{R}^m \quad (\mu_1 > 0),$$

$$\sum_{i,j=1}^{n} c_{ij}(y) \eta_i \eta_j \geq \mu_2 |\eta|^2 \quad \text{for all } (y, \eta) \in \overline{\Omega}_2 \times \mathrm{R}^n \quad (\mu_2 > 0).$$

$\mathbf{B}(x, D_x)$ and $\mathbf{D}(y, D_y)$ denote first-order linear differential operators with continuous coefficients:

$$\mathbf{B}(x, D_x) = \sum_{j=1}^{m} b_j(x) D_{x_j} + b_0(x), \quad \mathbf{D}(y, D_y) = \sum_{j=1}^{n} d_j(x) D_{y_j} + d_0(y).$$

$\tilde{\Phi}_1$ and $\tilde{\Phi}_2$ denote two continuous, linear functionals on $C(\overline{\Omega}_2; \mathrm{R})$ and $C(\overline{\Omega}_1; \mathrm{R})$, respectively, while Ψ stands for a continuous bilinear form acting on a Cartesian product $S_1 \times S_2$ of (possibly) infinitely dimensional Banach spaces. We assume also that $\tilde{\Phi}_1$, $\tilde{\Phi}_2$ and Ψ enjoy the following properties:

$$\tilde{\Phi}_1 \tilde{\Phi}_2[w] = \tilde{\Phi}_2 \tilde{\Phi}_1[w] \quad \text{for all } w \in W, \tag{2.73}$$

$$\mathbf{A}\Psi(u, v) = \Psi(\mathbf{A}u, v), \quad \mathbf{B}\Psi(u, v) = \Psi(\mathbf{B}u, v) \quad \text{for all } (u, v) \in U \times V, \tag{2.74}$$

$$\mathbf{C}\Psi(u, v) = \Psi(u, \mathbf{C}v), \quad \mathbf{D}\Psi(u, v) = \Psi(u, \mathbf{D}v) \quad \text{for all } (u, v) \in U \times V, \tag{2.75}$$

$$\tilde{\Phi}_1 \Psi(u, v) = \Psi\big(\Phi_1[u], v\big), \quad \tilde{\Phi}_2 \Psi(u, v) = \Psi\big(u, \Phi_2[v]\big) \quad \text{for all } (u, v) \in U \times V, \tag{2.76}$$

where

$$W = \big\{ z \in C^1(\overline{\Omega}_1 \times \overline{\Omega}_2; \mathrm{R}) : \mathbf{A}z \in C(\overline{\Omega}_1 \times \overline{\Omega}_2; \mathrm{R}),\ \mathbf{C}z \in C(\overline{\Omega}_1 \times \overline{\Omega}_2; \mathrm{R}) \big\},$$

$$U = \big\{ z \in C^1(\overline{\Omega}_1; S_1) : \mathbf{A}z \in C(\overline{\Omega}_1; S_1) \big\},$$

$$V = \big\{ z \in C^1(\overline{\Omega}_1; S_2) : \mathbf{C}z \in C(\overline{\Omega}_2; S_2) \big\},$$

and Φ_j $(j = 1, 2)$ is a continuous, linear operator from S_j to itself.

Finally, we note that $p : [0, T] \times \overline{\Omega}_1 \to \mathrm{R}$, $q : [0, T] \times \overline{\Omega}_2 \to \mathrm{R}$, $u_0 : \overline{\Omega}_1 \to S_1$ and $v_0 : \overline{\Omega}_2 \to S_2$ are prescribed functions satisfying the consistency conditions

$$p(0, x) = \Psi\big(u_0(x), \Phi_2[v_0]\big) \quad \text{for all } x \in \Omega_1, \tag{2.77}$$

$$q(0, y) = \Psi\big(\Phi_1[u_0], v_0(y)\big) \quad \text{for all } y \in \Omega_2, \tag{2.78}$$

$$\tilde{\Phi}_1\big[p(t, \cdot)\big] = \tilde{\Phi}_2\big[q(t, \cdot)\big] \quad \text{for all } t \in [0, T], \tag{2.79}$$

$$p(t,x) = 0 \quad \text{for all } (t,x) \in [0,T] \times \partial\Omega_1, \tag{2.80}$$

$$q(t,y) = 0 \quad \text{for all } (t,y) \in [0,T] \times \partial\Omega_2. \tag{2.81}$$

We conclude this section by noting that the seemingly restrictive initial condition (2.69) *can* be replaced by the usual one

$$w(0,x,y) = f(x,y) \quad (x,y) \in \Omega_1 \times \Omega_2. \tag{2.82}$$

This will be shown in section 6, where f is only assumed to be a (smooth) function satisfying the consistency conditions related to problem (2.68), (2.82), (2.70)–(2.72).

We look for a solution $w \in C^{1+\alpha}([0,T]; W)$ to problem (2.68)–(2.70) with separated variables of the form

$$w(t,x,y) = \Psi\big(u(t,x), v(t,y)\big), \tag{2.83}$$

where $u \in C^{1+\alpha}([0,T]; U)$, $v \in C^{1+\alpha}([0,T]; V)$ $(\alpha \in (0,1))$.

Consequently, equation (2.68) can be rewritten as

$$\begin{aligned} &\Psi\big(D_t u(t,x), v(t,y)\big) + \Psi\big(u(t,x), D_t v(t,y)\big) \\ &= \Psi\big(\mathbf{A}(x,D_x)u(t,x) + \lambda(t,x)\mathbf{B}(x,D_x)u(t,x), v(t.y)\big) \\ &\quad + \Psi\big(u(t,x), \mathbf{C}(y,D_y)v(t,y) + \mu(t,y)\mathbf{D}(y,D_y)v(t,y)\big). \end{aligned}$$

We now choose the pair (u,v) so as to satisfy the boundary value problems

$$D_t u(t.x) = \mathbf{A}(x,D_x)u(t,x) + \lambda(t,x)\mathbf{B}(x,D_x)u(t,x) \quad \text{in } (0,T) \times \Omega_1, \tag{2.84}$$

$$u(0,x) = u_0(x) \quad \text{in } \Omega_1, \tag{2.85}$$

$$u(t,x) = 0 \quad \text{on } (0,T) \times \partial\Omega_1 \tag{2.86}$$

and

$$D_t v(t.y) = \mathbf{C}(y,D_y)v(t,y) + \mu(t,y)\mathbf{D}(y,D_y)v(t,y) \quad \text{in } (0,T) \times \Omega_2, \tag{2.87}$$

$$v(0,y) = v_0(y) \quad \text{in } \Omega_2, \tag{2.88}$$

$$v(t,y) = 0 \quad \text{on } (0,T) \times \partial\Omega_2. \tag{2.89}$$

If the pair $(u.v)$ satisfies (2.84)–(2.87), it is easy to verify that the function w defined by (2.83) is the solution of direct problem (2.68)–(2.70). Moreover, if $\tilde{w}$ is any solution of (2.68)–(2.70). the the pair $\tilde{w} - w$ solves the linear problem (2.68)–(2.70) with homogeneous data. Hence, in a wide class of coefficients λ, μ and initial data u_0, v_0 , we get $\tilde{w} = w$ in $(0,T) \times \Omega_1 \times \Omega_2$.

Now rewrite equation (2.71) in the form

$$\Psi\big(u(t,x), \Phi_2[v(t,\cdot)]\big) = p(t,x). \tag{2.90}$$

Differentiating (2.90) with respect to t and using equation (2.84) we easily derive that λ solves

$$\begin{aligned} \mathbf{A}(x,D_x)p(t,x) &+ \lambda(t,x)\mathbf{B}(x,D_x)p(t,x) \\ &+ \Psi\big(u(t,x), D_t\Phi_2[v(t,\cdot)]\big) = D_t p(t,x) \quad (t,x) \in (0,T) \times \Omega_1. \end{aligned} \tag{2.91}$$

Assume

$$|\mathbf{B}(x,D_x)p(t,x)| \geq m_1 > 0 \quad \text{for all } (t,x) \in [0,T] \times \overline{\Omega}_1. \tag{2.92}$$

By virtue of (2.91)-(2.92) we can express λ in terms of the pair (u,v):

$$\begin{aligned} \lambda(t,x) &= \left[\mathbf{B}(x,D_x)p(t,x)\right]^{-1}\left[D_t p(t,x) - \mathbf{A}(x,D_x)p(t,x)\right] \\ &\quad -\left[\mathbf{B}(x,D_x)p(t,x)\right]^{-1}\Psi\big(u(t,x), D_t\Phi_2[v(t,\cdot)]\big) \\ &:= J_0(t,x) + J(u,D_t v)(t,x) \quad (t,x) \in (0,T)\times\Omega_1 \end{aligned} \tag{2.93}$$

Likewise, assuming

$$|\mathbf{D}(y,D_y)q(t,y)| \geq m_2 > 0 \quad \text{for all } (t,y) \in [0,T] \times \overline{\Omega}_2 \tag{2.94}$$

we can express μ in terms of the pair (u,v):

$$\begin{aligned} \mu(t,y) &= \left[\mathbf{D}(y,D_y)q(t,y)\right]^{-1}\left[D_t q(t,y) - \mathbf{C}(y,D_y)q(t,y)\right] \\ &\quad -\left[\mathbf{D}(y,D_y)q(t,y)\right]^{-1}\Psi\big(D_t\Phi_1[u(t,\cdot)], v(t,y)\big) \\ &:= K_0(t,x) + K(D_t u, v)(t,y) \quad (t,y) \in (0,T)\times\Omega_2. \end{aligned} \tag{2.95}$$

Consequently, the pair (u,v) solves the nonlinear system

$$D_t u(t,x) = \mathbf{A}(x,D_x)u(t,x) + \left[J_0(t,x) + J(u,D_t v)(t,x)\right]\mathbf{B}(x,D_x)u(t,x) \quad \text{in } (0,T)\times\Omega_1, \tag{2.96}$$

$$D_t v(t,y) = \mathbf{C}(y,D_y)v(t,y) + \left[K_0(t,x) + K(D_t u, v)(t,y)\right]\mathbf{D}(y,D_y)v(t,y) \quad \text{in } (0,T)\times\Omega_2, \tag{2.97}$$

$$u(0,x) = u_0(x) \quad \text{in } \Omega_1, \tag{2.98}$$

$$v(0,y) = v_0(y) \quad \text{in } \Omega_2, \tag{2.99}$$

$$u(t,x) = 0 \quad \text{on } (0,T)\times\partial\Omega_1, \tag{2.100}$$

$$v(t,y) = 0 \quad \text{on } (0,T)\times\partial\Omega_2. \tag{2.101}$$

Conversely, assume that (u,v) solves the boundary value problem (2.96)-(2.101) and introduce functions λ and μ defined by formulas (2.93)-(2.95). Then the triplet (w,λ,μ), where w is defined by (2.83), satisfies the identification problem (2.68)-(2.72). In fact, according to (2.93), (2.95)-(2.101), (2.74)-(2.75), we easily verify that (w,λ,μ) satisfies (2.68)-(2.70). In order to conclude the proof of our assertion, we need to show that the boundary conditions (2.71)-(2.72) are fulfilled as well. For this purpose, we introduce the functions

$$\tilde{p}(t,x) = \Psi\big(u(t,x), \Phi_2[v(t,\cdot)]\big) \quad \text{for all } (t,x) \in [0,T]\times\overline{\Omega}_1. \tag{2.102}$$

$$\tilde{q}(t,y) = \Psi\big(\Phi_1[u(t,\cdot)], v(t,y)\big) \quad \text{for all } (t,y) \in [0,T]\times\overline{\Omega}_2. \tag{2.103}$$

In order to show that $(\tilde{p},\tilde{q}) = (p,q)$, observe first that, according to (2.77), (2.102), we get the equation

$$\tilde{p}(0,x) = \Psi\big(u_0(x), \Phi_2[v_0]\big) = p(0,x) \quad \text{for all } x \in \overline{\Omega}_1.$$

Recalling that $u \in C^{1+\alpha}([0,T];U)$ and $v \in C^{1+\alpha}([0,T];V)$, from (2.92) and the continuity of $\mathbf{B}(x,D_x)p$, we deduce that there exists $T_0 \in (0,T]$ such that

$$|\mathbf{B}(x,D_x)\tilde{p}(t,x)| \geq \frac{1}{2}m_1 > 0 \quad \text{for all } (t,x) \in [0,T_0] \times \overline{\Omega}_1.$$

Then, we consider the equation (derived from (2.96))

$$\begin{aligned}\Psi(D_t u(t,x), \Phi_2[v(t,\cdot)]) &= \Psi(\mathbf{A}(x,D_x)u(t,x), \Phi_2[v(t,\cdot)]) \\ &+ [J_0(t,x) + J(u,D_t v)(t,x)]\Psi(\mathbf{B}(x,D_x)u(t,x), \Phi_2[v(t,\cdot)]) \quad \text{in } (0,T)\times\Omega_1.\end{aligned} \tag{2.104}$$

Solving (2.104) for $J_0 + J$, using definition (2.102) and comparing with (2.93), we easily derive the following equation involving the functions $\tilde{p}$ and p in $(0,T_0)\times\Omega_1$:

$$\begin{aligned}&[\mathbf{B}(x,D_x)\tilde{p}(t,x)]^{-1}\{D_t\tilde{p}(t,x) - \mathbf{A}(x,D_x)\tilde{p}(t,x) - \Psi(u(t,x), D_t\Phi_2[v(t,\cdot)])\} \\ &= [\mathbf{B}(x,D_x)p(t,x)]^{-1}\{D_t p(t,x) - \mathbf{A}(x,D_x)p(t,x) - \Psi(u(t,x), D_t\Phi_2[v(t,\cdot)])\}.\end{aligned}$$

Consequently, p and $\tilde{p}$ are both solutions of the following parabolic problem, where $f \in C([0,T_0]\times\overline{\Omega}_1)$ is a given function:

$$\begin{aligned}&D_t z(t,x) - \mathbf{A}(x,D_x)z(t,x) - f(t,x)\mathbf{B}(x,D_x)z(t,x) \\ &= \Psi(u(t,x), D_t\Phi_2[v(t,\cdot)]) \quad (t,x) \in [0,T]\times\Omega_1,\end{aligned} \tag{2.105}$$

$$z(0,x) = \Psi(u_0(x), \Phi_2[v_0]) \quad \text{for all } x \in \Omega_1, \tag{2.106}$$

$$z(t,x) = 0 \quad \text{for all } (t,x) \in [0,T]\times\partial\Omega_1. \tag{2.107}$$

By virtue of the uniqueness of the solution of problem (2.105)–(2.107), we deduce that $\tilde{p} = p$ in $[0,T_0]\times\overline{\Omega}_1$. Likewise, we can show that $\tilde{q} = q$ in $[0,T_0]\times\overline{\Omega}_2$.

We observe now that $\tilde{p}$ and $\tilde{q}$ satisfy in $[0,T_0]\times\overline{\Omega}_1$ bounds (2.92) and (2.94), respectively. Following the previous procedure, we can show that the equations $\tilde{p} = p$ and $\tilde{q} = q$ hold in the two larger cylinders $[0,T_1]\times\overline{\Omega}_1$ and $[0,T_0]\times\overline{\Omega}_2$, respectively, where $T_1 \in (T_0,T]$. This ensures the existence of a *maximal* time $T^* \in (T_0,T]$ for which

$$\tilde{p} = p \quad \text{in } [0,T^*)\times\overline{\Omega}_1, \quad \tilde{q} = q \quad \text{in } [0,T^*)\times\overline{\Omega}_2. \tag{2.108}$$

Assume, by contradiction, that $T^* < T$. Using the continuity of functions p, $\tilde{p}$ and q, $\tilde{q}$ in $[0,T]\times\overline{\Omega}_1$ and $[0,T]\times\overline{\Omega}_2$ respectively, from (2.108) we derive that such equations actually hold true in $[0,T^*]\times\overline{\Omega}_1$ and $[0,T^*]\times\overline{\Omega}_2$, respectively. Then, reasoning as above, we can show that equations $\tilde{p} = p$ and $\tilde{q} = q$ actually hold in $[0,\tilde{T}]\times\overline{\Omega}_1$ and $[0,\tilde{T})\times\overline{\Omega}_2$, respectively, for some $\tilde{T} \in (T^*,T)$. This contradicts our assumption: hence $T^* = T$.

We want to conclude this section by exhibiting the particularly simple case, when $S_1 = S_2 = \mathrm{R}$ and the bilinear functional Ψ has the form of a product:

$$\Psi(\xi,\eta) = \xi\eta.$$

First, we observe that $\tilde{\Phi}_1 = \Phi_1$, $\tilde{\Phi}_1 = \Phi_1$ and our data p, q, u_0, v_0 have to satisfy the following consistency conditions (cf. (2.77)–(2.81)):

$$p(0,x) = u_0(x)\Phi_2[v_0] \quad \text{for all } x \in \Omega_1, \tag{2.109}$$

$$q(0,y) = \Phi_1[u_0]v_0(y) \quad \text{for all } y \in \Omega_2, \tag{2.110}$$

$$\Phi_1\big[p(t,\cdot)\big] = \Phi_2\big[q(t,\cdot)\big] := w_0(t) \quad \text{for all } t \in [0,T], \tag{2.111}$$

$$p(t,x) = 0 \quad \text{for all } (t,x) \in [0,T] \times \partial\Omega_1, \tag{2.112}$$

$$q(t,y) = 0 \quad \text{for all } (t,y) \in [0,T] \times \partial\Omega_2. \tag{2.113}$$

We assume also that p, q satisfy the additional conditions

$$p(t,x) \neq 0 \quad \text{for all } (t,x) \in [0,T] \times \Omega_1, \tag{2.114}$$

$$q(t,y) \neq 0 \quad \text{for all } (t,y) \in [0,T] \times \Omega_2. \tag{2.115}$$

We look now for a solution of (2.68) of the form

$$w(t,x,y) = u(t,x)v(t,y),$$

where

$$u(t,x) = \frac{p(t,x)}{P(t)}, \quad v(t,y) = q(t,y).$$

and $P : [0,T] \to \mathrm{R}$ is an unknown function, *nowhere vanishing* in $[0,T]$.

Owing to consistency conditions (2.109)–(2.113) and assumptions (2.114)–(2.115), it is easy to verify that w satisfies conditions (2.69)–(2.72) if, and only if,

$$P(t) = w_0(t) \quad \text{for all } t \in [0,T].$$

Consequently, *a* solution of our identification problem (2.68)–(2.72) is given (in a closed form) by

$$w(t,x,y) = \frac{p(t,x)q(t,y)}{w_0(t)} \quad \text{for all } (t,x,y) \in [0,T] \times \Omega_1 \times \Omega_2, \tag{2.116}$$

$$\lambda(t,x) = \big[\mathbf{B}(x,D_x)p(t,x)\big]^{-1}\left[D_tp(t,x) - \mathbf{A}(x,D_x)p(t,x) - \frac{p(t,x)D_tw_0(t)}{w_0(t)}\right], \tag{2.117}$$

$$(t,x) \in [0,T] \times \overline{\Omega}_1,$$

$$\mu(t,y) = \big[\mathbf{D}(y,D_y)q(t,y)\big]^{-1}[D_tq(t,y) - \mathbf{C}(y,D_y)q(t,y)], \quad (t,x) \in [0,T] \times \overline{\Omega}_2. \tag{2.118}$$

REMARK 2.4. The previous procedure does not ensure the *uniqueness* of the solution assigned by (2.116)–(2.118).

However, we note that the existence of the solution, in the present case, is ensured whatever the pair $\tilde{\Phi}_1$ $\tilde{\Phi}_2$ may be.

REMARK 2.5. In the case where $S_1 = \mathrm{R}^\nu$, $S_2 = \mathrm{R}^\nu$ ($\mathrm{R}^\infty := \ell^2$),

$$\Psi(\xi,\eta) = \sum_{j=1}^{\nu} \xi_j\eta_j$$

and ν is (any) integer in $[2,+\infty]$, we cannot find any solution of problem (2.68)–(2.72) in a *closed form*. This justifies the theoretical approach developed in paper (Anikonov and Lorenzi, 1994).

2.7 FORMULAS IN PROBLEMS OF DETERMINATION OF SOURCES

The results of this section were obtained by Ayupova (1986).

Let **L** and **M** be differential operators such that

$$\mathbf{LM} = \mathbf{ML}. \tag{2.119}$$

Consider an equation

$$\mathbf{L}u = f, \tag{2.120}$$

where f satisfies a condition

$$\mathbf{M}f = 0. \tag{2.121}$$

There exist initial and boundary conditions for function u depending on operators **L** and **M**. The problem is posed: to find the source f by information for u.

In this section the formulas for determination of f for the concrete operators **L** and **M** are given.

The algorithm of the source determination is based on the structure method of the solution of inverse problems which is determined below. Let us apply the operator **M** to both sides of the equation (2.120). By (2.119) and (2.121) we obtain

$$\mathbf{LM}u = 0.$$

Denoting $\mathbf{M}u = w$, we have

$$\mathbf{L}w = 0.$$

Solving consequently equations $\mathbf{L}w = 0$ and $\mathbf{M}u = w$, we find u. Substituting u in equation (2.120), we obtain f. Further, everywhere

$$\Delta = \sum_{i=1}^{n} \frac{\partial^2}{\partial x_i^2}, \quad \Delta_{n-1} = \sum_{j=1, j\neq i}^{n} \frac{\partial^2}{\partial x_j^2}$$

$$\tilde{\xi} = (\xi_1, \ldots, \xi_{i-1}, \xi_{i+1}, \ldots, \xi_n), \quad \tilde{x} = (x_1, \ldots, x_{i-1}, x_{i+1}, \ldots, x_n)$$

Example 2.6.

$$\mathbf{L}u = u_t - \Delta u = f(x,t),$$
$$\mathbf{M}f = f_{x_i} - \Delta_{n-1} f = 0,$$
$$t \in \mathrm{R}^+, \quad x_i \in \mathrm{R}^+, \quad \tilde{x} \in \mathrm{R}^{n-1},$$
$$u|_{t=0} = u_0(x), \quad u|_{x_i=0} = p_0(\tilde{x}, t),$$

$$f(x,t) = \frac{1}{(2\sqrt{\pi x_i})^{n-1}} \int\limits_{\mathrm{R}^{n-1}} \Big[p_{0t}(\tilde{\xi}, t) - \Delta_{n-1} p_0(\tilde{\xi}, t) - w_{\xi_i}(\tilde{\xi}, 0, t)$$
$$- \Delta_{n-1}\Delta_{n-1} p_0(\tilde{\xi}, t) - \Delta_{n-1} w(\tilde{\xi}, 0, t)\Big] e^{-\frac{|\tilde{x}-\tilde{\xi}|^2}{4x_i}} \, d\tilde{\xi},$$

where $w(x,t)$ is a solution of the problem

$$w_t - \Delta w = 0,$$

$$w\big|_{t=0} = u_{0x_i} - \Delta_{n-1}u_0.$$

Example 2.7.

$$\mathbf{L}u = u_t - \Delta u = f(x,t),$$
$$\mathbf{M}f = f_{x_ix_i} + \omega^2 f = 0,$$
$$t \in \mathrm{R}^+, \quad x \in \mathrm{R}^n,$$
$$u\big|_{t=0} = u_0(x), \quad u\big|_{x_i=0} = p_0(\tilde{x},t), \quad u_{x_i}\big|_{x_i=0} = p_1(\tilde{x},t),$$
$$f(x,t) = a(\tilde{x},t)\cos\omega x_i + b(\tilde{x},t)\sin\omega x_i,$$

where

$$a(\tilde{x},t) = -w(\tilde{x},0,t) + p_{0t}(\tilde{x},t) - \Delta_{n-1}p_0(\tilde{x},t) + \omega^2 p_0(\tilde{x},t),$$

$$b(\tilde{x},t) = -\frac{1}{\omega}w_{x_i}(\tilde{x},0,t) + \frac{1}{\omega}p_{1t}(\tilde{x},t) - \frac{1}{\omega}\Delta_{n-1}p_1(\tilde{x},t) + \omega p_1(\tilde{x},t),$$

w is a solution of the problem

$$w_t - \Delta w = 0,$$
$$w\big|_{t=0} = u_{0x_ix_i} + \omega^2 u_0.$$

Example 2.8.

$$\mathbf{L}u = u_t - \Delta u = f(x,t),$$
$$\mathbf{M}f = f_{x_ix_i} - f_{x_jx_j} = 0,$$
$$t \in \mathrm{R}^+, \quad x_i \in \mathrm{R}^+, \quad \tilde{x} \in \mathrm{R}^{n-1},$$
$$u\big|_{t=0} = u_0(x), \quad u\big|_{x_i} = p_0(\tilde{x},t), \quad u_{x_i}\big|_{x_i} = p_1(\tilde{x},t),$$

$$\begin{aligned} f(x,t) = {} & \frac{1}{2}\int\limits_{x_j-x_i}^{x_j+x_i} \left[p_{1t}(x',\xi_j,t) - \Delta_{n-2}p_1(x',\xi_j,t) - w_{\xi_j}(x',0,\xi_j,t)\right] \mathrm{d}\xi_j \\ & + \frac{1}{2}\left[p_{0t}(x',x_j - x_i,t) + p_{0t}(x',x_j + x_i,t)\right] \\ & - \frac{1}{2}\left[\Delta_{n-2}p_0(x',x_j - x_i,t) + \Delta_{n-2}p_0(x',x_j + x_i,t)\right] \\ & - \frac{1}{2}\left[w(x'.0,x_j + x_i,t) + w(x',0,x_j - x_i,t)\right] \\ & - \left[p_{0x_j}(x'.x_j + x_i,t) - p_{0x_j}(x',x_j - x_i,t)\right] \\ & - \left[p_{0x_jx_j}(x',x_j + x_i,t) + p_{0x_jx_j}(x',x_j - x_i,t)\right], \end{aligned}$$

where $x' = (x_1, \dots, x_{i-1}, x_{i+1}, \dots, x_{j-1}, x_{j+1}, \dots, x_n)$, $\Delta_{n-2} = \sum\limits_{\substack{k\neq i\\ k\neq j}} \dfrac{\partial^2}{\partial x_k^2}$ and w is a solution of the problem

$$w_t - \Delta w = 0,$$

$$w\big|_{t=0} = u_{0x_ix_i} + u_{0x_jx_j}.$$

Example 2.9.

$$\mathbf{L}u = u_t - \Delta u = f(x,t),$$

$$\mathbf{M}f = \Delta u = 0,$$

$$t \in \mathrm{R}^+, \quad x \in \mathrm{R}^n,$$

$$u\big|_{t=0} = u_0(x), \quad u\big|_{\partial S} = p_0(x,t).$$

It should be $u_0\big|_{\partial S} = p_0(x,0)$, S is sphere of the radius R in the space R^n.

$$f(x,t) = \int\limits_{|\xi|=R} P_R(x,\xi) p_{0t}(\xi,t)\, dS_\xi$$

$$- \int\limits_{|\xi|<R} G_R(x,\xi) \left[-\frac{n}{(2\sqrt{\pi})^n 2(\sqrt{t})^{n+2}} \int\limits_{\mathrm{R}^n} \Delta u_0(\eta)\, e^{-\frac{|\xi-\eta|}{4t}}\, d\eta \right.$$

$$\left. + \frac{1}{(2\sqrt{\pi t})^n 4t^2} \int\limits_{\mathrm{R}^n} |\xi - \eta|\, \Delta u_0(\eta)\, e^{-\frac{|\xi-\eta|}{4t}}\, d\eta \right] d\xi - \frac{1}{(2\sqrt{\pi t})^n} \int\limits_{\mathrm{R}^n} \Delta u_0(\xi)\, e^{-\frac{|x-\xi|}{4t}}\, d\xi$$

for $x \in S$, where

$$P_R(x,\xi) = \frac{R^2 - |x|^2}{\sigma_n R\, |x-\xi|^n},$$

$$G_R(x,\xi) = \frac{1}{\sigma_n} \left(\frac{1}{|x-\xi|^{n-2}} - \frac{(R/|x|)^{n-2}}{|(R/|x|)^2 x - \xi|^{n-2}} \right),$$

$\sigma_n = n\pi^{n/2}/\Gamma(n/2+1)$ is square of the unit sphere in the space R^n.

Example 2.10.

$$\mathbf{L}u = u_{tt} - \Delta u = f(x,t),$$

$$\mathbf{M}f = f_{x_i} - \Delta_{n-1} f = 0,$$

$$t \in \mathrm{R}^+, \quad x_i \in \mathrm{R}^+, \quad \tilde{x} \in \mathrm{R}^n,$$

$$u\big|_{t=0} = u_0(x), \quad u_t\big|_{t=0} = u_1(x), \quad u\big|_{x_i} = p_0(\tilde{x},t)$$

$$f(x,t) = \frac{1}{(2\sqrt{\pi x_i})^{n-1}} \int\limits_{\mathrm{R}^{n-1}} \Big[p_{0tt}(\tilde{\xi},t) - \Delta_{n-1} p_0(\tilde{\xi},t) - w_{\xi_i}(\tilde{\xi},0,t)$$

$$- \Delta_{n-1}\Delta_{n-1} p_0(\tilde{\xi},t) - \Delta_{n-1} w(\tilde{\xi},0,t) \Big]\, e^{-\frac{|\tilde{x}-\tilde{\xi}|^2}{4x_i}}\, d\tilde{\xi},$$

where $w(x,t)$ is a solution of the problem

$$w_{tt} - \Delta w = 0,$$

$$w\big|_{t=0} = u_{0x_i} - \Delta_{n-1}u_0, \quad w_t\big|_{t=0} = u_{1x_i} - \Delta_{n-1}u_1.$$

Example 2.11.

$$\mathbf{L}u = u_{tt} - \Delta u = f(x,t),$$
$$\mathbf{M}f = f_{x_ix_i} + \omega^2 f = 0,$$
$$t \in \mathrm{R}^+, \quad x \in \mathrm{R}^n,$$
$$u\big|_{t=0} = u_0(x), \quad u\big|_{x_i=0} = p_0(\tilde{x},t),$$
$$u_t\big|_{t=0} = u_1(x), \quad u_{x_i}\big|_{x_i=0} = p_1(\tilde{x},t),$$
$$f(x,t) = a(\tilde{x},t)\cos\omega x_i + b(\tilde{x},t)\sin\omega x_i,$$

where

$$a(\tilde{x},t) = -w(\tilde{x},0,t) + p_{0tt}(\tilde{x},t) - \Delta_{n-1}p_0(\tilde{x},t) + \omega^2 p_0(\tilde{x},t),$$

$$b(\tilde{x},t) = -\frac{1}{\omega}w_x(\tilde{x},0,t) + \frac{1}{\omega}p_{1tt}(\tilde{x},t) - \frac{1}{\omega}\Delta_{n-1}p_1(\tilde{x},t) + \omega p_1(\tilde{x},t),$$

w is a solution of the problem

$$w_{tt} - \Delta w = 0,$$
$$w\big|_{t=0} = u_{0x_ix_i} + \omega^2 u_0,$$
$$w_t\big|_{t=0} = u_{1x_ix_i} + \omega^2 u_1.$$

Example 2.12.

$$\mathbf{L}u = u_{tt} - \Delta u = f(x,t),$$
$$\mathbf{M}f = f_{x_ix_i} - f_{x_jx_j} = 0,$$
$$t \in \mathrm{R}^+, \quad x_i \in \mathrm{R}^+, \quad \tilde{x} \in \mathrm{R}^{n-1},$$
$$u\big|_{t=0} = u_0(x), \quad u\big|_{x_i=0} = p_0(\tilde{x},t),$$
$$u_t\big|_{t=0} = u_1(x) \quad u_{x_i}\big|_{x_i=0} = p_1(\tilde{x},t),$$

As in Example 2.8

$$f(x,t) = \frac{1}{2}\int_{x_j-x_i}^{x_j+x_i} \left[p_{1tt}(x',\xi_j,t) - \Delta_{n-2}p_1(x',\xi_j,t) - w_{\xi_j}(x',0,\xi_j,t)\right] d\xi_j$$

$$+\frac{1}{2}\left[p_{0tt}(x',x_j-x_i,t) + p_{0tt}(x',x_j+x_i,t)\right]$$

$$-\frac{1}{2}\left[\Delta_{n-2}p_0(x',x_j-x_i,t) + \Delta_{n-2}p_0(x',x_j+x_i,t)\right]$$

$$-\frac{1}{2}\left[w(x',0,x_j+x_i,t)+w(x',0,x_j-x_i,t)\right]$$

$$-\left[p_{0x_j}(x',x_j+x_i,t)-p_{0x_j}(x',x_j-x_i,t)\right]$$

$$-\left[p_{0x_jx_j}(x',x_j+x_i,t)+p_{0x_jx_j}(x',x_j-x_i,t)\right],$$

where $x'=(x_1,\dots,x_{i-1},x_{i+1},\dots,x_{j-1},x_{j+1},\dots,x_n)$, $\Delta_{n-2}=\sum\limits_{\substack{k\neq i\\ k\neq j}}\frac{\partial^2}{\partial x_k^2}$ and $w(x,t)$ is a solution of the problem

$$w_{tt}-\Delta w=0,$$

$$w\big|_{t=0}=u_{0x_ix_i}-u_{0x_jx_j},\quad w_t\big|_{t=0}=u_{1x_ix_i}-u_{1x_jx_j}.$$

Example 2.13.

$$\mathbf{L}u=u_{tt}-\Delta u=f(x,t),$$

$$\mathbf{M}f=\Delta u=0,$$

$$t\in\mathrm{R}^+,\quad x\in\mathrm{R}^n,$$

$$u\big|_{t=0}=u_0(x),\quad u\big|_{\partial S}=p_0(x,t),$$

$$u_t\big|_{t=0}=u_1(x)$$

It should be $u_0|_{\partial S}=p_0(x,0)$, S is sphere of the radius R in the space R^n.

$$f(x,t)=\int\limits_{|\xi|=R}P_R(x,\xi)p_{0tt}(\xi,t)\,\mathrm{d}S_\xi-\int\limits_{|\xi|<R}G_R(x,\xi)w_{tt}\,\mathrm{d}\xi-w(x,t)$$

for $x\in S$, where $P_R(x,\xi)$, $G_R(x,\xi)$ are the same as in Example 2.9, $w(x,t)$ is a solution of the problem

$$w_{tt}-\Delta w=0,$$

$$w\big|_{t=0}=\Delta u_0(x),\quad w_t\big|_{t=0}=\Delta u_1(x).$$

Example 2.14.

$$\mathbf{L}u=u_t-u_{xx}=f(x,t),$$

$$\mathbf{M}f=f_{xx}+\omega^2f=0,$$

$$t\in\mathrm{R}^+,\quad x\in\mathrm{R},$$

$$u\big|_{t=0}=u_0(x),\ u\big|_{x=0}=p_0(t),\ u_x\big|_{x=0}=p_1(t),$$

$$f(x,t)=a(t)\cos\omega x+b(t)\sin\omega x,$$

where

$$a(t) = -w(0,t) + p_0'(t) + \omega^2 p_0(t),$$

$$b(t) = -\frac{1}{\omega} w_x(0,t) + \frac{1}{\omega} p_1'(t) + \omega p_1(t),$$

w is a solution of the problem

$$w_t - w_{xx} = 0,$$
$$w\big|_{t=0} = u_{0xx} + \omega^2 u_0.$$

Example 2.15.

$$\mathbf{L}u = u_{tt} - u_{xx} = f(x,t),$$
$$\mathbf{M}f = f_{xx} + \omega^2 f = 0,$$
$$t \in \mathrm{R}^+, \quad x \in \mathrm{R},$$
$$u\big|_{t=0} = u_0(x),\ u\big|_{x=0} = p_0(t),$$
$$u_t\big|_{t=0} = u_1(x),\ u_x\big|_{x=0} = p_1(t),$$
$$f(x,t) = a(t)\cos\omega x + b(t)\sin\omega x,$$

where

$$a(t) = -w(0,t) + p_0''(t) + \omega^2 p_0(t),$$

$$b(t) = -\frac{1}{\omega} w_x(0,t) + \frac{1}{\omega} p_1''(t) + \omega p_1(t),$$

w is a solution of the problem

$$w_{tt} - w_{xx} = 0,$$
$$w\big|_{t=0} = u_{0xx} + \omega^2 u_0,$$
$$w_t\big|_{t=0} = u_{1xx} + \omega^2 u_1.$$

CHAPTER 3

Formulas in inverse problems for the kinematic problems in seismology

This chapter deals with constructive methods of studying multidimensional inverse kinematic problems in seismic surveying. Various formulas and inequalities and, in particular, generalization of the Herglotz formula are given.

Much attention is given to the derivation of integro-differential relations between the hodograph and the velocity which are used as a basis for the construction of the algorithms for solving inverse kinematic problems. The methods and general results are contained in (Anikonov, 1973, 1978, 1979, 1980, 1995a, b, d).

3.1 INVERSE KINEMATIC PROBLEM. SOME RELATIONS

Let D be a domain in the Euclidean space R^n of variables $x = (x^1, \ldots, x^n)$, $n \geq 2$, with the smooth boundary ∂D. We consider in the domain D the conformal Euclidean metric

$$ds^2 = \lambda^2(x)\,|dx|^2, \tag{3.1}$$

where the function $\lambda(x)$ has a sufficient number of derivatives. Let the length of geodesics $\gamma(y, z)$ connecting y and z of the metric (3.1) be known for all points pairs $y \in \partial D$, $z \in \partial D$, i. e. the function

$$\tau(y, z) = \int\limits_{\gamma(y,z)} \lambda(x)\,|dx|, \quad y \in \partial D, \quad z \in \partial D. \tag{3.2}$$

be known. The problem of determining $\lambda(x)$, $x \in D$, by this equation is called the inverse kinematic problem or the nonlinear tomography problem.

The function $\lambda(x)$ in geodesics is reciprocal to the velocity of refracted waves with a particular polarization, and the multivalued function $\tau(y, z)$ is the time of propagation along the ray (geodesic) $\gamma(y, z)$. In other problems of tomography $\lambda(x)$ is the attenuation.

Constructive methods of studying multidimensional inverse kinematic problems are especially interesting from the viewpoint of applications.

We first give equations for extremals, various forms of equations for the geodesics of the conformal Euclidean metric $ds^2 = \lambda^2(x)\,|dx|^2$, and eikonal equations.

The systems of equations

$$\frac{\partial}{\partial x^i}\mathcal{L} = \frac{d}{dt}\frac{\partial \mathcal{L}}{\partial \dot{x}^i}, \quad i = 1, \ldots, n \tag{3.3}$$

are called the equations for extremals $x(t)$. Here $\dot{x} = \mathrm{d}x/\mathrm{d}t$, $\mathcal{L}$ is a Lagrangian; t is a parameter. For the case $\mathcal{L}(x, \dot{x}) = \lambda(x)\,|\dot{x}|$ and the special choice of the parameter t in (3.3) we obtain the frequently used equations for the geodesics γ of the conformal Euclidean metric:

$$\frac{\mathrm{d}^2 x^i}{\mathrm{d}t^2} = |\dot{x}|\,\frac{\partial \ln \lambda}{\partial x^i} - 2\sum_{j=1}^{n} \frac{\partial \ln \lambda}{\partial x^j}\dot{x}^j \dot{x}^i, \quad \lambda(x)\,|\dot{x}| = 1,$$

$$\frac{\mathrm{d}^2 x^i}{\mathrm{d}t^2} = \frac{\partial \lambda}{\partial x^i} - \sum_{j=1}^{n} \frac{\partial \lambda}{\partial x^j}\dot{x}^j \dot{x}^i, \quad |\dot{x}| = 1,$$

$$\frac{\mathrm{d}^2 x^i}{\mathrm{d}t^2} = \frac{1}{2}\frac{\partial \lambda^2}{\partial x^i}, \quad |\dot{x}| = \lambda(x).$$

Let

$$w(y,z) = \int_{\gamma(y,z)} \lambda(x)\,|\mathrm{d}x|,$$

where $\gamma(y,z)$ is the geodesic of the metric $\mathrm{d}s^2 = \lambda^2(x)\,|\mathrm{d}x|^2$ which connects the points $y \in D$ and $z \in D$. The parameter t on the geodesic is chosen according to the condition $|\dot{x}| = \lambda(x)$. Then

$$w(y,z) = \int_{\gamma(y,z)} \lambda^2(x(t))\,\mathrm{d}t,$$

$$\frac{\partial w}{\partial y^i} = -\left.\frac{\mathrm{d}x^i}{\mathrm{d}t}\right|_{x=y}, \quad \frac{\partial w}{\partial z^i} = \left.\frac{\mathrm{d}x^i}{\mathrm{d}t}\right|_{x=z}, \quad i = 1,\dots,n, \tag{3.4}$$

$$|\mathrm{grad}_y\, w(y,z)| = \lambda(y), \quad |\mathrm{grad}_z\, w(y,z)| = \lambda(z).$$

The equations

$$|\mathrm{grad}_y\, w(y,z)| = \lambda(y), \quad |\mathrm{grad}_z\, w(y,z)| = \lambda(z).$$

are called the eikonal equations.

3.2 FORMULAS FOR THE RAY AND TIME

In this section, for the special functions $\lambda(x)$ which specify the metrics $\mathrm{d}s^2 = \lambda^2(x)\,|\mathrm{d}x|^2$, $x \in D \subset \mathrm{R}^n$, we give the explicit formulas for rays and time which correspond to these metrics. The derived formulas for time:

$$w(y,z) = \int_{\gamma(y,z)} \lambda(x)\,|\mathrm{d}x|, \quad y \in D, \quad x \in D$$

may be substantially extended in the plane case by conformal transformations and used in direct and inverse tomography problems.

We consider two problems.

Problem 3.1. Determine the ray that issues from a given point in a given direction, i. e. solve the Cauchy problem:

$$\ddot{x} = \lambda(x)\,\mathrm{grad}\,\lambda, \quad |\dot{x}| = \lambda(x), \quad x(0) = y, \quad \dot{x}(0) = p,$$

and calculate the time

$$w(y,p,t) = \int_0^t \lambda^2\big(x(y,p,\tau)\big)\,d\tau$$

as an explicit function of the parameter t and the Cauchy data.

Problem 3.2. Determine the ray that connects two given points, i. e. the boundary value problem:

$$\ddot{x} = \lambda(x)\,\mathrm{grad}\,\lambda, \quad |\dot{x}| = \lambda(x), \quad x(0) = y, \quad x(T) = z$$

and calculate the time in an explicit form:

$$w(y,z) = \int_{\gamma(y,z)} \lambda(x)\,|dx|$$

depending on the pair of points y, z.

We consider several cases.

1. $\lambda^2(x) = \lambda_0^2 + 2(\mathbf{a}, x)$; λ_0 is a constant number, and $\mathbf{a}$ is a constant vector. The solution of Problem 3.1 is

$$x(t) = \frac{1}{2}\mathbf{a}t^2 + pt + y, \quad |p|^2 = \lambda_0^2 + 2(\mathbf{a}, x), \quad t \geq 0,$$

$$w(y,p,t) = \frac{1}{3}|\mathbf{a}|^2 t^3 + \left[\lambda_0^2 + (\mathbf{a}, p)\right]t^2 + 2(\mathbf{a}, y)t.$$

The solution of Problem 3.2 is

$$x(t) = \frac{(T-t)y + tz}{T} - \frac{1}{2}t(T-t), \quad 0 \leq t \leq T,$$

where

$$T^2 = \frac{2}{|\mathbf{a}|^2}\left[\bar{\lambda}^2 - \sqrt{\bar{\lambda}^4 - |\mathbf{a}|^2|y-z|^2}\right], \quad \bar{\lambda}^2 = \lambda_0^2 + (\mathbf{a}, y+z),$$

$$w(y,z) = \frac{2\bar{\lambda}^2 + \sqrt{\bar{\lambda}^4 - |\mathbf{a}|^2|y-z|^2}}{3|\mathbf{a}|}\sqrt{2\left(\bar{\lambda}^2 - \sqrt{\bar{\lambda}^4 - |\mathbf{a}|^2|y-z|^2}\right)}.$$

2. $\lambda^2(x) = \lambda_0^2 + a^2|x|^2$, a is a constant. The solution of Problem 3.1 is

$$x(t) = y\cosh at + \frac{1}{a}p\sinh at, \quad |p|^2 = \lambda_0^2 + a^2|x|^2, \quad t \geq 0,$$

$$w(y,p,t) = \frac{1}{2}\lambda_0^2 t + \frac{\lambda_0^2 + 2a^2|x|^2}{4a}\sinh 2at + (y,p)\sinh^2 at.$$

The solution of Problem 3.2 is:

$$x(t) = y\cosh at + \frac{z - My}{\sqrt{M^2-1}}\sinh at, \quad 0 \le t \le \frac{1}{a}\ln(M + \sqrt{M^2-1}).$$

where

$$M = \frac{1}{\lambda_0^4}\left[a^4(y,z)^2 + \lambda^2(\lambda_0^2 + a^2|y|^2 + a^2|z|^2) - a^2(y,z)\right],$$

$$w(y,z) = \frac{\lambda_0^2}{2a}\left[N\sqrt{M^2-1} + \ln\left(M + \sqrt{M^2-1}\right)\right].$$

Here,

$$N = \frac{1}{\lambda_0^2}\left[\sqrt{a^2(y,z)^2 + \lambda_0^2(\lambda_0^2 + a^2|y|^2 + a^2|z|^2)} + a^2(y,z)\right].$$

3. $\lambda^2(x) = \lambda_0^2 - a^2|y|^2$, $a > 0$. The solution of Problem 3.1 is:

$$x(t) = y\cos at + \frac{1}{a}p\sin at, \quad |p|^2 = \lambda_0^2 - a^2|x|^2, \quad t \ge 0,$$

$$w(y,p,t) = \frac{1}{2}\lambda_0^2 t + \frac{\lambda_0^2 - 2a^2|y|^2}{4a}\sin 2at - (y,p)\sin^2 at.$$

The solution of Problem 3.2 is:

$$x(t) = y\cos at - \frac{z - My}{\sqrt{1-M^2}}\sin at, \quad 0 \le t \le \frac{1}{a}\arccos M, \quad 0 \le M \le \pi,$$

where

$$M = \frac{1}{\lambda_0^2}\left[\sqrt{a^4(y,z)^2 + \lambda_0^2(\lambda_0^2 - a^2|y|^2 - a^2|z|^2)} + a^2(y,z)\right],$$

$$w(y,z) = \frac{\lambda_0^2}{2a}\left[\arccos M + N\sqrt{1-M^2}\right].$$

Here,

$$N = \frac{1}{\lambda_0^2}\left[\sqrt{a^4(y,z)^2 + \lambda_0^2(\lambda_0^2 - a^2|y|^2 - a^2|z|^2)} - a^2(y,z)\right].$$

Since the form of the metric $ds^2 = \lambda^2(x)|dx|^2$ is invariant with respect to conformal transformations, we may extend the formulas obtained, i. e. we may use $\varphi(y)$ and $\varphi(z)$ instead of y and z, where $\varphi(q)$ is an analytic function of the variable $q = q_1 + i q_2$.

4. Let $f(q)$, $q = q_1 + i q_2$, be a complex analytic function in the domain $|q| < 1$, $|f(q)| < 1$, and let

$$ds^2 = \frac{|f'(q)|}{1 - |f^2(q)|^2}|dq|^2 = \lambda^2(q)|dq|^2$$

be a hyperbolic metric. We set:

$$w(y,z) = \int\limits_{\gamma(y,z)} \lambda(q)\,|dq|,$$

where $\gamma(y,z)$ is a geodesic of the metric $ds^2 = \lambda^2(q)|dq|^2$ which connects the points y and z; $|y| < 1$; $|z| < 1$. Then

$$w(y,z) = \frac{1}{2}\ln\frac{|1 - f(y)\bar{f}(z)| + |f(y) - f(z)|}{|1 - f(y)\bar{f}(z)| - |f(y) - f(z)|}.$$

5. Let

$$\tau(\alpha,\varphi)=\int\limits_{\gamma(\alpha,\varphi)}\lambda(x,y)\sqrt{dx^2+dy^2},$$

where $\gamma(\alpha,\varphi)$ is the geodesic of the metric $ds^2=\lambda^2(x,y)(dx^2+dy^2)$ which connects the points $(\cos\alpha,\sin\alpha)$ and $(\cos\varphi,\sin\varphi)$ of the circle $x^2+y^2=1$.

Then

$$\iint\limits_{x^2+y^2\le 1}\lambda^2(x,y)\,dx\,dy=\left|\frac{1}{2\pi}\int\limits_0^{2\pi}\int\limits_0^{2\pi}\frac{\partial\tau(\alpha,\varphi)}{\partial\alpha}\frac{\partial\tau(\alpha,\varphi)}{\partial\varphi}\,d\alpha\,d\varphi\right|.$$

Let

$$\mu(x)=\big(u(x),\operatorname{grad}\lambda(x)\big)+\lambda(x)\frac{1}{n}\operatorname{sp}\left(\frac{\partial u_i}{\partial x^j}\right)+\beta\lambda(x),$$

where β is a constant, $n>2$,

$$u(x)=a_0x+Ax-\mathbf{b}\,|x|^2+2x(\mathbf{b},x)+\mathbf{c},$$

a_0 is a constant, $\mathbf{b}$ and $\mathbf{c}$ are constant vectors, A is a constant skew-symmetric matrix. The equality

$$\int\limits_{\gamma(y,z)}\mu(x)\,|dx|=\big(u(z),\operatorname{grad}_z w(y,z)\big)+\big(u(y),\operatorname{grad}_y w(y,z)\big)+\beta w(y,z)$$

holds, where

$$w(y,z)=\int\limits_{\gamma(y,z)}\lambda(x)\,|dx|.$$

Here, $\gamma(y,z)$ is the geodesic of the metric $ds^2=\lambda^2(x)\,|dx|^2$ which connects the points $y\in D$ and $z\in D$.

6. Let

$$\lambda^2(x)=\sum_{i=1}^{n}\lambda_i^2(x^i),$$

where $\lambda_i(x^i)$ is a function dependent only on the variable x^i, $i=,\dots,n$. The metrics

$$ds^2=\sum_{i=1}^{n}\lambda_i^2(x^i)\,|dx|^2$$

are called the Liouville metrics. The equations for geodesics in this case lead to the quadratures:

$$t=\int\frac{dx^i}{\sqrt{\lambda_i^2(x^i)+p_i}},\qquad\sum_{i=1}^{n}p_i=0.$$

7. Let

$$\lambda^2(x)=\sum_{i,j=1}^{n}a_{ij}x^ix^j+\sum_{i=1}^{n}b_ix^i+c,$$

where a_{ij}, b_i, and c are constants. If we choose the parameter according to the condition $|\dot{x}| = \lambda(x)$, the system of equations for geodesics is linear:

$$\frac{d^2x^i}{dt^2} = \sum_{i=1}^{n} a_{ij}x^i + b_i, \quad \sum_{i=1}^{n} \left(\frac{dx^i}{dt}\right)^2 = \lambda^2(x).$$

For $b = 0$ and $A = (a_{ij})$, the solution of the system of equations $d^2x/dt^2 = Ax$ is written as:

$$x(t) = x_0 \cos\sqrt{A}t + \frac{p}{\sqrt{A}} \sin\sqrt{A}t, \quad |p|^2 = \sum_{i,j=1}^{n} a_{ij}x_0^i x_0^j$$

and the following equalities hold:

$$x(0) = x_0, \quad \left.\frac{dx}{dt}\right|_{t=0} = p, \quad x_0 \in R^n, \quad p \in R^n.$$

3.3 INTEGRO-DIFFERENTIAL RELATIONS AND THE FIRST INTEGRALS

One of the ways to construct solutions of an inverse kinematic problem is to obtain and use integro-differential identities of the type:

$$\begin{aligned} (\mathbf{a}, \operatorname{grad}_y w(y,z)) + (\mathbf{a}, \operatorname{grad}_z w(y,z)) &= \int_{\gamma(y,z)} (\mathbf{a}, \operatorname{grad} \lambda(x))\, |dx|, \\ (\mathbf{a}, \operatorname{grad}_y w(y,z))^2 - (\mathbf{a}, \operatorname{grad}_z w(y,z))^2 &= \int_{\gamma(y,z)} (\mathbf{a}, \operatorname{grad} \lambda^2(x)) \sum_{i=1}^{n} a_i\, dx^i, \end{aligned}$$

where

$$w(y,z) = \int_{\gamma(y,z)} \lambda(x)\, |dx|.$$

a is a constant vector; $\gamma(y,z)$ is the geodesic of the metric $ds^2 = \lambda^2(x)\,|dx|^2$ which connects the points y and z. We shall now construct the more general relations.

Let $u = \left(u_{i_1 \dots i_m}(x)\right)$ be a symmetric tensor field of the order m, $m \geq 0$, such that

$$du = \boldsymbol{\sigma}(\delta \otimes v), \tag{3.5}$$

where $d = \boldsymbol{\sigma}\nabla$; $\nabla = (\partial/\partial x^1, \dots, \partial/\partial x^n)$; $\boldsymbol{\sigma}$ is a symmetrization operator; $\delta = (\delta_{ij})$ is the Kronecker delta; $\otimes$ denotes tensor multiplication; $v(x)$ is a symmetric tensor field of the order $m-1$ [note that $v(x)$ is expressed in terms of the derivatives of $u(x)$]. For example for $m = 1$:

$$(du)_{ij} = \frac{1}{2}\left(\frac{\partial u_i}{\partial x^j} + \frac{\partial u_j}{\partial x^i}\right), \quad v(x) = \frac{1}{n} \operatorname{div} u(x). \tag{3.6}$$

The identity

$$\begin{aligned}\mathbf{L}^u w(y,z) &= \sum_{i_1,\dots,i_m}^{n} u_{i_1\dots i_m}(z) w_{z^{i_1}} \dots w_{z^{i_m}} + (-1)^{m-1} u_{i_1\dots i_m}(y) w_{y^{i_1}} \dots w_{y^{i_m}} \\ &= \int\limits_{\gamma(y,z)} \left[\lambda^2(x) v_{i_1\dots i_m}(x) + m\lambda(x) \sum_{i_m=1}^{n} u_{i_1\dots i_m}(x)\lambda_{x^{i_m}}\right] \dot{x}^{i_1} \dots \dot{x}^{i_{m-1}}\, dt\end{aligned} \tag{3.7}$$

holds for any field that satisfies (3.5) and any function $\lambda(x)$. In the following we imply summation from 1 to n over the same subscript and superscript. The above identity and the identity

$$\frac{d}{dt}\left(u_{i_1\dots i_m}(x)\dot{x}^{i_1}\dots\dot{x}^{i_m}\right) = (\tilde{I}\lambda)_{i_1\dots i_{m-1}}(x)\dot{x}^{i_1}\dots\dot{x}^{i_{m-1}},$$

where

$$(\tilde{I}\lambda)_{i_1\dots i_{m-1}} = \lambda^2 v_{i_1\dots i_{m-1}} + m\lambda \sum_{i_m=1}^{n} u_{i_1\dots i_m}(x)\lambda_{x^{i_m}}$$

lead to the following assertion (Anikonov, 1995a).

Theorem 3.1. *Let $\lambda(x)$ be a smooth function and $u(x)$ a smooth tensor field that satisfies (3.5). Then the following conditions are equivalent:*

1. *The equations for rays admit the first integral $u_{i_1\dots i_m}(x)\dot{x}^{i_1}\dots\dot{x}^{i_m}$, i. e.*

$$\frac{d}{dt}\left(u_{i_1\dots i_m}(x)\dot{x}^{i_1}\dots\dot{x}^{i_m}\right) = 0.$$

2. *The function $\lambda(x)$ satisfies the system of equations*

$$(\tilde{I}\lambda)_{i_1\dots i_{m-1}} = 0, \quad i_1\dots i_{m-1} = (\overline{1,n}), \quad x \in D. \tag{3.8}$$

3. *The function $w(y,z)$ satisfies the equation*

$$\mathbf{L}^u w(y,z) = 0, \quad y \in D, \quad z \in D.$$

In inverse problems, the relations $\mathbf{L}w(y,z) = 0$, where $\mathbf{L}$ is a differential operator, may generally be checked only at the boundary ∂D of the domain D, i. e. when $y \in \partial D$, $z \in \partial D$. Therefore, the question is of particular importance. Let the equality

$$\mathbf{L}^{u^0} w(y,z) = 0, \quad y \in \partial D, \quad z \in \partial D$$

hold for a tensor field u^0 which is specified at the boundary ∂D of the domain D. The question is whether the extension $\mathbf{L}^u w(y,z) = 0$, $y \in D$, $z \in D$, of the differential relation from the boundary ∂D to the domain D exists and is unique when the condition (3.5) is satisfied. According to Anikonov and Pestov(1993) and Anikonov (1995a) the answer is positive under rather general constraints.

Because of this result and Theorem 3.1, the relations $\mathbf{L}^u w(y,z) = 0$, $y \in \partial D$, $z \in \partial D$, lead to the differential relations $\tilde{I}\lambda(x) = 0$, $x \in D$ which, in turn, characterize the structure of the Riemannian space, which is important in applications. Sections 3.4 and 3.5 give more concrete results.

3.4 LINEAR FIRST INTEGRAL

We consider the system of equations:

$$\begin{aligned} &\frac{\partial u_i}{\partial x^j} + \frac{\partial u_j}{\partial x^i} = 0, && i \neq j, \\ &\frac{\partial u_i}{\partial x^i} = \frac{\partial u_j}{\partial x^j}, && i,j = 1,\dots,n. \end{aligned} \tag{3.9}$$

For $n = 2$, it is the Cauchy-Riemann system; and its general solution for $n > 2$ is given by the formula:

$$u(x) = a_0 x + Ax - \mathbf{b}\,|x|^2 + 2(\mathbf{b}, x) + \mathbf{c},$$

where $a_0 = \text{const}$; $\mathbf{b}$ and $\mathbf{c}$ are constant vectors; A is a constant skew-symmetric matrix. Note that the trajectories of the vector field $u(x)$ that satisfies (3.9) generate a group of conformal transformations of R^n.

Identity (3.7) (Section 3.3) in this case has the form:

$$\big(u(y), \operatorname{grad}_y w(y,z)\big) + \big(u(z), \operatorname{grad}_z w(y,z)\big) = \int\limits_{\gamma(y,z)} \left[(u, \operatorname{grad}\lambda) + \frac{1}{n}\lambda \operatorname{div} u\right] |\mathrm{d}x|\,.$$

The more general identity

$$I_\beta w(y,z) = \int\limits_{\gamma(y,z)} \tilde{I}_\beta \lambda(x)\, |\mathrm{d}x| \tag{3.10}$$

holds, where

$$\begin{aligned} I_\beta w(y,z) &= \big(u(y), \operatorname{grad}_y w(y,z)\big) + \big(u(z), \operatorname{grad}_z w(y,z)\big) + \beta w(y,z), \\ \tilde{I}_\beta \lambda(x) &= \big(u(y), \operatorname{grad}\lambda(x)\big) + \frac{1}{n}\lambda(x) \operatorname{div} u(x) + \beta\lambda(x), \end{aligned}$$

and β is a constant.

Let B be a solid sphere of unit radius in R^n, and the metric $\mathrm{d}s^2 = \lambda^2(x)\,|\mathrm{d}x|^2$ is such that there exists a single geodesic in B which connects the points y and z.

Theorem 3.2. *The following relations hold:*

1. $\displaystyle \left\| \frac{\tilde{I}_\beta \lambda}{\lambda} \right\|_{C(B)} \geq \left| \frac{I_\beta w(y,z)}{w(y,z)} \right|$ *for* $|y| = 1,\ |z| = 1;$

2. $\displaystyle \int\limits_B \lambda^{n-2} \left|\tilde{I}_\beta \lambda\right|^2 \mathrm{d}x \leq \frac{\Gamma(n/2)}{(n-1)(2\pi)^{n/2}} \left| \int\limits_{\partial B}\int\limits_{\partial B} M w(y,z)\, \mathrm{d}s_y\, \mathrm{d}s_z \right|,$

where Γ is a gamma function; ds_y *and* ds_z *are elements of the area of the sphere* $\partial B = \{x : |x| = 1\}$ *at points* $y \in \partial B$ *and* $z \in \partial B$. *The determinant* Mw *is given by:*

$$Mw(y,z) = \begin{vmatrix} 0 & 0 & \dfrac{\partial}{\partial z^1} I_\beta w & \dots & \dfrac{\partial}{\partial z^n} I_\beta w \\ 0 & 0 & \dfrac{z^1}{|z|} & \dots & \dfrac{z^n}{|z|} \\ \dfrac{\partial}{\partial y^1} I_\beta w & \dfrac{y^1}{|y|} & \dots & \dots & \dots \\ \vdots & \vdots & & \dfrac{\partial^2}{\partial y^i \partial z^j} I_\beta w & \\ \dfrac{\partial}{\partial y^n} I_\beta w & \dfrac{y^n}{|y|} & & & \end{vmatrix}$$

Equation (3.10), in particular, implies that if $\mathbf{L}_\beta w(y,z) = 0$ only for $|y| = 1$, $|z| = 1$, then $\tilde{I}_\beta \lambda(x) = 0$, $x \in B$, and $I_\beta w(y,z) = 0$, $y \in B$, $z \in B$, which is very important in applications. Thus, we have an extension of the differential relation $I_\beta w(y,z) = 0$ from the boundary $\partial B \times \partial B$ to the domain $B \times B$. Therefore, we are interested in a more concrete form of the function in view of the relation

$$I_\beta w(y,z) = 0, \quad |y| = 1, \quad |z| = 1.$$

We consider only the case $\beta = 0$ and examine the equation

$$\tilde{I}_\beta \lambda = \big(u(y), \operatorname{grad} \lambda\big) + \frac{1}{n} \lambda \operatorname{div} u = 0.$$

Let $n = 3$ and the vector field $u(x)$ have the form:

$$u(x) = a_0 x + Ax - \mathbf{b}\,|x|^2 + 2(\mathbf{b}, x) + \mathbf{c}.$$

The general solutions of the equations $\tilde{I}\lambda = 0$ for particular $u(x)$ are:

$u(x)$, $\qquad v(x) = 1/\lambda(x)$,

$u = (0,0,1) = k$, $\qquad v = F(x_1, x_2)$,

$u = [k, x]$, $\qquad v = F\left(x_1^2 + x_2^2, x_3\right)$,

$u = x$, $\qquad v = |x|\, F\left(\dfrac{x}{|x|}\right)$,

$u = a_0 x + [k, x]$, $\qquad v = x_3 F\left(\dfrac{\sqrt{x_1^2 + x_2^2}}{x_3}, \sqrt{x_1^2 + x_2^2}\, e^{-a_0 \arctan(x_2/x_1)}\right)$,

$u = [k, x] + \alpha k$, $\qquad v = F\left(\sqrt{x_1^2 + x_2^2}, x_3 - \alpha \arctan \dfrac{x_2}{x_1}\right)$,

$$u = -k|x|^2 + 2x(k,x), \qquad v = |x|^2 F\left(\frac{x_1}{|x|^2}, \frac{x_2}{|x|^2}\right),$$

$$u = a_0 - k|x|^2 + 2x(k,x), \qquad v = \sqrt{x_1^2+x_2^2}F\left(\frac{x_2}{x_1}, \frac{a_0x_3+|x|^2}{\sqrt{x_1^2+x_2^2}}\right),$$

$$u = 2[a,x] - k|x|^2 + 2x(k,x), \quad v = \sqrt{(x_1+a_2)^2+(x_2-a_1)^2}$$

$$a_3 = 0, \qquad \times F\left(\frac{x_2-a_1}{x_1+a_2}, \frac{|x|^2 - a_2x_1 + a_1x_2}{\sqrt{(x_1+a_2)^2+(x_2-a_1)^2}}\right),$$

$$u = \alpha[k,x] - k|x|^2 + 2x(k,x), \quad v = \sqrt{x_1^2+x_2^2}F\left(\frac{\sqrt{x_1^2+x_2^2}}{|x|}, \alpha\frac{x_3}{|x|^2} + \arctan\frac{x_2}{x_1}\right),$$

$$u = -k|x|^2 + 2x(k,x) + \alpha k, \quad v = \sqrt{x_1^2+x_2^2}F\left(\frac{x_2}{x_1}, \frac{|x|^2+\alpha}{\sqrt{x_1^2+x_2^2}}\right).$$

3.5 QUADRATIC FIRST INTEGRAL

This section gives the results analogous to those of Section 3.4 and associated with the quadratic first integral.

As before, let the components $u_i(x)$, $i =, \dots, n$, of the vector function $u(x)$ satisfy the system of equations:

$$\begin{aligned} &\frac{\partial u_i}{\partial x^j} + \frac{\partial u_j}{\partial x^i} = 0, && i \neq j \\ &\frac{\partial u_i}{\partial x^i} = \frac{\partial u_j}{\partial x^j}, && i,j = 1,\dots,n, \quad n \geq 2. \end{aligned} \tag{3.11}$$

For $n = 2$, as noted above, it is a system of Cauchy-Riemann equations; for $n > 2$ its general solution is given by the formula

$$u(x) = Ax - \mathbf{b}|x|^2 + 2(\mathbf{b}, x) + \mathbf{c}, \tag{3.12}$$

where A is a constant matrix with elements such that $A_{ij} = -A_{ji}$, $i \neq j$, $A_{ii} = A_0$, $\mathbf{b}$ and $\mathbf{c}$ are constant vectors; $(\mathbf{b}, x)$ is a scalar product. It is obvious that there is a substantial difference between the classes of vector functions $u(x)$ for $n = 2$ and $n > 2$.

Let $u^k(x) = (u_1^k, \dots, u_n^k)$ and $v^k(x) = (v_1^k, \dots, v_n^k)$ be vector functions that satisfy (3.11) for any k, $k = 1, \dots, n$. We set:

$$a_{ij}(x) = \frac{1}{2}\sum_{k=1}^{n}\left(u_i^k v_j^k + u_j^k v_i^k\right)$$

and

$$Q = \sum_{k=1}^{n}\left(u^k, \dot{x}\right)\left(v^k, \dot{x}\right) = \sum_{i,j=1}^{n} a_{ij}(x)\dot{x}^i\dot{x}^j.$$

Along any geodesic γ the equality

$$\frac{\mathrm{d}}{\mathrm{d}t}Q = \sum_{k=1}^{n} (v_i^k \tilde{\mathbf{L}}_k^u \lambda + u_i^k \tilde{\mathbf{L}}_k^v \lambda)\dot{x}^i$$

holds, where

$$\tilde{\mathbf{L}}_k^u \lambda = \lambda\big(u^k, \operatorname{grad}\lambda\big) + \frac{\lambda^2}{n}\operatorname{div} u^k,$$
$$\tilde{\mathbf{L}}_k^v \lambda = \lambda\big(v^k, \operatorname{grad}\lambda\big) + \frac{\lambda^2}{n}\operatorname{div} v^k, \qquad k = 1,\dots,n, \quad |x| \le 1. \tag{3.13}$$

If this equality holds along γ, then

$$\frac{\mathrm{d}}{\mathrm{d}t}(Q - \varphi(x)) = 0.$$

We consider the differential expression:

$$\mathbf{L}w = \sum_{k=1}^{n} \left[\left(u^k(z), \frac{\partial w}{\partial z}\right)\left(v^k(z), \frac{\partial w}{\partial z}\right) - \left(u^k(y), \frac{\partial w}{\partial y}\right)\left(v^k(y), \frac{\partial w}{\partial y}\right)\right],$$

where

$$w(y,z) = \int\limits_{\gamma(y,z)} \lambda(x)\,|\mathrm{d}x|,$$

$$\frac{\partial w}{\partial y} = \left(\frac{\partial w}{\partial y^1},\dots,\frac{\partial w}{\partial y^n}\right), \quad \frac{\partial w}{\partial z} = \left(\frac{\partial w}{\partial z^1},\dots,\frac{\partial w}{\partial z^n}\right),$$

u^k and v^k are vector functions that satisfy (3.11).

Theorem 3.3. *In order for* $\mathbf{L}w = \varphi(z) - \varphi(y)$, $|y| = 1$, $|z| = 1$, *to hold for a continuously differentiable function* $\varphi(x)$, $|x| \le 1$, *it is necessary and sufficient that*

$$\frac{\partial \varphi}{\partial x^i} = \sum_{k=1}^{n} (v_i^k I_k^u \lambda + u_i^k \mathbf{L}_k^v \lambda), \quad i = 1,\dots,n, \quad |x| \le 1.$$

We consider the form of functions $\lambda(x)$, whose equations for geodesics admit the quadratic first integral for $n = 3$:

$$I(x,\dot{x}) = \big(u(x),\dot{x}\big)^2 - \varphi(x), \quad |\dot{x}| = \lambda(x), \tag{3.14}$$

where $\varphi(x)$ is a function, and the vector field $u(x)$ is specified in (3.12).

In order for a function $I(x,\dot{x})$ to be the first integral it is necessary that the field $u(x)$ satisfy the additional relation $(u, \operatorname{curl} u) = 0$. The vector fields $u(x)$ that satisfy this condition are reduced by a linear conformal transformation to one of the forms:

1. $u = k = (0,0,1)$;
2. $u = [k, x]$;
3. $u = x$;
4. $u = -k\,|x|^2 + 2(k,x) + \alpha k$.

(3.15)

The system of equations for $\lambda(x)$ transforms to:

$$\operatorname{curl}\left\{\left[\lambda(u,\operatorname{grad}\lambda)+\frac{1}{3}\lambda^2\operatorname{div}u\right]u\right\}=0.$$

The general solutions of this system for $u(x)$ of the form (3.15) are given below:

$$\begin{array}{ll}
u(x), & \lambda^2(x) \\
u=k=(0,0,1), & \lambda^2=F(x_1,x_2)+G(x_3) \\
u=[k,x], & \lambda^2=\dfrac{1}{\sqrt{x_1^2+x_2^2}}\left[F(x_1^2+x_2^2.x_3)+G\left(\dfrac{x_2}{x_1}\right)\right] \\
u=x, & \lambda^2=\dfrac{1}{|x|^2}\left[F\left(\dfrac{x}{|x|}\right)+G(|x|)\right] \\
u=-|x|^2+2x(k,x)+\alpha k, & \lambda^2=\dfrac{1}{(|x|^2-\alpha)^2+4\alpha x_3^2} \\
 & \times\left[F\left(\dfrac{\sqrt{x_1^2+x_2^2}}{|x|^2+\alpha},\dfrac{x_2}{x_1}\right)+G\left(\dfrac{x_3}{\sqrt{x_1^2-x_2^2}}\right)\right].
\end{array}$$

Thus the equations for geodesics in R^3 admit the first integral of the form (3.14) if and only if the function $\lambda^2(x)$ has the above form up to a linear conformal transformation. The function:

$$w(y,z)=\int_{\gamma(y,z)}\lambda(x)\,|dx|,\quad y\in\partial D,\quad x\in\partial D$$

is a solution of the equation:

$$\mathrm{L}w=\left(u(z),\frac{\partial w}{\partial z}\right)^2-\left(u(y),\frac{\partial w}{\partial y}\right)^2+\varphi(y)-\varphi(z)=0,\quad y\in\partial D,\quad z\in\partial D$$

and this equation is extended to $y\in D$, $z\in D$.

3.6 GENERALIZATION OF THE HERGLOTZ FORMULA

According to the Sections 3.4 and 3.5, the integro-differential identities allow us to determine only the general form of the sought function in an inverse kinematic problem. In order to eliminate arbitrariness, further investigations are necessary, which folow in this section.

Let E^{n+1} be the $(n+1)$-dimensional Euclidean space of variables (x,y), $x=(x_1,\ldots,x_n)$, $-\infty<y<\infty$, $n\geq 1$. We denote by $\{\lambda\}$ a set of twice continuously differentiable functions $\lambda(y)$ that are given in the domain $y\geq 0$ and satisfy the conditions:

$$\lambda(y)>0,\quad \lambda'(y)<0,\quad \lambda(0)=1$$

and by $\{a\}$ a set of triply continuously differentiable functions $a(x)$, $x \in E^n$, such that $a(x) \geq 0$, $a(0) = 0$. We specify in the half-space $y \geq 0$ the set of metrics ds:

$$ds^2 = (\lambda^2(y) + a^2(x))(dx^2 + dy^2), \quad \lambda \in \{\lambda\}, \quad a \in \{a\}, \quad dx^2 = \sum_{i=1}^{n} dx_i^2. \tag{3.16}$$

These metrics correspond to the quadratic first integral for $\bar{u}(\bar{x}) = (0, 0, 1)$ (see Section 3.5). There exist for each metric (3.16) numbers $r > 0$, $R > 0$, $r \leq R$, such that any two points ξ and η in the hyperplane $y = 0$, $|\xi| < r$, $|\eta| < R$ are connected with a single geodesic $\gamma(y, z)$ of the metric (3.16).

We consider the following problem. The function

$$\tau(\xi, \eta) = \int\limits_{\gamma(\xi,\eta)} \sqrt{(\lambda^2(y) + a^2(x))(dx^2 + dy^2)}$$

is given in the domain $|\xi| < r$, $|\eta| < R$. It is required that the functions $\lambda(y) \in \{\lambda\}$, $a \in \{a\}$ be found. This problem is, generally speaking, nonlinear because the geodesics $\gamma(\xi, \eta)$ are substantially dependent on the sought functions (λ, a). Note that the function $\tau(\xi, \eta)$ for $a(x) \equiv 0$ depends only on $|\xi - \eta|$, and the inversion formula derived by Herglotz holds true. Namely, if we set:

$$f(\alpha) = \tau|_{\xi=0, \eta=(0,0,\ldots,\alpha)}, \quad 0 \leq \alpha < R$$

then, according to Herglotz,

$$\lambda^{-1}(q) = \frac{1}{\pi} \int\limits_{q}^{q} \frac{f'^{-1}(h)\, dh}{\sqrt{h^2 - q^2}}, \tag{3.17}$$

where $\lambda^{-1}(h)$ and $f'^{-1}(h)$ are inverse functions to $\lambda(z)$ and $f'(\alpha) = df/d\alpha$ respectively.

In our case it turns out that

$$\sum_{i=1}^{n} \left(\frac{\partial \tau}{\partial \eta_i}\right)^2 - \left(\frac{\partial \tau}{\partial \xi_i}\right)^2 = a^2(\eta) - a^2(\xi)$$

and the function $a(x)$ is determined by $\tau(\xi, \eta)$ and the formula:

$$a(x) = \sqrt{\sum_{i=1}^{n} \left(\frac{\partial \tau}{\partial \eta_i}\right)^2 - \left(\frac{\partial \tau}{\partial \xi_i}\right)^2 \Big|_{\eta=x, \xi=0}}$$

and (3.17) holds for the function $\lambda^{-1}(q)$, where instead of $f'^{-1}(h)$, a function $d(h)$ appears which is determined (implicitly) by the functions $f'(\alpha)$ and $a(x)$. Before we formulate the result exactly, we introduce some additional information.

Let x^0 be a fixed vector of unit length which belongs to the space E^n. We introduce functions $f(\alpha)$ and $\tilde{a}(x)$, assuming

$$f(\alpha) = \tau(0, \alpha x^0), \quad 0 \leq \alpha < R,$$

$$\tilde{a}(x) = \sqrt{\sum_{i=1}^{n} \left(\frac{\partial \tau}{\partial \eta_i}\right)^2 - \left(\frac{\partial \tau}{\partial \xi_i}\right)^2 \Big|_{\eta=x, \xi=0}}, \quad |x| < R.$$

Let $x(t;p)$ be a solution of the Cauchy problem:

$$\frac{d^2\tilde{x}}{dt^2} = \tilde{a}\,\mathrm{grad}\,\tilde{a}, \quad \tilde{x}|_{t=0} = 0, \quad \left.\frac{d\tilde{x}}{dt}\right|_{t=0} = p, \quad \tilde{x} \in E^n, \quad p \in E^n. \tag{3.18}$$

We form the system of equations for the parameter t and the vector p:

$$\tilde{x}(t;p) = \alpha x^0, \quad \left(\frac{d\tilde{x}}{dt}, x^0\right) = f'(\alpha), \quad 0 \le \alpha < R. \tag{3.19}$$

The brackets on the left-hand side of the second equation denote a scalar product. Note that (3.18) has a solution $\tilde{x} = tp$ for $\tilde{a}(x) \equiv 0$, and (3.19) yields

$$tp = \alpha x^0, \quad (p, x^0) = f'(\alpha).$$

Hence

$$t = \frac{\alpha}{f'(\alpha)}, \quad p = f'(\alpha) x^0.$$

If $n = 1$, we obtain for the sought functions $t(\alpha)$ and $p(\alpha)$ the expressions

$$t(\alpha) = \int_0^\alpha \frac{dq}{\sqrt{\tilde{u}^2(q) + p^2(\alpha)}}, \quad p^2(\alpha) = f'^2(\alpha) - \tilde{u}^2(\alpha).$$

We fail to obtain the formulas for $t(\alpha)$ and $p(\alpha)$ for the general case.

Theorem 3.4. *Let $(\lambda(y), a(x))$, $\lambda(y) \in \{\lambda\}$, $a \in \{a\}$, be a solution of the problem. Then,*

1. *the function $\tau(\xi, \eta)$ satisfies the equation:*

$$\sum_{i=1}^n \left(\frac{\partial \tau}{\partial \eta_i}\right)^2 - \left(\frac{\partial \tau}{\partial \xi_i}\right)^2 = a^2(\eta) - a^2(\xi);$$

2. $a(x) = \tilde{a}(x) = \sqrt{\sum_{i=1}^n \left(\frac{\partial \tau}{\partial \eta_i}\right)^2 - \left(\frac{\partial \tau}{\partial \xi_i}\right)^2}\Bigg|_{\xi=0,\eta=x}$;

3. *the system of equation (3.19) has the unique solution $t(\alpha)$, $p(\alpha)$ which is determined on an interval $0 \le \alpha \le \alpha_0$, α_0, the function $h(\alpha) = |p(\alpha)|$ strictly monotonically decreases and $h(0) = 1$;*

4. *the following formulas hold true*

$$\lambda^{-1}(q) = \frac{1}{\pi} \int_q^1 \frac{d(h)\,dh}{\sqrt{h^2 - q^2}},$$

$$d(q) = q t\big(h^{-1}(q)\big), \quad h(\alpha_0) \le q \le 1.$$

PROOF. Let $ds^2 = Q^2(r)\,|dr|^2$, $r = (r_1, \ldots, r_m)$, $m \geq 2$ be a conformally Euclidean metric; let $\gamma(\bar{a}, \bar{b})$ be its geodesic connecting the points $\bar{a}$ and $\bar{b}$ in the space. We choose the parameter t on the geodesic in the form

$$t = \int_{\gamma} \frac{|dr|}{Q(r)}.$$

Then the equations for the geodesic γ, as noted above, have the form:

$$\frac{d^2 r}{dt^2} = Q \operatorname{grad} Q, \quad \left|\frac{dr}{dt}\right|^2 = Q^2(r)$$

and if

$$w(\bar{a}, \bar{b}) = \int_{\gamma(\bar{a},\bar{b})} Q(r)\,|dr|^2, \quad \bar{a} \in E^m, \quad \bar{b} \in E^m,$$

then

$$\left(\frac{\partial w}{\partial b_i}\right)^2 = \left(\frac{\partial r_i}{\partial t}\right)^2\bigg|_{r=\bar{b}}, \quad \left(\frac{\partial w}{\partial a_i}\right)^2 = \left(\frac{\partial r_i}{\partial t}\right)^2\bigg|_{r=\bar{a}}, \quad i = 1, \ldots, m.$$

Hence, in particular, we have for $r = (x, y) = Q^2 = \lambda^2(y) + a^2(x)$,

$$\frac{d^2 y}{dt^2} = \lambda\lambda'(y), \quad \frac{d^2 x}{dt^2} = a \operatorname{grad} a$$

$$\left(\frac{dy}{dt}\right)^2 + \left|\frac{dx}{dt}\right|^2 = \lambda^2(y) + a^2(x). \tag{3.20}$$

If we set $\bar{a} = (\xi, 0)$, $\bar{b} = (\eta, 0)$, then $w(\bar{a}, \bar{b}) = \tau(\xi, \eta)$,

$$\sum_{i=1}^{n} \left(\frac{\partial \tau}{\partial \xi_i}^2\right) = \left|\frac{dx}{dt}\right|^2_{x=\xi}, \quad \sum_{i=1}^{n} \left(\frac{\partial \tau}{\partial \eta_i}^2\right) = \left|\frac{dx}{dt}\right|^2_{x=\eta}. \tag{3.21}$$

On the other hand, we have from (3.20):

$$\left|\frac{dx}{dt}\right|^2_{x=\xi} = \lambda^0(0) + a^2(\xi) - \left(\frac{dy}{dt}\right)^2_{y=0},$$

$$\left|\frac{dx}{dt}\right|^2_{x=\eta} = \lambda^0(0) + a^2(\eta) - \left(\frac{dy}{dt}\right)^2_{y=0}.$$

Hence

$$\left|\frac{dx}{dt}\right|^2_{x=\eta} - \left|\frac{dx}{dt}\right|^2_{x=\xi} = a^2(\eta) - a^2(\xi).$$

Using (3.21), we get:

$$\sum_{i=1}^{n} \left[\left(\frac{\partial \tau}{\partial \eta_i}^2\right) - \left(\frac{\partial \tau}{\partial \xi_i}^2\right)\right] = a^2(\eta) - a^2(\xi). \tag{3.22}$$

Thus, proposition 1 of the theorem is proved.

If we assume $\xi = 0$, $\eta = x$ in this equation and recall that $a(0) = 0$, we obtain the formula for $a(x)$:

$$a(x) = \sqrt{\sum_{i=1}^{n} \left(\frac{\partial \tau}{\partial \eta_i}\right)^2 - \left(\frac{\partial \tau}{\partial \xi_i}\right)^2 \Bigg|_{\eta = x, \xi = 0}}. \tag{3.23}$$

Let us prove propositions 3 and 4 of the theorem. When proving them we assume the variable α to be sufficiently small unless otherwise stated. We denote by $\gamma_\lambda(\alpha)$ the geodesic of the metric

$$ds^2 = (\lambda^2(y) + a^2(x))(dy^2 + dx^2)$$

which connects the origin of coordinates with the point αx^0; the function $a(x)$ is specified here by (3.23) and $\lambda(y) \in \{\lambda\}$. We first note that whatever the function $\lambda(y) \in \{\lambda\}$, the functions

$$f(\alpha) = \int_{\gamma_\lambda} \sqrt{\lambda^2 + a^2}\sqrt{dx^2 + dy^2}, \quad t(\alpha) = \int_{\gamma_\lambda} \frac{\sqrt{dx^2 + dy^2}}{\sqrt{\lambda^2 + a^2}}$$

are continuously differentiable, and

$$f'(0)\sqrt{\lambda^2(0) + a^2(0)} = 1, \quad t'(0) = \frac{1}{\sqrt{\lambda^2(0) + a^2(0)}} = 1.$$

Hence, in particular, $f(\alpha)$ and $t(\alpha)$ are strictly monotonically increasing functions on an interval.

We show that the determination of $\lambda(y)$ by $f'(\alpha)$ is reduced to the problem:

find the vector function $x(t, \alpha)$ and the functions $t(\alpha)$ and $\lambda(y)$ such that

(I) $\dfrac{d^2x}{dt^2} = a \operatorname{grad} a, \; x|_{t=0} = 0, \; x|_{t=t(\alpha)} = \alpha x^0, \; \left(\dfrac{dx}{dt}, x^0\right)\Big|_{t=t(\alpha)} = f'(\alpha);$

(II) $\dfrac{d^2y}{dt^2} = \lambda\lambda', \; y|_{t=0} = 0, \; y|_{t=t(\alpha)} = 0, \; \left(\dfrac{dy}{dt}\right)^2 + \left|\dfrac{dx}{dt}\right|^2 = \lambda^2(y) + a^2(x).$

This system of equations arises, as shown above, for

$$t = \int_{\gamma} \frac{\sqrt{dx^2 + dy^2}}{\sqrt{a^2 + \lambda^2}}.$$

Besides, according to the condition of the problem, the geodesic γ_λ must satisfy the boundary data:

$$x|_{t=0} = 0, \quad x|_{t=t(\alpha)} = \alpha x^0, \quad y|_{t=0} = 0, \quad y|_{t=t(\alpha)} = 0.$$

Therefore, it is enough to show that

$$\left(\frac{dx}{dt}, x^0\right)\Big|_{t=t(\alpha)} = f'(\alpha).$$

We do this using the function

$$\tau(x) = \int_{\gamma_\lambda} \sqrt{(\lambda^2 + a^2)(dx^2 + dy^2)},$$

where γ_λ is the geodesic of the metric (3.16) connecting the origin of coordinates in the space E^{n+1} with the point x in the hyperplane $y = 0$. Using the relation,

$$\operatorname{grad}\tau = x'\frac{\sqrt{\lambda^2(0)+a^2(x)}}{\sqrt{x'^2+y'^2}}, \quad \lambda^2+a^2 = x'^2+y'^2, \quad f(\alpha) = \tau(\alpha x^0)$$

we obtain the equalities

$$f'(\alpha) = (\operatorname{grad} w, x^0)\big|_{x=\alpha x^0} = \left(\frac{\mathrm{d}x}{\mathrm{d}t}, x^0\right)\bigg|_{t=t(\alpha)}.$$

We see that condition (I) of the above (reduced) problem uniquely specifies the function $d(q)$. Then, condition (II) leads to a formula for the function $\lambda^{-1}(q)$ inverse to the sought one. Let us prove this.

Let $\tilde{x}(t,p)$ be a solution of the Cauchy problem:

$$\frac{\mathrm{d}^2\tilde{x}}{\mathrm{d}t^2} = \tilde{a}\operatorname{grad}\tilde{a}, \quad \tilde{x}|_{t=0} = 0, \quad \frac{\mathrm{d}\tilde{x}}{\mathrm{d}t}\bigg|_{t=0} = p.$$

According to the condition, the function $a(x)$ is triply continuously differentiable; therefore, the solution $\tilde{x}(t,p)$ of this problem exists and is unique for some t. Moreover, $\tilde{x}(t,p)$ is at least twice continuously differentiable with respect to p. We derive the system of equations for p and t,

$$\tilde{x}(t,p) = \alpha x^0, \quad \left(\frac{\mathrm{d}x}{\mathrm{d}t}, x^0\right)\bigg|_{t=t(\alpha)} = f'(\alpha). \tag{3.24}$$

If we show that this system has the unique solution $p(\alpha)$ and $t(\alpha)$, the vector function $x(t,x) = \tilde{x}(t,p(\alpha))$ and the function $t(\alpha)$ are the unique solution of the problem

$$\frac{\mathrm{d}^2x}{\mathrm{d}t^2} = a\operatorname{grad}a, \quad x|_{t=0} = 0, \quad x|_{t=t(\alpha)} = \alpha x^0, \quad 0 \le \alpha \le t(\alpha).$$

We represent the solution $\tilde{x}(t,p)$ of the Cauchy problem as

$$\tilde{x}(t,p) = pt + \varphi(t,p).$$

Using this notation, we rewrite (3.24) as

$$pt + \varphi = \alpha x^0, \quad (pt + \varphi', x^0) = f'(\alpha). \tag{3.25}$$

Since $u(0) = 0$ and $\mathrm{d}\tilde{x}/\mathrm{d}t|_{t=0} = p$,

$$\frac{\mathrm{d}\varphi}{\mathrm{d}t}\bigg|_{t=0} = \frac{\mathrm{d}^2\varphi}{\mathrm{d}t^2}\bigg|_{t=0} = 0.$$

We set $t = q\alpha$, $\psi = \varphi/\alpha$, $0 \le q \le 1$. Using this notation, we rewrite (3.25) as

$$pt + \varphi = x^0, \quad (pt + \psi', x^0) = f'(\alpha). \tag{3.26}$$

Since

$$\frac{\mathrm{d}\varphi}{\mathrm{d}t}\bigg|_{t=0} = \frac{\mathrm{d}^2\varphi}{\mathrm{d}t^2}\bigg|_{t=0} = 0, \quad f'(0) = 1, \quad |x^0| = 1$$

the solution $q(\alpha)$, $p(\alpha)$ of (3.26) for $\alpha = 0$ is $q = 1$, $p = x^0$. The determinant of the Jacobi matrix (3.26) at the point $q = 1$, $p = x^0$, $\alpha = 0$ equals:

$$\begin{vmatrix} x_1^0 & 1 & & & 0 \\ x_2^0 & & 1 & & \\ \vdots & & 0 & \ddots & \\ x_n^0 & & & & 1 \\ 0 & x_1^0 & x_2^0 & \dots & x_n^0 \end{vmatrix} = (-1)^{n+1}$$

hence, by the theorem of implicit functions, (3.23) has the unique solution $q(\alpha)$, $p(\alpha)$, $q(0) = 1$, $p(0) = x^0$. System of equations (I) in the above (reduced) problem has the unique solution $x(t,\alpha)$, $t(\alpha)$, too.

We now show that there exists a number α_0 such that the function $h(\alpha) = |p(\alpha)|$, $0 \le \alpha \le \alpha_0$, strictly monotonically decreases, and

$$d(q) = qt(h^{-1}(q)) = 2 \int_0^{\lambda^{-1}(q)} \frac{q\,\mathrm{d}z}{\sqrt{\lambda^2(z) - q^2}}. \tag{3.27}$$

We first consider the problem

$$\frac{\mathrm{d}^2 y}{\mathrm{d}t^2} = \lambda(y)\lambda'(y), \quad y|_{t=0} = y|_{t=t(\alpha)} = 0. \quad 0 \le t \le t(\alpha)$$

where $t(\alpha)$ is a solution of (3.24), $\lambda(y) \in \{\lambda\}$ is a fixed function.

For any function $\lambda(y) \in \{\lambda\}$ the solution $y(t,\alpha)$ of this problem satisfies the conditions as a function of the variables t:

1. $y(t,\alpha)$ is a strictly convex function, and $y > 0$, $0 < t < t(\alpha)$;
2. $(\mathrm{d}y/\mathrm{d}t)^2 = \lambda^2 - c(\alpha)$, where $c(\alpha)$ is a strictly monotonically decreasing function dependent only on α;
3. the graph of $y(t,\alpha)$ is symmetric about the straight line $t = t(\alpha)/2$.

Indeed, convexity results from the inequalities $\lambda(y) > 0$, $\lambda'(y) < 0$ which also lead to $y(t,\alpha) > 0$, $0 < t < t(\alpha)$. Then, $(\mathrm{d}y/\mathrm{d}t)^2 = \lambda^2 - c(\alpha)$ is the first integral of the equation:

$$\frac{\mathrm{d}^2 y}{\mathrm{d}t^2} = \lambda(y)\lambda'(y).$$

Property 3 follows, for example, from property 2. We are to show that the function $c(\alpha)$ is rigorously monotonically decreasing. Indeed, taking into account properties 1–3, we have:

$$\lambda^2\left(y\left(\frac{t(\alpha)}{2}, \alpha\right)\right) = c(\alpha).$$

Since $\lambda(y)$ is monotonic, $c(\alpha)$ is also monotonic provided that

$$y\left(\frac{t(\alpha_1)}{2}, \alpha_1\right) > y\left(\frac{t(\alpha_2)}{2}, \alpha_2\right)$$

for $\alpha_1 > \alpha_2$. We even show that for any α_1 and α_2, $\alpha_1 > \alpha_2$, and any t, $0 < t \leq t(\alpha_2)$,

$$y(t, \alpha_1) > y(t, \alpha_2).$$

We suppose that this inequality does not hold. We have provided that $t(\alpha_1) > t(\alpha_2)$. Both solutions $y_i = y(t/\alpha_i)$, $i = 1, 2$, have properties 1–3. Therefore, there exists at least a single point t_0 such that

$$y_1\big|_{t=t_0} = y_2\big|_{t=0} = y^0.$$

Since y_1 and y_2 are solutions of the same equation $y'' = \lambda'(y)\lambda(y)$ and

$$y_i\big|_{t=0} = 0, \quad y_i\big|_{t=t_0} = z^0.$$

then $y_1 \equiv y_2$, $0 \leq t \leq t(\alpha_2)$. We write

$$z\big(t(\alpha_1), \alpha_1\big) = z\big(t(\alpha_2), \alpha_2\big) = 0, \quad t(\alpha_1) > t(\alpha_2).$$

The latter equality contradicts property 1. Thus, properties 1–3 are proved.

Finally, let the function $\lambda(y) \in \{\lambda\}$ satisfy condition (II) in the reduced problem. We have:

$$\left(\frac{\mathrm{d}y}{\mathrm{d}t}\right)^2 = \lambda^2 - c(\alpha), \quad \left(\frac{\mathrm{d}y}{\mathrm{d}t}\right)^2 + \left|\frac{\mathrm{d}x}{\mathrm{d}t}\right|^2 = \lambda^2(y) + a^2(x). \tag{3.28}$$

Here $x(t, \alpha)$ is a solution of condition (II).

Since

$$\left|\frac{\mathrm{d}x}{\mathrm{d}t}\right|^2_{t=0} = |p(\alpha)|^2 = h^2(\alpha), \quad a(0) = 0$$

we obtain the equality $c(\alpha) = h^2(\alpha)$ from (3.28). We have proved that $h(0) = |p(0)| = |x^0| = 1$. Taking into account $c(\alpha) = h^2(\alpha)$, we obtain from the first equality in (3.28):

$$\frac{\mathrm{d}t}{\mathrm{d}y} = \pm\frac{1}{\sqrt{\lambda^2(q) - h^2(\alpha)}}.$$

Integrating this relation, we obtain the expression for $t(\alpha)$:

$$t(\alpha) = 2\int_0^{y(\alpha)} \frac{\mathrm{d}y}{\sqrt{\lambda^2(q) - h^2(\alpha)}}. \tag{3.29}$$

Let $\lambda^{-1}(q)$ and $h^{-1}(q)$ be inverse functions to $\lambda(y)$ and $h(x)$, respectively. Assuming $h(x) = q$ in (3.29), we have:

$$d(q) = qt\big(h^{-1}(q)\big) = 2\int_0^{\lambda^{-1}(q)} \frac{q\,\mathrm{d}z}{\sqrt{\lambda^2(z) - q^2}}.$$

Thus, (3.27) is established. We consider it as an equation for the function $\lambda(y)$. Let us show that the inversion formula holds:

$$\lambda^{-1}(q) = \frac{1}{\pi}\int_q^1 \frac{d(r)\,\mathrm{d}r}{\sqrt{r^2 - q^2}}.$$

Indeed

$$\frac{1}{\pi}\int_q^1 \frac{d(r)\,dr}{\sqrt{r^2-q^2}} = \frac{2}{\pi}\int_q^1\int_0^{\lambda^{-1}(q)} \frac{r\,dz\,dr}{\sqrt{(\lambda^2(z)-r^2)(r^2-q^2)}}$$

$$= -\frac{2}{\pi}\int_0^{\lambda^{-1}(q)} dz \int_q^{\lambda(z)} \frac{d}{dr}\arctan\sqrt{\frac{\lambda^2(z)-z^2}{z^2-q^2}}\,dz = \lambda^{-1}(q).$$

This completes the proof of the theorem.

It is simpler to find the solution of a two-dimensional inverse kinematic problem. We give some examples.

Example 3.1. $\lambda^2(x,y) = f^2(x) + g^2(y)$, $f(0) = 0$, $g(0) = 1$, $g'(y) > 0$. The data are given on the line $y = 0$.

$$\tau(x_1,x_2) = \int_{\gamma(x_1,x_2)} \sqrt{f^2(x)+g^2(y)}\sqrt{dx^2+dy^2},$$

where $\gamma(x_1,x_2)$ is the geodesic connecting the points $(x_1,0)$ and $(x_2,0)$. Then $f^2(x)$ is determined by the formula:

$$f^2(x) = \tau_x^2(0,x) - \tau_y^2(x,y)\big|_{y=0}.$$

Let $f(x)$ be known and the hodograph

$$T(\alpha) = \int_{\gamma(0,\alpha)} \sqrt{f^2(x)+g^2(y)}\sqrt{dx^2+dy^2}$$

be given; $\gamma(0,\alpha)$ is the geodesic connecting the points $(0,0)$ and $(\alpha,0)$. Then $g(y)$ is determined by the formula:

$$g^{-1}(p) = \frac{1}{\pi}\int_p^1 \frac{qt(\alpha(q))}{\sqrt{q^2-p^2}}\,dq,$$

where

$$t(\alpha) = \int_0^\alpha \frac{dx}{\sqrt{f^2(x)+q^2(\alpha)}}, \quad q^2(\alpha) = T'^2(\alpha) - f^2(\alpha),$$

$\alpha(q)$ is the inverse function to $q(\alpha)$.

Example 3.2.

$$\lambda^2(x,y) = \frac{1}{x^2+y^2}\left(g^2\left(\arctan\frac{y}{x}\right) + f^2\left(\sqrt{x^2+y^2}\right)\right),$$

$$f(p_0) = 0, \quad g(0) = 1, \quad g'(\varphi) > 0.$$

The data are given on the line $y = 0$:

$$\tau(x_1,x_2) = \int_{\gamma(x_1,x_2)} \frac{1}{p}\sqrt{g^2(\varphi)+f^2(p)}\sqrt{dp^2+p^2\,d\varphi^2}, \quad x_1 > 0, \quad x_2 > 0,$$

where $\gamma(x_1, x_2)$ is the geodesic connecting the points $(x_1, 0)$ and $(x_2, 0)$. Then $f^2(p)$ is determined by the formula:

$$f^2(p) = p^2 \tau_p^2(p_0, p) - p_0^2 \tau_r^2(p, r)\big|_{r=p_0}.$$

Let $f(p)$ be known and the hodograph

$$T(\alpha) = \int\limits_{\gamma(p_0, \alpha)} \frac{1}{p} \sqrt{g^2(\varphi) + f^2(p)} \sqrt{\mathrm{d}p^2 + p^2\, \mathrm{d}\varphi^2}$$

be given; $\gamma(p_0, \alpha)$ is the geodesic connecting the points (p_0, α) and $(\alpha, 0)$. Then $g(\varphi)$ is determined by the formulas:

$$g^{-1}(p) = \frac{1}{\pi} \int\limits_p^1 \frac{qt(\alpha(q))}{\sqrt{q^2 - p^2}}\, \mathrm{d}q,$$

where

$$t(\alpha) = \int\limits_{p_0}^{\alpha} \frac{\mathrm{d}p}{p\sqrt{f^2(p) + q^2(\alpha)}}, \quad q^2(\alpha) = \alpha T'^2(\alpha) - f^2(\alpha),$$

$\alpha(q)$ is the inverse function to $q(\alpha)$.

Example 3.3.

$$\lambda^2(x, y) = \frac{1}{x^2 + y^2} \left(f^2 \left(\arctan \frac{y}{x} \right) + g^2 \left(\sqrt{x^2 + y^2} \right) \right),$$

$$f(0) = 0, \quad g(1) = 1, \quad g'(p) > 0.$$

The data are given on the circle $x^2 + y^2 = 1$

$$\tau(\alpha, \beta) = \int\limits_{\gamma(\alpha, \beta)} \frac{1}{p} \sqrt{f^2(\varphi) + g^2(p)} \sqrt{\mathrm{d}p^2 + p^2\, \mathrm{d}\varphi^2},$$

where $\gamma(\alpha, \beta)$ is the geodesic connecting the points $(\cos\alpha, \sin\alpha)$ and $(\cos\beta, \sin\beta)$. Then $f^2(\varphi)$ is determined by the formula:

$$f^2(\varphi) = \tau_\varphi^2(0, \varphi) - \tau_\beta^2(\varphi, \beta)\big|_{\beta=0}.$$

Let $f(\varphi)$ be known and the hodograph

$$T(\alpha) = \int\limits_{\gamma(0, \alpha)} \frac{1}{p} \sqrt{f^2(\varphi) + g^2(p)} \sqrt{\mathrm{d}p^2 + p^2\, \mathrm{d}\varphi^2}$$

be given; $\gamma(0, \alpha)$ is the geodesic connecting the points $(1, 0)$ and $(\cos\alpha, \sin\alpha)$. Then $g(p)$ is determined by the formula:

$$g^{-1}(p) = \exp\left[\frac{1}{\pi} \int\limits_p^1 \frac{qt(\alpha(q))}{\sqrt{q^2 - p^2}}\, \mathrm{d}q \right],$$

where

$$t(\alpha) = \int\limits_{p_0}^{\alpha} \frac{\mathrm{d}\varphi}{\sqrt{f^2(\varphi) + q^2(\alpha)}}, \quad q^2(\alpha) = T'^2(\alpha) - f^2(\alpha),$$

$\alpha(q)$ is the inverse function to $q(\alpha)$.

We now consider the two-dimensional Riemannian metric with the linear element $\mathrm{d}s$:

$$\mathrm{d}s^2 = \lambda^2(y)\,\mathrm{d}x^2 + 2b(\lambda^2(y))\,\mathrm{d}x\,\mathrm{d}y + c^2(\lambda^2(y))\,\mathrm{d}y^2 \tag{3.30}$$

where b and c are known functions, and $\lambda(y)$ is an unknown function. We suppose that for a family of geodesics we know the lengths of the geodesics in the metric (3.30).

We denote by $\gamma(x,0)$ the geodesic of the metric (3.30) which is located in the half-space $y \geq 0$, originates at the point $(0,0)$ and ends at $(x,0)$. We know the function:

$$\tau(x) = \int\limits_{\gamma(x)} \sqrt{a^2\,\mathrm{d}x^2 + 2b\,\mathrm{d}x\,\mathrm{d}y + c^2\,\mathrm{d}y^2}.$$

We must specify the function $\lambda(y)$ for $y \geq 0$. We assume that the following conditions are satisfied:

(a) $\lambda(y)$ is a positive, continuously differentiable, rigorously decreasing function, and $\lambda(0) = 1$;

(b) $b(\xi)$, $c(\xi)$, $0 \leq \xi \leq 1$, are continuous functions such that for $\xi > 0$

$$\xi c^2(\xi) - b^2(\xi) > 0, \quad c(1) > 0, \quad \alpha = \frac{b(1)}{c(1)} \geq 0;$$

(c) the curve in the plane (x,t) specified by the equation $t = t(x)$ [$t(x)$ is generally a many-valued function] is continuously differentiable, and if $p(x)$ is the angle between this curve and the x-axis, then $p(x)$ is a strictly decreasing function.

Let all the three conditions be satisfied. If $y(p)$, $x(p)$ are inverse functions to $\lambda(y)$ and $p(x)$, respectively, then

$$y(p) = -\int\limits_{p}^{1} \frac{B'(q)}{\varphi(q^2)}\,\mathrm{d}q,$$

where

$$B'(q) = \frac{1}{\pi}\frac{\mathrm{d}}{\mathrm{d}q}\int\limits_{q}^{1} \frac{x(p)}{\sqrt{p^2 - q^2}}\,\mathrm{d}p, \quad \varphi(q^2) = \frac{1}{q^2}\sqrt{q^2c^2(q^2) - b^2(q^2)}.$$

The Herglotz formula is obtained for $b = 0$ and $c(\lambda) = \lambda$. The condition of monotonicity for the unknown functions is substantial.

3.7 INTEGRO-DIFFERENTIAL IDENTITIES IN THE PLANE

3.7.1

Before proceeding to the problems, we give the lemma of analytic functions which will be useful in the following.

Let E^n be the Euclidean space of variables $(x_1, \dots, x_n)$; D_0 be the half-space $x_n > 0$; $\omega(r,t)$ the intersection D_0 of with the open n-dimensional sphere $s(r,t)$ of radius $r > 0$ with its centre at the point t that belongs to the hyperplane $x_n = 0$. The intersection $\omega(r,t)$ is called the hemisphere.

We consider the set $\{\omega\}$ of hemispheres $\omega(r,t)$, $0 < r < a$, $|t| < b$, where $a > 0$, $b > 0$ are fixed numbers.

Lemma 3.1. *Let real functions $\lambda_1(x)$ and $\lambda_2(x)$ be analytic in the half-space $x_n \geq 0$ and*

$$\lambda_1(x) \not\equiv \lambda_2(x).$$

Then there exists a hemisphere $\omega_0 = \omega(t_0, r_0) \in \{\omega\}$ such that for all $x \in \omega_0$:

$$|\lambda_1(x) - \lambda_2(x)| \geq c x_n^\alpha, \tag{3.31}$$

where $c > 0$ and $\alpha > 0$ are constants.

PROOF. It is sufficient to prove (3.31) only when

$$\lambda_1\big|_{x_n=0} = \lambda_2\big|_{x_n=0}.$$

Let $\lambda(x)$ be an arbitrary analytic function that is given in the half-space $x_n \geq 0$, not identically equal to zero and such that $\lambda\big|_{x_n=0}$. We first show that there exists a hemisphere $\omega_1 \in \{\omega\}$ for which the inequality $|\lambda(x)| > 0$ holds.

We suppose that this is not the case. Then each intersection $\omega \in \{\omega\}$ contains a point $x_\omega = (x_1^\omega, \dots, x_n^\omega)$ such that

$$\lambda(x_\omega) = 0. \tag{3.32}$$

We show that this equality, in particular, yields

$$\left.\frac{\partial^m \lambda}{\partial x_n^m}\right|_{x_n=0} = 0, \tag{3.33}$$

where m is any integer.

The assumed analyticity of the function λ implies the equality $\lambda(x) = 0$ in spite of the condition $\lambda(x) \not\equiv 0$.

Equality (3.33) for $m = 0$ follows from (3.32). Let ω be an arbitrary hemisphere and x_ω a point at which $\lambda(x_\omega) = 0$. We set $t_\omega = (x_1^\omega, x_2^\omega, \dots, x_{n-1}^\omega, 0)$ and consider the function

$$\lambda(\xi) = \lambda(\xi t_\omega + (1-\xi)x_\omega), \quad 0 \leq \xi \leq 1.$$

The definition of the function $\tilde{\lambda}(\xi)$ leads to the equalities:

$$\tilde{\lambda}(0) = \tilde{\lambda}(1) = 0.$$

Therefore, there exists a point ξ_0, $0 < \xi_0 < 1$, such that $\mathrm{d}\tilde{\lambda}/\mathrm{d}\xi = 0$ for $\xi = \xi_0$.

On the other hand, the definition of the point t_ω implies:

$$\frac{\partial \lambda}{\partial x_n} = -\frac{\partial \tilde{\lambda}}{\partial \xi}.$$

The above arguments show that there exists a point x_ω^1 at which the function $\lambda^1(x) = \partial\lambda/\partial x_n$ vanishes. Because of the arbitrariness of ω we conclude: whatever the hemisphere ω, there exists a point x_ω^1 such that $\lambda^1(x_\omega^1)$. Hence, in particular:

$$\lambda^1\Big|_{x_n=0} = \frac{\partial \lambda}{\partial x_n}\Big|_{x_n=0} = 0.$$

Using the same reasonings, we find the point

$$x_\omega^2 : \quad \left\{\lambda^2(x_\omega^2) = \frac{\partial^2 \lambda}{\partial x_n^2} = 0\right\}$$

in any hemisphere ω. Hence $\dfrac{\partial^2 \lambda}{\partial x_n^2}\Big|_{x_n=0} = 0$, etc, $\dfrac{\partial^m \lambda}{\partial x_n^m}\Big|_{x_n=0} = 0$.

Thus, we have shown that there exists a hemisphere $\omega_1 = \omega(t_0, r_1)$ for which the inequality $|\lambda(x)| > 0$ holds.

Now let $\omega_0 = \omega(t_0, r_0)$, where $r_0 = r_1/4$. Since the function $\lambda(x)$ is zero, when $x_n = 0$ and $\lambda \neq 0$ for $x \in \omega_0$, because of the Loiasiewicz inequality we have:

$$|\lambda(x)| \geq c x_n^\alpha, \quad x \in \omega_0, \tag{3.34}$$

where $c > 0$ and $\alpha > 0$ are constants. Assuming $\lambda = \lambda_1 - \lambda_2$ in (3.34) we obtain (3.31).

REMARK 3.1. The assumed analycity of both functions λ_1 and λ_2 in the closed domain x_n is substantial. Indeed, (3.31) is obviously invalid for the functions $\lambda_1 = 0$ and $\lambda_2(0) = 0$, $\lambda_2(x) = \mathrm{e}^{-1/x_n^2} \sin(1/x_n)$. The function $\lambda_2(x)$ is analytic only in the open domain $x_n > 0$.

REMARK 3.2. The proof of the lemma implies that if λ_1 and λ_2 , $\lambda_1 \neq \lambda_2$, are infinitely differentiable and analytic in x_n, including the hyperplane $x_n = 0$, there exists a hemisphere $\omega_0 \in \{\omega\}$ such that the inequality holds:

$$|\lambda_1(x) - \lambda_2(x)| > 0, \quad x \in \omega_0.$$

3.7.2

Let D be a simply connected domain in the plane E^2 of variables $x = (x_1, x_2)$ with the smooth boundary

$$\Gamma = \{\varphi(t), \quad \varphi = (\varphi_1, \varphi_2), \quad |t| < t_0\}.$$

We suppose that the Riemannian metric is given in the domain D:

$$\mathrm{d}s^2 = \lambda^2(x_1, x_2)(\mathrm{d}x_1^2 + \mathrm{d}x_2^2) = \lambda^2(x)\,|\mathrm{d}x|^2 \tag{3.35}$$

such that any two points $\varphi(a)$ and $\varphi(b)$ on a part of the curve Γ ($|a \leq r|$, $|b| \leq r$) may be connected with the single geodesic $\gamma(a, b)$ of the metric (3.35).

We consider the function $\tau(a,b)$ in the domain $|a| \le r$, $|b| \le r$:

$$\tau(a,b) = \int\limits_{\gamma(a,b)} \lambda(x)\,|\mathrm{d}x|\,.$$

We show here that if the function $\tau(a,b)$ satisfies a differential equation, we uniquely find a differential equation for the function $\lambda(x)$, and *vice versa*. This leads, in particular, to the formulas for the function λ.

Let us proceed to the exact formulations.

Let $u(x)$ and $v(x)$ be harmonic conjugate functions given in the closed domain D such that $u + \mathrm{i}\,v$ is an analytic function of the complex variable $x_1 + \mathrm{i}\,x_2$. We set

$$u_0(t) = u\big(\varphi(t)\big), \quad v_0(t) = v\big(\varphi(t)\big)$$

and let

$$f(t) = u_0\frac{\mathrm{d}\varphi_1}{\mathrm{d}t} + v_0\frac{\mathrm{d}\varphi_2}{\mathrm{d}t}, \quad g(t) = u_0\frac{\mathrm{d}\varphi_2}{\mathrm{d}t} - v_0\frac{\mathrm{d}\varphi_1}{\mathrm{d}t}.$$

We consider the first-order differential operators $\mathbf{L}\tau$ and $\tilde{\mathbf{L}}\lambda$:

$$\begin{aligned}\mathbf{L}\tau \;=\;& f(a)\frac{\partial\tau}{\partial a} + f(b)\frac{\partial\tau}{\partial b} + g(a)\sqrt{\left(\max_b\frac{\partial\tau}{\partial a}\right)^2 - \left(\frac{\partial\tau}{\partial a}\right)^2}\\ &+g(b)\sqrt{\left(\max_a\frac{\partial\tau}{\partial b}\right)^2 - \left(\frac{\partial\tau}{\partial b}\right)^2} + \beta\tau(a,b), \quad |a| \le r, \quad |b| \le r,\end{aligned}$$

where β is a constant,

$$\tilde{\mathbf{L}}\lambda = u(x)\frac{\partial\lambda}{\partial x_1} + v(x)\frac{\partial\lambda}{\partial x_2} + \frac{1}{2}\left(\frac{\partial u}{\partial x_1} + \frac{\partial u}{\partial x_2}\right)\lambda + \beta\lambda, \quad x \in D.$$

In the following, we suppose that the function $\tilde{\mathbf{L}}\lambda$ is analytic in the closed domain D, and the function $\tau(a,b)$ satisfies the inequality

$$\tau(a,b) < \tau(a,b_1) + \tau(b_1,b), \quad a < b_1 < b.$$

Theorem 3.5. *If* $\mathbf{L}\tau = 0$, *then* $\tilde{\mathbf{L}}\lambda = 0$, *and vice versa.*

PROOF. Let y and z be points in the domain

$$D\big(y = (y_1, y_2);\; z = (z_1, z_2)\big)$$

and $\gamma(y,z)$ be the geodesic of the metric (3.35) which connects these points. We consider the function:

$$w(y,z) = \int\limits_{\gamma(y,z)} \lambda(x)\,|\mathrm{d}x|\,,$$

where we integrate from y to z. If $u_1(x)$ and $u_2(x)$ are differentiable functions and $x(\xi)$ is the parametrization of γ, then

$$\int_{\gamma(y,z)} \left[\sum_{i=1}^{2} u_i \frac{\partial \lambda}{\partial x_i} + \frac{\lambda}{|x'_\xi|^2} \int_{i=1}^{2} \sum_{j=1}^{2} \frac{\partial u_i}{\partial x_j} \frac{\mathrm{d}x_i}{\mathrm{d}\xi} \frac{\mathrm{d}x_j}{\mathrm{d}\xi} + \beta\lambda \right] |\mathrm{d}x|$$
$$= \sum_{i=1}^{2} \frac{\partial w}{\partial z_i} u_i(z) + \frac{\partial w}{\partial y_i} u_i(y) + \beta w(y,z), \quad \beta = \text{const} . \tag{3.36}$$

Indeed, first, let

$$\xi = \int_\gamma \frac{|\mathrm{d}x|}{\lambda}.$$

This choice of the parameter ξ inplies that the system of Euler equations specifying the geodesic γ has the form:

$$\frac{\mathrm{d}^2 x_i}{\mathrm{d}\xi^2} = \lambda \frac{\partial \lambda}{\partial x_i}, \quad i = 1, 2.$$

Using these equalities, we have:

$$\sum_{i=1}^{2} \frac{\mathrm{d}}{\mathrm{d}\xi} \left(u_i(x) \frac{\mathrm{d}x_i}{\mathrm{d}\xi} \right) = \sum_{i=1}^{2} \lambda \left(u_i \frac{\mathrm{d}\lambda}{\mathrm{d}x_i} \right) + \sum_{i=1}^{2} \sum_{j=1}^{2} \frac{\partial u_i}{\partial x_j} \frac{\mathrm{d}x_i}{\mathrm{d}\xi} \frac{\mathrm{d}x_j}{\mathrm{d}\xi}. \tag{3.37}$$

Note that

$$\left. \frac{\mathrm{d}x_i}{\mathrm{d}\xi} \right|_{x=y} = -\frac{\partial w}{\partial y_i}, \quad \left. \frac{\mathrm{d}x_i}{\mathrm{d}\xi} \right|_{x=z} = \frac{\partial w}{\partial z_i}. \tag{3.38}$$

If we integrate (3.37) along γ and use (3.38), we obtain (3.36). We now assume $u_1 = u(x)$ and $u_2 = u(x)$, where $u(x)$ and $v(x)$ are harmonic conjugate functions. This assumption leads to

$$\frac{\lambda}{|x'_\xi|^2} \sum_{i=1}^{2} \sum_{j=1}^{2} \frac{\partial u_i}{\partial x_j} \frac{\mathrm{d}x_i}{\mathrm{d}\xi} \frac{\mathrm{d}x_j}{\mathrm{d}\xi} = \lambda \frac{1}{2} \left(\frac{\partial u}{\partial x_1} + \frac{\partial u}{\partial x_2} \right)$$

This identity and (3.36) yield:

$$\int_{\gamma(y,z)} \tilde{\mathbf{L}}\lambda \, |\mathrm{d}x| = \mathbf{L}w = \frac{\partial w}{\partial z_1} u(z) + \frac{\partial w}{\partial z_2} v(z) + \frac{\partial w}{\partial y_1} u(y) + \frac{\partial w}{\partial y_2} v(y) + \beta w. \tag{3.39}$$

The definition of the function $w(y, z)$ implies that the equality $w(\varphi(a), \varphi(b)) = \tau(a, b)$ holds for $y = \varphi(a)$, $z = \varphi(b)$.

Hence, in particular,

$$\frac{\partial \tau}{\partial b} = \sum_{i=1}^{2} \left. \frac{\partial w}{\partial z_i} \frac{\mathrm{d}\varphi_i}{\mathrm{d}t} \right|_{z=\varphi(b), t=b}, \quad \frac{\partial \tau}{\partial a} = \sum_{i=1}^{2} \left. \frac{\partial w}{\partial y_i} \frac{\mathrm{d}\varphi_i}{\mathrm{d}t} \right|_{y=\varphi(a), t=a}. \tag{3.40}$$

Let $(\cos\alpha_1, \cos\alpha_2)$ and $(\cos\theta_1, \cos\theta_2)$ be unit vectors tangent to the geodesic $\gamma(a, b)$ at points $\varphi(a)$ and $\varphi(b)$, respectively. We have:

$$\frac{\partial w}{\partial z_1} = \lambda(\varphi(b)) \cos\alpha_1, \quad \frac{\partial w}{\partial y_1} = \lambda(\varphi(b)) \cos\theta_1, \quad i = 1, 2. \tag{3.41}$$

Equalities (3.40) and (3.41) yield:

$$\left(\max_a \frac{\partial \tau}{\partial b}\right)^2 = \lambda^2(\varphi(b))\,|\varphi'_t|^2, \quad \left(\max_a \frac{\partial \tau}{\partial a}\right)^2 = \lambda^2(\varphi(a))\,|\varphi'_t|^2,$$

$$\frac{\partial w}{\partial z_1} = \frac{\partial \tau}{\partial b}\frac{\partial \varphi_1}{\partial b} + \sqrt{\left(\max_a \frac{\partial \tau}{\partial b}\right)^2 - \left(\frac{\partial \tau}{\partial b}\right)^2}\frac{\mathrm{d}\varphi_2}{\mathrm{d}b},$$

$$\frac{\partial w}{\partial z_2} = \frac{\partial \tau}{\partial b}\frac{\partial \varphi_2}{\partial b} + \sqrt{\left(\max_a \frac{\partial \tau}{\partial b}\right)^2 - \left(\frac{\partial \tau}{\partial b}\right)^2}\frac{\mathrm{d}\varphi_2}{\mathrm{d}b}.$$

Similar formulas hold for $\partial w/\partial y_i$, $i = 1, 2$, as well. Substituting $\partial w/\partial z_i$, $\partial w/\partial y_i$, $i = 1, 2$, into (3.39), we derive the principal identity:

$$\mathbf{L}\tau = \int_{\gamma(a,b)} \tilde{\mathbf{L}}\lambda\,|\mathrm{d}x|.$$

Hence, if $\tilde{\mathbf{L}}\lambda = 0$, $x \in D$, then $\mathbf{L}\tau = 0$ in the domain $|a| \le r$, $|b| \le r$. Let us prove otherwise. If $\mathbf{L}\tau = 0$,

$$\int_{\gamma(a,b)} \tilde{\mathbf{L}}\lambda\,|\mathrm{d}x| = 0 \tag{3.42}$$

along each geodesic $\gamma(a,b)$, $|a| \le r$, $|b| \le r$. Under the above assumptions the integrand is analytic in the closed domain D.

Locally, geodesics are the shortest. Because of the inequality:

$$\tau(a,b) < \tau(a,b_1) + \tau(b_1,b), \quad a < b_1 < b$$

any interior point of the shortest $\gamma(a,b)$ cannot belong to the boundary Γ. Consequently any shortest $\gamma(a,b)$, except for the ends $\varphi(a)$ and $\varphi(b)$, completely belongs to the domain D.

Let $\{\omega\}$ be a set of all intersections of the domain D with open circles whose centres belong to the boundary Γ. The integrand $\tilde{\mathbf{L}}\lambda = \lambda_1(x)$ is assumed to be analytic in the closed domain D. Therefore, if $\tilde{\mathbf{L}}\lambda \ne 0$, by Lemma 3.1 there exists $\omega_0 \in \{\omega\}$ such that $|\lambda_1(x)| > 0$ for $x \in \omega_0$. Consequently, we obtain

$$\int_{\gamma(a,b)} \tilde{\mathbf{L}}\lambda\,|\mathrm{d}x| \ne 0$$

for the shortest $\gamma_0 \subset \omega_0$, which contradicts (3.42). This completes the proof of the theorem.

Thus, the problem of determining the function $\lambda(x)$ by $\tau(a,b)$ provided that $\mathbf{L}\tau = 0$ splits into two stages:

1. determine the analytic function $u(x) + \mathrm{i}\,v(x)$ by the boundary data:

$$f(t) = u_0\frac{\mathrm{d}\varphi_1}{\mathrm{d}t} + v_0\frac{\mathrm{d}\varphi_2}{\mathrm{d}t}, \quad g(t) = u_0\frac{\mathrm{d}\varphi_2}{\mathrm{d}t} - v_0\frac{\mathrm{d}\varphi_1}{\mathrm{d}t}, \quad \mathbf{L}\tau = 0, \quad |a| \le r, \quad |b| \le r;$$

2. investigate the equation $\tilde{\mathbf{L}}\lambda = 0$, $x \in D$.

We give two examples to illustrate possible applications.

Example 3.4. Let a domain be the half-space $x_2 > 0$, and $\varphi_1 = t = x_1$, $\varphi_2 = 0$, $a \le b$. If

$$\mathbf{L}\tau = \frac{\partial \tau}{\partial a} + \frac{\partial \tau}{\partial b} - \alpha \left[\sqrt{\left(\max_a \frac{\partial \tau}{\partial a}\right)^2 - \left(\frac{\partial \tau}{\partial b}\right)^2} + \sqrt{\left(\max_a \frac{\partial \tau}{\partial b}\right)^2 - \left(\frac{\partial \tau}{\partial b}\right)^2} \right] + \beta\tau = 0$$

where α and β are constants, $\alpha \neq 0$, then

$$\lambda(x) = \lambda_0(x_1 - x_2/\alpha)\,e^{-(\beta/\alpha)x_2}, \quad \lambda_0(q) = \max_a \left.\frac{\partial \tau}{\partial b}\right|_{b=q}.$$

Indeed, in this case $u_0(t) = 1$, $v_0(t) = \alpha$. Because of this and the uniqueness of the analytic extension, we have $u(x) = 1$ and $v(x) = \alpha$, $x \in D$. The equation $\tilde{\mathbf{L}}\lambda = 0$ for these u and v is written as

$$\tilde{\mathbf{L}}\lambda = \frac{\partial \lambda}{\partial x_1} + \alpha \frac{\partial \lambda}{\partial x_2} + \beta\lambda = 0.$$

This equality and the equality $\lambda_0(x_1) = \lambda(x_1, 0)$ lead to the desired assertion.

Example 3.5. Let Γ be a closed curve and its parametrization $\varphi(t)$, $|t| \le t_0$ be given. We assume that the numbers $u_0(a_0)$ and $v_0(a_0)$ are known, where a_0 is a point on the segment $|t| \le t_0$. Under this assumption, the equality $\mathbf{L}\tau = 0$ in the domain $|t| \le t_0$, $a = a_0$ may be written as

$$u_0(b)A_1(b) + v_0(b)A_2(b) = A_3(b), \tag{3.43}$$

where $A_i(b)$ are given functions specified by $\tau(a,b)$ and $\varphi(t)$.

To find the analytic function $u + \mathrm{i}\,v$ by using the boundary condition (3.43) is the Riemann-Hilbert problem. Therefore, the determination of the differential equation $\tilde{\mathbf{L}}\lambda = 0$ for the function $\lambda(x)$ is reduced to this problem.

3.7.3

Our aim is to give the necessary and sufficient conditions under which the function $\lambda(x,y)$ satisfies the second-order differential equation:

$$\tilde{\mathbf{L}}\lambda = \frac{\partial^2}{\partial x \partial y}\left[u\lambda^2\right] + v\left[\frac{\partial^2 \lambda^2}{\partial y^2} - \frac{\partial^2 \lambda^2}{\partial x^2}\right] + \frac{\partial v}{\partial y}\frac{\partial \lambda^2}{\partial y} - \frac{\partial v}{\partial x}\frac{\partial \lambda^2}{\partial x} + \beta\Delta \ln \lambda = 0,$$

where $u(x,y)$ and $v(x,y)$ are harmonic conjugate functions, β is a constant, Δ is the Laplace operator.

The above conditions are formulated in terms of the harmonic functions $u(x,y)$ and $v(x,y)$ for $y = 0$. We designate:

1. $u_0(x) = u(x,0)$, $v_0(x) = v(x,0)$;

2. $\lambda_0(x) = \lambda(x,0) = \left.\dfrac{\partial \tau}{\partial b}\right|_{a=b=x}$;

3. $B(a,b) = \sqrt{\lambda_0^2(b) - \left(\dfrac{\partial \tau}{\partial b}\right)^2}$, $A(a,b) = \sqrt{\lambda_0^2(a) - \left(\dfrac{\partial \tau}{\partial a}\right)^2}$;

4. $\lambda_1(x) = \left.\dfrac{\partial\lambda}{\partial y}\right|_{y=0} = -\lim\limits_{b\to x, a\to x} \dfrac{A+B}{b-a};$

5. $k(a,b) = \arctan \dfrac{B}{\partial\tau/\partial b} + \arctan \dfrac{A}{\partial\tau/\partial a} + \displaystyle\int_a^b \dfrac{\lambda_1(t)}{\lambda_0(t)}\, dt.$

Note that the functions $\lambda_0(x)$, $\lambda_1(x)$, $A(a,b)$, $B(a,b)$, $k(a,b)$ are specified only by the function $\tau(a,b)$ according to the above formulas.

We consider the expression:

$$\begin{aligned} \mathbf{L}\tau &= u_0(b)\left(\frac{\partial\tau}{\partial b}\right)^2 - u_0(a)\left(\frac{\partial\tau}{\partial a}\right)^2 + u_0(a)\lambda_0^2(a) - u_0(b)\lambda_0^2(b) \\ &\quad -2\left[v_0(b)\frac{\partial\tau}{\partial b}B + v_0(a)\frac{\partial\tau}{\partial a}A\right] - 2\int_a^b u_0\lambda_0\lambda_1\, dx - \beta k(a,b). \end{aligned}$$

This expression includes as coefficients the values $u_0(x)$ and $v_0(x)$ of the harmonic functions $u(x,y)$ and $v(x,y)$ at the boundary $y=0$.

Theorem 3.6. *If $\tilde{\mathbf{L}}\lambda$ is an analytic function in the domain $y \geq 0$, and $\tau(a,b)$ satisfies the inequality:*

$$\tau(a,b) < \tau(a,b_1) + \tau(b_1,b), \quad a < b_1 < b,$$

then $\tilde{\mathbf{L}}\lambda = 0$ if and only if $\tilde{\mathbf{L}}\tau = 0$.

Proof. Let $D(a,b)$ be a domain in the half-plane $y \geq 0$, which is bounded by the geodesic $\gamma(a,b)$ and the segment $[a,b]$ of the straight line $y=0$. We first show that the principal identity holds:

$$\int_{D(a,b)} \tilde{\mathbf{L}}\lambda\, dx\, dy = \mathbf{L}\tau. \tag{3.44}$$

As before, let

$$t = \int_\gamma \frac{\sqrt{dx^2 + dy^2}}{\lambda(x,y)}$$

then

$$\frac{d^2x}{dt^2} = \lambda\frac{\partial\lambda}{\partial x}, \quad \frac{d^2y}{dt^2} = \lambda\frac{\partial\lambda}{\partial y}. \tag{3.45}$$

Using (3.45) and the relation

$$x'^2 + y_t'^2 = \lambda^2$$

we check by direct calculation that

$$\begin{aligned} &\frac{d}{dt}\left[u(x'^2 - y'^2) + 2vx'y' + \beta\arctan\frac{y'}{x'}\right] \\ &= \left[\frac{\partial}{\partial x}(u\lambda^2) + v\frac{\partial}{\partial y}\lambda^2 + \beta\frac{\partial}{\partial y}(\ln\lambda)\right]x' + \left[-\frac{\partial}{\partial y}(u\lambda^2) + v\frac{\partial}{\partial x}\lambda^2 - \beta\frac{\partial}{\partial x}(\ln\lambda)\right]y' \end{aligned} \tag{3.46}$$

along γ. Note that

$$x'_t|_{x=b,y=0} = \frac{\partial\tau}{\partial b}, \quad x'_t|_{x=a,y=0} = -\frac{\partial\tau}{\partial a},$$
$$y'_t|_{x=b,y=0} = -B(a,b), \quad y'_t|_{x=a,y=0} = -A(a,b). \tag{3.47}$$

If we integrate (3.46) along $\gamma(a,b)$, add to the obtained relation the integral:

$$-\int_a^b \left[\frac{\partial}{\partial x}(u_0\lambda_0^2) + 2v_0\lambda_0\lambda_1 + \beta\frac{\lambda_1}{\lambda_0}\right] dx$$

and use the Green formula and (3.47) we obtain (3.44).

Equation (3.44) implies that if $\mathbf{L}\tau = 0$, the assumed analyticity of the function $\tilde{\mathbf{L}}\lambda$ leads to $\tilde{\mathbf{L}}\lambda = 0$, similar to the proof of Theorem 3.5. This completes the proof of Theorem 3.6.

This theorem shows that in order to find the differential equation $\tilde{\mathbf{L}}\lambda = 0$, i. e. the coefficients $u(x,y)$ and $v(x,y)$, we must determine $u_0(x)$, $v_0(x)$, β from the equality $\mathbf{L}\tau = 0$ and continue them analytically.

We shall give some corollaries. Let $\mathbf{L}\tau = 0$.

1. if $u_0 = 0$, $v_0 = 0$, $\beta \neq 0$, we have the equation $\Delta \ln \lambda = 0$ and the conditions
$$\lambda\big|_{y=0} = \lambda_0(x), \quad \left.\frac{\partial\lambda}{\partial y}\right|_{y=0} = \lambda_1(x), \quad |x| \leq r; \tag{3.48}$$
2. if $u_0 = 1$, $v_0 = 0$, $\beta = 0$, then
$$\frac{\partial^2\lambda^2}{\partial x\partial y} = 0$$
and, therefore, $\lambda^2 = f(x) + g(y)$, where f and g are some functions;
3. if $u_0 = 0$, $v_0 = 1$, $\beta = 0$, then
$$\frac{\partial^2\lambda^2}{\partial y^2} - \frac{\partial^2\lambda^2}{\partial x^2} = 0$$
and, consequently, $\lambda^2 = f(x+y) + g(x-y)$, f and g being uniquely determined from (3.48).

3.7.4

We shall give the general procedure of reducing a two-dimensional inverse kinematic problem to the investigation of systems of differential equations and inequalities. As before, our results may be formulated as follows: it is a hodograph as a function of two variables satisfies a differential equation (or an inequality), we find a differential equation (or an inequality) for the unknown velocity, and *vice versa*. These results allow us to investigate the structure of velocity level curves and find various formulas for the velocity.

Let $\{\lambda\}$ be a set of real analytic functions $\lambda(x,y)$, $-\infty < x < \infty$, $y \geq 0$, such that $\lambda > 0$, $\partial\lambda/\partial y < 0$. There exists for any function $\lambda \in \{\lambda\}$ a number $\alpha > 0$, such that there exists the unique geodesic $\gamma(a,b)$ of the metric $ds^2 = \lambda^2(x,y)(dx^2 + dy^2)$ which connects the points a and b, on the straight line $y = 0$, whatever the numbers a and b, $-\alpha < a < b < \alpha$, may be.

We consider the functions:

1. $\lambda_0 = \lim_{a\to b=x} \dfrac{\partial\tau}{\partial b} = \lambda(x,0)$;

2. $\tau(a,b) = \displaystyle\int_{\gamma(a,b)} \lambda(x,y)\sqrt{dx^2+dy^2}, \ -\alpha < a < b < \alpha$;

3. $u_k(x,y)$, $v_k(x,y)$ are a solution of the system of equations:

$$\begin{cases} \dfrac{\partial u_k}{\partial y} = -\dfrac{\partial v_k}{\partial x} - 2(k+1)\lambda\dfrac{\partial\lambda}{\partial y}u_{k+1} \\ \dfrac{\partial v_k}{\partial y} = \dfrac{\partial u_k}{\partial x} - u_{k+1}\lambda\dfrac{\partial\lambda}{\partial x} - \dfrac{\partial u_{k+1}}{\partial x}\lambda^2 - (2k+3)\lambda\dfrac{\partial\lambda}{\partial y}v_{k+1}; \end{cases} \tag{3.49}$$

4. $R(x,y,x',y') = x'\sum_{k=0}^{\infty} u_k(x,y)y'^{2k} + \sum_{k=0}^{\infty} v_k(x,y)y'^{2k+1}$,
   where $u_k(x,y)$, $v_k(x,y)$ are a solution of (3.49);

5. we define the differential operators $\mathbf{L}\tau$ and $\tilde{\mathbf{L}}\lambda$:

$$\mathbf{L}\tau = R\left(b,0,\frac{\partial\tau}{\partial b},-\sqrt{\lambda_0^2(b)-\left(\frac{\partial\tau}{\partial b}\right)^2}\right) - R\left(a,0,\frac{\partial\tau}{\partial a},-\sqrt{\lambda_0^2(a)-\left(\frac{\partial\tau}{\partial a}\right)^2}\right),$$

$$\tilde{\mathbf{L}}\lambda = u_0(x,y)\frac{\partial\lambda}{\partial x} + v_0(x,y)\frac{\partial\lambda}{\partial y} + \frac{\partial u_0}{\partial x}\lambda,$$

$u_0 = u_k$ for $k=0$, $v_0 = v_k$ for $k=0$.

Theorem 3.7.

(a) *For any function $\lambda(x,y)\in\{\lambda\}$ and any number s, there exists an analytic solution $u_k(x,y)$, $v_k(x,y)$, $k=0,1,2,\ldots$, of the system of equations (3.49) such that*

$$\mathbf{L}\tau = s\tau, \quad \tilde{\mathbf{L}}\lambda = s\lambda;$$

(b) *in order for $\mathbf{L}\tau = s\tau$ to hold, it is necessary and sufficient that $\tilde{\mathbf{L}}\lambda = s\lambda$;*

(c) *there exists a semicircle*

$$\omega = \{(x,y) : (x-x_0)^2 + y^2 \le r^2, \quad y>0\}$$

such that if $\mathbf{L}\tau \le s\tau$ ($\mathbf{L}\tau \ge s\tau$), then $\tilde{\mathbf{L}}\lambda \le s\lambda$ ($\tilde{\mathbf{L}}\lambda \ge s\lambda$) and, conversely, if $\tilde{\mathbf{L}}\lambda \le s\lambda$ ($\tilde{\mathbf{L}}\lambda \ge s\lambda$), then $\mathbf{L}\tau \le s\tau$ ($\mathbf{L}\tau \ge s\tau$).

In particular, this theorem reduces the inverse kinematic problem to the investigation of the system of equations (3.49) and

$$\tilde{\mathbf{L}}\lambda = s\lambda$$

with the boundary conditions

$$\mathbf{L}\tau = s\tau, \quad \lambda(x,0) = \lambda_0(x).$$

The proof of Theorem 3.7 is analogous to that of Theorems 3.5 and 3.6.

REMARK 3.3. If $s = 0$ and $\tilde{\mathbf{L}}\lambda = 0$, the function

$$R(x,y,x',y') = x'\sum_{k=0}^{\infty} u_k(x,y)y'^{2k} + \sum_{k=0}^{\infty} v_k y'^{2k+1},$$

where $u_k(x,y)$, $v_k(x,y)$ are a solution of (3.49), is the first integral of the geodesics of the metric $ds^2 = \lambda^2(x,y)(dx^2 + dy^2)$, and

$$R(x,y,x',y') = R\left(b, 0, \frac{\partial\tau}{\partial b}, -\sqrt{\lambda_0^2(b) - \left(\frac{\partial\tau}{\partial b}\right)^2}\right)$$

along $\gamma(a,b)$. At the vertex (x^*, y^*) of the geodesic $\gamma(a,b)$ we have

$$\frac{dx}{dt} = \lambda(x^*, y^*), \quad \frac{dy}{dt} = 0.$$

Therefore,

$$\lambda(x^*, y^*)u_0(x^*, y^*) = R\left(b, 0, \frac{\partial\tau}{\partial b}, -\sqrt{\lambda_0^2(b) - \left(\frac{\partial\tau}{\partial b}\right)^2}\right)$$

that, for a one-dimensional case, generalizes the formula for the so-called apparent velocities: $n = 1$, $\lambda = \lambda(y)$, $R = \partial\tau/\partial b$.

3.8 DETERMINATION OF THE RIEMANN METRIC OF A SPECIAL FORM

We consider the problem of specifying the function $\lambda(x,y)$ in the half-plane $y \geq 0$ if the function is given in the domain $0 \leq a \leq b \leq 1$:

$$\tau(a,b) = \int_{\gamma(a,b)} \lambda(x,y)\sqrt{dx^2 + dy^2},$$

where $\gamma(a,b)$ is the geodesic of the metric $ds^2 = \lambda^2(dx^2 + dy^2)$ which connects the points a and b of the straight line $y = 0$.

In the following, we suppose that the given function is represented as

$$\tau(a,b) = \tau_0\big(h(b) - h(a)\big), \tag{3.50}$$

where τ_0 and $h(t)$ are strictly monotonically increasing functions, and

$$\frac{d^2\tau_0}{dt^2} < 0, \quad h(0) = 0, \quad h(1) = 1, \quad \left.\frac{d\tau_0}{dt}\right|_{t=0} = 1.$$

These functions $\tau(a,b)$, as shown below, correspond to the functions that admit the representation

$$\lambda(x,y) = \lambda_0\big(v(x,y)\big)\sqrt{\left(\frac{\partial v}{\partial x}\right)^2 + \left(\frac{\partial v}{\partial y}\right)^2},$$

where $\lambda_0(x)$ is a function of a single variable, and $v(x,y)$ is a harmonic function that vanished for $y = 0$, $0 \le x \le 1$. When (3.50) is satisfied, the formulas are given for the unknown functions $\tau_0(t)$, $h(t)$, and $\lambda(x,y)$. If $h(t) = t$, the formula for the function λ is identical to the Herglotz formula.

We denote by $h^{-1}(t)$, $\varphi(t)$ the inverse function to $h(t)$ and $\tau_0'(t)$ respectively, and by $g(t)$ the function specified by the equality:

$$g(t) = \left.\frac{\partial \tau}{\partial b}\right|_{a=b=t}.$$

Theorem 3.8. *The following equalities hold:*

$$h(t) = \int_0^t g(\xi)\, d\xi,$$

$$\tau_0(t) = \tau\big(0, h^{-1}(t)\big),$$

$$\lambda(x,y) = \lambda_0\big(v(x,y)\big)\sqrt{\left(\frac{\partial v}{\partial x}\right)^2 + \left(\frac{\partial v}{\partial y}\right)^2},$$

where $v(x,y)$ is a harmonic function that satisfies the conditions:

$$v|_{y=0} = 0, \quad \left.\frac{\partial v}{\partial y}\right|_{y=0} = -h'(x), \quad 0 \le x \le 1$$

and the function $\lambda_0(t)$ is inverse to:

$$t(p) = \frac{1}{\pi}\int_p^1 \frac{\varphi(q)\, dq}{\sqrt{q^2 - p^2}}.$$

PROOF. We firstly specify $\tau_0(t)$ and $h(t)$ by the function $\tau(a,b)$. We consider a function $b(a,\alpha)$, $0 \le a \le b_0$, $\alpha > 0$, such that

$$\tau\big(a, b(a,\alpha)\big) = \alpha.$$

We have from (3.50):

$$h\big(b(a,\alpha)\big) - h(a) = \tau_0^{-1}(\alpha). \tag{3.51}$$

Since $\tau(a,\alpha) = 0$, then $b(a,0) = 0$. Therefore, we have from (3.51):

$$\left.\frac{d}{d\alpha}\tau_0^{-1}(\alpha)\right|_{\alpha=0} = 1. \tag{3.52}$$

The definition of the function $b(a,\alpha)$ implies

$$\left.\frac{\partial b}{\partial \alpha}\frac{\partial \tau}{\partial b}\right|_{\alpha=0} = 1.$$

Using (3.52) and this equality, we get

$$h'(t) = \frac{\partial \tau}{\partial b}\bigg|_{a=b=t} = g(t), \quad 0 \le t \le 1.$$

Hence, taking into account the data $h(0) = 0$, $h(1) = 1$, we obtain for $h(t)$:

$$h(t) = \int_0^t g(\xi)\,d\xi. \tag{3.53}$$

We now find the function $\tau_0(t)$. Assuming $a = 0$ in (3.50), we have

$$\tau(0, b) = \tau_0\big(h(b)\big),$$

where $h(t)$ is uniquely specified by (3.53). We find $\tau_0(t)$ from the last equality:

$$\tau_0(t) = \tau\big(0, h^{-1}(t)\big). \tag{3.54}$$

Let us now find the function $\lambda(x, y)$. We consider the conformal transformation $u = u(x, y)$, $v = v(x, y)$ of the half-plane $y \ge 0$ into the half-plane $v \ge 0$, which is specified by the boundary conditions:

$$u\big|_{y=0} = h(x), \quad v\big|_{y=0} = 0, \quad 0 \le x \le 1.$$

The points a and b on the straight line $y = 0$ map into the points $h = h_1(a)$ and $h = h_2(b)$ on the straight line $v = 0$, and the geodesic maps γ into $\hat{\gamma}$. Because of (3.50) we have:

$$\tau_0(h_2 - h_1) = \int_{\hat{\gamma}(h_1, h_2)} \hat{\lambda}(u, v)\sqrt{du^2 + dv^2}, \tag{3.55}$$

where

$$\hat{\lambda}(u, v) = \lambda\big(x(u, v), y(u, v)\big)\sqrt{\left(\frac{\partial x}{\partial u}\right)^2 + \left(\frac{\partial y}{\partial u}\right)^2},$$

$\hat{\gamma}(h_1, h_2)$ is the geodesic of the metric

$$ds^2 = \hat{\lambda}^2(u, v)(du^2 + dv^2)$$

which connects the points h_1 and h_2, $h_1 = h(a)$, $h_2 = h(b)$.

Since the function τ_0 is dependent only on the difference $h_2 - h_1$, the solution of (3.55) is the function λ_0 of a single variable v specified by the Herglotz formula:

$$t(p) = \frac{1}{\pi}\int_p^1 \frac{\varphi(q)\,dq}{\sqrt{q^2 - p^2}},$$

where $t(p)$ is an inverse function to $\lambda_0(t)$, and $\varphi(q)$ is an inverse function to $\tau_0'(t)$. Since the functions $h(t)$, $\tau_0(t)$, $\lambda(t)$, $v(x, y)$ are specified uniquely we derive from the equality

$$\lambda_0^2(v)(du^2 + dv^2) = \lambda(x, y)(dx^2 + dy^2)$$

the formula

$$\lambda(x,y) = \lambda_0(v(x,y))\sqrt{\left(\frac{\partial v}{\partial x}\right)^2 + \left(\frac{\partial v}{\partial y}\right)^2}.$$

This completes the proof of the theorem.

REMARK 3.4. The function

$$g(t) = \left.\frac{\partial \tau}{\partial b}\right|_{a=b=t}$$

is identical to the function $\lambda(t,0)$. Therefore, the above theorem implies that if the functions $\tau(0,b)$ and $\lambda(t,0)$ are given, the function

$$\lambda(x,y) = \lambda_0(v(x,y))\sqrt{\left(\frac{\partial v}{\partial x}\right)^2 + \left(\frac{\partial v}{\partial y}\right)^2}$$

is uniquely specified by (3.50)–(3.52).

REMARK 3.5. A similar procedure is used for investigating the case in which

$$\lambda^2(x,y) = \left[\lambda_0(v) + a^2(u)\right]\sqrt{\left(\frac{\partial v}{\partial x}\right)^2 + \left(\frac{\partial v}{\partial y}\right)^2},$$

$v = v(x,y)$ and $u = u(x,y)$ are harmonic conjugate functions, $\lambda_0(t)$ and $a(t)$ are functions of a single variable.

3.9 THE PROBLEM ASSOCIATED WITH AN EQUATION FOR GEODESICS OF THE LIOUVILLE METRIC

We consider the problem: find twice continuously differentiable non-negative functions $a(y)$ and $b(x)$ in the domain $0 \le x \le h$, $y \ge 0$ such that

1. there exists a unique solution $y(x,z)$ of the boundary value problem:

$$\sqrt{a^2+b^2}\frac{\mathrm{d}^2 y}{\mathrm{d}x^2} = \left[a\frac{\mathrm{d}a}{\mathrm{d}y} - b\frac{\mathrm{d}b}{\mathrm{d}x}\frac{\mathrm{d}y}{\mathrm{d}x}\right]\left[1 + \left(\frac{\mathrm{d}y}{\mathrm{d}x}\right)^2\right],$$

$$y(0,z) = 0, \quad y(h,z) = z, \quad 0 \le x \le h, \quad 0 \le z \le z_0;$$

2. the functions

$$\varphi(z) = \left.\frac{\mathrm{d}y}{\mathrm{d}x}\right|_{x=0}, \quad \tau(z) = \int_0^h \sqrt{a^2(y(x,z)) + b^2(x)}\sqrt{1 + \left(\frac{\mathrm{d}y}{\mathrm{d}x}\right)^2}\,\mathrm{d}x$$

are given in the domain $0 \le z \le z_0$.

The problem considered has geometric and physical interpretations.

The curve $y = y(x,z)$ is the geodesic of the Liouville metric $\mathrm{d}s^2 = (a^2+b^2)(\mathrm{d}x^2+\mathrm{d}y^2)$ which connects the points $(0,0)$ and (h,z);

$\arctan\varphi(z)$ is the angle between this geodesic at the point $(0,0)$ and the x-axis;

$\tau(z)$ is the distance in the metric ds between the points $(0,0)$ and (h,z).

On the other hand, if we suppose that the strip $0 \le x \le h$, $y \ge 0$ is an elastic inhomogeneous medium and a perturbation occurs at the origin of coordinates, the function $v = 1/\sqrt{(a^2+b^2)}$ is considered as the velocity of the extended perturbation and the function $\tau(z)$ as the propagation time of the perturbation; the time of arrival of this perturbation is recorded at the point $(0,0)$. Thus the above problem is to determine the velocity $v = 1/\sqrt{(a^2+b^2)}$ by using time $\tau(z)$ and the slope angle $\varphi(z)$ of the tangent to the geodesic $\gamma(z)$ at the point $(0,0)$. In the following we assume that the function $b(x)$ rigorously monotonically decreases and the function $a(y)$ monotonically increases, and $b(0) = 1$, $a(0) = 0$.

We set

$$f(z) = \frac{\varphi^2(z)}{1+\varphi^2(z)},$$

$$g(z) = \left(\frac{d\tau}{dz}\right)^2 - f(z).$$

Theorem 3.9. *Let the functions $f(z)$ and $g(z)$ be non-negative and the function $f(z)$, besides, strictly monotonically increase, and $f(0) = 0$. Then the formulas*

1. $a^2(z) = g(z)$, $0 \le z \le z_0$;

2. $\displaystyle\int_0^h \frac{d\xi}{b^{2k+1}(\xi)} = t_k$, $k = 0, 1, 2, \ldots$;

hold, where

$$t_k = \frac{2^k}{(2k+1)!!}\frac{d^k}{dp^k}\int_0^{z(p)} \frac{d\xi}{\sqrt{g(\xi)-p}}\Bigg|_{p=0}$$

and the function $z(p)$ is inverse to $f(z)$.

PROOF. Let $(x(t), y(t))$ be a parametrization of the geodesic $\gamma = \{y = y(x,z)\}$, where

$$t = \int_\gamma \frac{\sqrt{dx^2+dy^2}}{v}.$$

Because of this choice of the parameter t, the system of the Euler equations

$$\begin{aligned}
\frac{\partial}{\partial x}\left[\frac{1}{v}\sqrt{x_t'^2+y_t'^2}\right] &= \frac{d}{dt}\left[\frac{1}{v}\frac{x_t'}{\sqrt{x_t'^2+y_t'^2}}\right],\\
\frac{\partial}{\partial y}\left[\frac{1}{v}\sqrt{x_t'^2+y_t'^2}\right] &= \frac{d}{dt}\left[\frac{1}{v}\frac{y_t'}{\sqrt{x_t'^2+y_t'^2}}\right]
\end{aligned} \tag{3.56}$$

that specifies the geodesic γ has the form:

$$\begin{aligned}
\frac{d^2x}{dt^2} &= \frac{1}{2}\frac{\partial}{\partial x}\left(\frac{1}{v^2}\right),\\
\frac{d^2y}{dt^2} &= \frac{1}{2}\frac{\partial}{\partial y}\left(\frac{1}{v^2}\right).
\end{aligned} \tag{3.57}$$

For

$$\frac{1}{v^2} = a^2(y) + b^2(x)$$

the system (3.57) splits into:

$$\frac{d^2x}{dt^2} = b\frac{db}{dx},$$

$$\frac{d^2y}{dt^2} = a\frac{da}{dy}.$$

Hence, because of the above remark, its solution is found in the form:

$$t = \int_0^x \frac{d\xi}{\sqrt{b^2(\xi) - p}}, \tag{3.58}$$

$$t = \int_0^y \frac{d\xi}{\sqrt{a^2(\xi) + p}}. \tag{3.59}$$

Here $p \geq 0$ is a constant.

Let us show that

$$\left(\frac{d\tau}{dz}\right)^2 = \left(\frac{dy}{dt}\right)^2\bigg|_{x=h,y=z}, \tag{3.60}$$

$$p = \frac{\varphi^2(z)}{1+\varphi^2(z)} = f(z). \tag{3.61}$$

Since $t = \int_\gamma \frac{1}{v}\, ds$, equation (3.60) follows from the relation:

$$\left(\frac{d\tau}{dz}\right)^2 = \left(\frac{1}{v}\frac{y'_t}{\sqrt{x_t'^2 + y_t'^2}}\right)^2\Bigg|_{x=h,y=z} = \left(\frac{dy}{dt}\right)^2\bigg|_{x=h,y=z}.$$

Further, we have from (3.58) and (3.59):

$$\frac{dx}{dt}\bigg|_{t=0} = \sqrt{1-p}, \quad \frac{dy}{dt}\bigg|_{t=0} = \sqrt{p}.$$

Hence,

$$\varphi(z) = \frac{y'_t}{x'_t} = \frac{\sqrt{p}}{\sqrt{1-p}}.$$

This equality leads to (3.61).

We now prove the first assertion of the theorem. If we differentiate (3.59) and assume $x = h$, we get

$$\frac{dy}{dt}\bigg|_{x=h} = \sqrt{a^2(z) + p}.$$

Using (3.60) and (3.61), we obtain the sought formula:

$$a^2(z) = \left(\frac{d\tau}{dz}\right)^2 - \frac{\varphi^2(z)}{1+\varphi^2(z)} = g(z).$$

Now let $z(p)$ be the inverse function to $f(z)$. Note that if $x = h$, then $y = z(p)$.

If we assume $y = z(p)$ in (3.59), we find the function

$$t(p) = \int_0^{z(p)} \frac{d\xi}{\sqrt{a^2(\xi) + p}}, \quad p \in [0, f(z_0)]$$

and, thus, we obtain for the function $b(x)$:

$$t(p) = \int_0^h \frac{d\xi}{\sqrt{b^2(\xi) - p}}.$$

If we differentiate both sides of this equation k times with respect to p and assume $p = 0$, we get

$$\int_0^h \frac{d\xi}{b^{2k+1}(\xi)} = \frac{2^k}{(2k+1)!!} \frac{d^k}{dp^k} t(p)\bigg|_{p=0} = t_k, \quad k = 0, 1, 2, \ldots.$$

This completes the proof of the theorem.

3.10 DETERMINATION OF A HOMOGENEOUS FUNCTIONAL

Let E^n be the n-dimensional Euclidean space of variables $x = (x_1, \ldots, x_n)$, $n \geq 2$; $F(x, x')$ be a function, given in $E^n \times E^n$, $x \in E^n$, $x' \in E^n$, with the properties:

1. $F(x, x')$ is a non-negative and triply continuously differentiable function for $x' \neq 0$;
2. $F(x, x')$ is a positive homogeneous function, of degree 1 with respect to the variable x', even and strictly convex with respect to the same variable.

These functions appear in variational calculus. The curve $x = x(\xi)$ specified by the system of equations:

$$\frac{\partial F}{\partial x_i} = \frac{d}{d\xi} \frac{\partial F}{\partial x_i}, \quad i = 1, \ldots, n, \quad x' = \frac{dx}{d\xi}, \quad 0 \leq \xi \leq \xi_0$$

is called the extremal and denotes as γ in the nomenclature of variational calculus.

Let $F(x, x') = \lambda(x)\Phi(x, x')$, where Φ is fixed. We assume $\lambda(x)$ and $\Phi(x, x')$ to be homogeneous functions with respect to the variable x:

$$\lambda(tx) = t^k \lambda(x), \quad \Phi(tx, x') = t^l \Phi(x, x'), \quad \lambda(0) = 0, \quad t > 0.$$

We consider the more general kinematic problem: find the function $\lambda(x)$ if

1. there exists a unique extremal $\gamma(a)$ which connects the points $x = 0$ and $x = (a, h)$, where $a = (a_1, a_2, \ldots, a_{n-1})$, $|a| < r$, and $h > 0$, $r > 0$ are fixed constants;
2. the function is given in the domain $|a| < r$:

$$\tau(a) = \int_{\gamma(a)} \lambda(x)\Phi(x, x')\, d\xi.$$

Note that for the case

$$\Phi(x,x') = \sqrt{\sum_{i,j}^{n} g_{ij}(x)x_i'x_j'}$$

this problem may be considered as an inverse problem for a hyperbolic equation, namely: find $\lambda(x)$ if there exists a unique solution $u(x,\tau)$ of the problem

$$\lambda^2(x)\frac{\partial^2 u}{\partial\tau^2} = \sum_{i,j}^{n} c_{ij}(x)\frac{\partial^2 u}{\partial x_i \partial x_j} + \delta(\tau),$$

$$u\big|_{\tau=0} = 0, \quad \frac{\partial u}{\partial\tau}\bigg|_{\tau=0} = 0.$$

Here δ is the delta function, $\big(c_{ij}(x)\big)$ is the matrix inverse to $\big(d_{ij}(x)\big)$, and the function $\tau(a) = \big\{u(a,\tau) = 0 \text{ for all } \tau \le \tau(a)\big\}$ is given in the domain $|a| < r$.

Indeed

$$\tau(a) = \int_{\gamma(a)} \lambda(x)\sqrt{\sum_{i,j}^{n} g_{ij}(x)x_i'x_j'}\,d\xi.$$

Theorem 3.10. *The formula*

$$\lambda(x) = x_n^k\varphi\left(\frac{x_1}{x_n}, \frac{x_2}{x_n}, \dots, \frac{x_{n-1}}{x_n}\right), \quad 0 < x_n < h$$

holds with

$$\varphi(a) = H\left(a, h, \frac{\partial\tau}{\partial a_1}, \dots, \frac{\partial\tau}{\partial a_{n-1}}, \frac{(k+l+1)\tau(a) - \sum\limits_{i=1}^{n-1}\frac{\partial\tau}{\partial a_i}a_i}{h}\right),$$

$$H(x,p) = \sup_{x'}\frac{\sum\limits_{i=1}^{n} p_i x_i'}{\Phi(x,x')}.$$

Proof. Let $F(x,x')$ be an arbitrary function that obeys condition 1 and 2 and is positive homogeneous, of degree s with respect to the variable $x \in E^n$. We consider a domain $D \subset E^n$ such that any two points $y \in D$ and $z \in D$ may be connected with a single extremal $\gamma(y,z)$ corresponding to the function $F(x,x')$. We set

$$w(y,z) = \int_{\gamma(y,z)} F(x,x')\,d\xi,$$

where we integrate from y to z, and show that the function $w(y,z)$ is positive homogeneous, of degree $s+1$, i. e.

$$w(ty,tz) = t^{s+1}w(y,z).$$

Indeed, we have

$$x_i \frac{\partial F}{\partial x_i} = x_i \frac{\mathrm{d}}{\mathrm{d}\xi}\frac{\partial F}{\partial x_i'} = \frac{\mathrm{d}}{\mathrm{d}\xi}\left(x_i \frac{\partial F}{\partial x_i'}\right) - \frac{\partial F}{\partial x_i'} x_i', \quad i = 1, \dots, n$$

along $\gamma(y, z)$. Hence, because of the equality

$$\sum_{i=1}^{n} \frac{\partial F}{\partial x_i'} x_i' = F(x, x')$$

we obtain

$$\sum_{i=1}^{n} x_i \frac{\partial F}{\partial x_i} + F = \sum_{i=1}^{n} \frac{\mathrm{d}}{\mathrm{d}\xi}\left(x_i \frac{\partial F}{\partial x_i'}\right).$$

If we integrate this equality over $\gamma(y, z)$ and take into account the relations

$$\left.\frac{\partial F}{\partial x_i'}\right|_{x=z} = \frac{\partial w}{\partial z_i}, \quad \left.\frac{\partial F}{\partial x_i'}\right|_{x=y} = \frac{\partial w}{\partial y_i}, \quad i = 1, \dots, n$$

we obtain:

$$\int\limits_{\gamma(y,z)} \left[\sum_{i=1}^{n} x_i \frac{\partial F}{\partial x_i} - sF\right] \mathrm{d}\xi = \sum_{i=1}^{n}\left(\frac{\partial w}{\partial z_i} + \frac{\partial w}{\partial y_i}\right) - (s+1) w(y, z).$$

Because of the assumed homogeneity of the function $F(x, x')$ of degree s the integrand here is zero. Therefore, $w(ty, tz) = t^{s+1} w(y, z)$.

Let $F(x, x') = \lambda(x)\Phi(x, x')$. According to the condition of the problem, the function F is homogeneous of degree $k + l$. As proved above, the function

$$w(y, z) = \int\limits_{\gamma(y,z)} \lambda(x)\Phi(x, x')\, \mathrm{d}\xi$$

is homogeneous, of degree $k + l + 1$.

Note that since $\lambda(0) = 0$, then

$$\left.\frac{\partial w}{\partial y_i}\right|_{y_i=0} = -\left.\frac{\partial F}{\partial x_i'}\right|_{x=0} = \lambda(0)\frac{\partial \Phi}{\partial x_i'} = 0.$$

The equation

$$\sum_{i=1}^{n}\left(\frac{\partial w}{\partial z_i} + \frac{\partial w}{\partial y_i}\right) = (k + l + 1) w(y, z)$$

for $y = 0$, $z_i = a_i$, $i = 1, 2, \dots, n-1$, $z_n = h$ takes the form:

$$\sum_{i=1}^{n-1} \frac{\partial \tau}{\partial a_i} + \frac{\partial w}{\partial z_n} h = (k + l + 1)\tau(a). \tag{3.62}$$

We consider the function

$$H(x, p) = \sup_{x'} \frac{\sum_{i=1}^{n} p_i x_i'}{\Phi(x, x')}.$$

We show that

$$H\left(z, \frac{\partial w}{\partial z_1}, \frac{\partial w}{\partial z_2}, \ldots, \frac{\partial w}{\partial z_n}\right) = \lambda(z). \tag{3.63}$$

Indeed, if

$$\tilde{H}(x, p) = \sup_{x'} \frac{\sum_{i=1}^{n} p_i x'_i}{\lambda(x)\Phi(x, x')}$$

is a Hamiltonian function, then

$$\tilde{H}\left(z, \frac{\partial w}{\partial z_1}, \frac{\partial w}{\partial z_2}, \ldots, \frac{\partial w}{\partial z_n}\right) = 1.$$

Since $\tilde{H} = H/\lambda$, hence (3.63). If we set $z_i = a_i$ and $z_n = h$ in (3.63) we obtain

$$H\left(a, h, \frac{\partial \tau}{\partial a_1}, \ldots, \frac{\partial \tau}{\partial a_{n-1}}, \frac{\partial w}{\partial z_n}\right).$$

If we substitute here $\partial w/\partial z_n$ from (3.62), we get

$$\lambda(a) = H\left(a, h, \frac{\partial \tau}{\partial a_1}, \ldots, \frac{\partial \tau}{\partial a_{n-1}}, \frac{(k+l+1)\tau(a) - \sum_{i=1}^{n-1} \frac{\partial \tau}{\partial a_i} a_i}{h}\right). \tag{3.64}$$

Since the function $\lambda(x)$ is assumed to be homogeneous of degree k, (3.64) leads to the assertion of the theorem.

CHAPTER 4

Formulas in integral geometry and tomography

4.1 TOMOGRAPHY AND INVERSE PROBLEMS FOR KINETIC EQUATIONS

X-ray tomography is described by the transport equation:

$$\frac{\partial w}{\partial t} + (\mathrm{grad}_x\, w, n) + \lambda(x, E)w = \int\limits_{E_1}^{E_2} \int\limits_{\Omega} I(t, n, n', x, E')w \, \mathrm{d}n' \, \mathrm{d}E' + F(t, n, x)$$

Here $x \in D \subset \mathrm{R}^n$, $\alpha \le t \le \beta$, $|n| = 1$, $w(x, n, t)$ is flow of photons, E is energy, F is function of source, I is indicatrix, and λ is a coefficient.

If function w does not depend on variable t and $I = F = 0$, we have the equation

$$(\mathrm{grad}_x\, w, n) + \lambda(x, E)w = 0.$$

This equation is connected with problems of integral geometry (Anikonov, 1995a). The problem of tomography is an inverse problem for the transport equation.

Now, we show one result which is contained in Anikonov *et al.* (1993). We consider the stationary transport equation

$$(\mathrm{grad}_x\, w, n) + \lambda(x)w(x, n) = \frac{1}{4\pi} \int\limits_{\Omega} I(x, n, n')w \, \mathrm{d}n' + F, \tag{4.1}$$

where $x \in D \subset \mathrm{R}^3$ and D is a convex domain, $|n| = 1$, $n \in \Omega$, and ∂D is the smooth boundary D of the class C^2. We denote the length of intersection of the ray $(x + tn, t > 0)$ with domain D by $\sigma(x, n)$ and denote by Γ^- the set of points $(x, n) \in \partial D \times \Omega$, $x \in \partial D$, $n \in \Omega$, for which $x = x_0 - \sigma(x_0, -n)n$, $x_0 \in D$ and by Γ^+ the set of points $(x, n) \in \partial D \times \Omega$ for which $x = x_0 + \sigma(x_0, -n)n$, $x_0 \in D$. We add the boundary conditions to equation (4.1)

$$h(x, n) = w(x, n), \quad (x, n) \in \Gamma^-, \tag{4.2}$$

$$H(x, n) = w(x, n), \quad (x, n) \in \Gamma^+. \tag{4.3}$$

The inverse problem is to determine the coefficient $\lambda(x)$, $x \in D$ from equation (4.1) and the boundary conditions (4.2), (4.3) if the functions $h(x, n)$, $H(x, n)$ are known.

This inverse problem is the real problem of tomography when the indicatrix of scattering $I \not\equiv 0$. Assume that the functions $\lambda(x)$, $I(x,n,n')$, $F(x,n)$ belong to the classes of singular continuously differential functions.

Let the vector $n \in \Omega$ be characterized by spherical angles, (θ,γ), $0 \leq \theta \leq \pi$, $0 \leq \gamma < 2\pi$, i.e. $n = n(\theta,\gamma)$. The angle $\theta = \frac{\pi}{2}$ corresponds to the horizontal plane. We assume that the function $h(x,n(\theta,\gamma))$ is continuous except in the plane $\theta = \frac{\pi}{2}$ where the external radiation intensity has a jump. This implies that

$$\lim_{\varepsilon\to 0}\left|h(x,n(\frac{\pi}{2}+\varepsilon,\gamma)) - h(x,n(\frac{\pi}{2}-\varepsilon,\gamma))\right| > 0 \tag{4.4}$$

for all $(x,n) \in \Gamma^-$.

Theorem 4.1. *If (4.4) is satisfied, then the formula*

$$\begin{aligned}\int_0^{\sigma(x,-n(\frac{\pi}{2},\gamma))} \lambda(x-\tau n(\frac{\pi}{2},\gamma))\,d\tau \;=\; & -\ln\lim_{\varepsilon\to 0}\Big[\big(H(x,n(\frac{\pi}{2}+\varepsilon,\gamma)) - H(x,n(\frac{\pi}{2}-\varepsilon,\gamma))\big)^{-1} \\ & \times\big(h(x-\sigma(x,-n(\frac{\pi}{2}+\varepsilon,\gamma))n(\frac{\pi}{2}+\varepsilon,\gamma),n(\frac{\pi}{2}+\varepsilon,\gamma)) \\ & -h(x-\sigma(x,-n(\frac{\pi}{2}-\varepsilon,\gamma))n(\frac{\pi}{2}-\varepsilon,\gamma),n(\frac{\pi}{2}-\varepsilon,\gamma))\big)\Big]\end{aligned}$$

holds for all $(x,n) \in \Gamma^+$.

From this theorem it follows that the inverse problem of tomography is reduced to the Radon transform.

4.2 INVERSION FORMULAS IN INTEGRAL GEOMETRY

The problems of determing unknown objects (functions, differential forms, tensor fields), their integral known on manifolds (curves, surfaces), are called the problems of integral geometry.

The problems of integral geometry are connected with differential equations, inverse problems, and group representations and one can find them in geophysics, astronomy, or medicine.

One general problem of integral geometry is to find the function

$$f(x) \in C^\infty(\mathrm{R}^n), \quad \sup_{x\in\mathrm{R}^n}(1+|x|^m)\left|\frac{\partial^\alpha f}{\partial x_1^{\alpha_1},\ldots,\partial x_n^{\alpha_n}}\right| < \infty, \quad \forall m \geq 0$$

if we know the function

$$\hat{f}(y) = \int_{B(y)} f(x)d\mu, \quad y \in \mathrm{R}^n,$$

where $B(y) \subset \mathrm{R}^n$ is a smooth manifold, $y \in \mathrm{R}^n$, and μ is a measure on $B(y)$ (Gel'fand *et al.*, 1966). The Radon transform is

$$\hat{f}(\omega,p) = \int\limits_{(\omega,x)=p} f(x)d\mu, \quad y=(\omega,p),$$

$$B(y) = \{x : (\omega,x)=p\}, \quad \omega \in \mathrm{R}^n, \quad |\omega|=1, \quad p \in \mathrm{R}, \quad x \in \mathrm{R}^n.$$

The inversion formulas in problems of the integral geometry play a very important role in both theory and applications.

Theorem 4.2. *(Radon, 1917; Gel'fand* et al.*, 1966; Helgason, 1980). Let a function $\hat{f}(\omega,p)$ be the Radon transform*

$$\hat{f}(\omega,p) = \int\limits_{(x,\omega)=p} f(x)\,d\nu.$$

Then the function $\hat{f}(\omega,p)$, $\omega \in \Omega$, $\Omega = \{\omega, |\omega|=1\}$, $p \in \mathrm{R}$, satisfies the conditions:

1. $\hat{f}(\omega,p) \in C^\infty(\Omega \times \mathrm{R})$,

2. $\sup\limits_{p\in\mathrm{R}}(1+|p|^m)\left|\dfrac{\partial^\alpha \hat{f}}{\partial p^\alpha}\right|, \quad \forall m \geq 0, \quad \alpha \geq 0,$

3. $\int\limits_{\mathrm{R}} \hat{f}(\omega,p)p^k\,dp$ *is a homogeneous polynomial of the degree k depending on $\omega = (\omega_1,\ldots,\omega_n)$ and the inversion Radon formula*

$$f(x) = \frac{(-4\pi)^{(n-1)/2}\Gamma(\frac{n}{2})}{\Gamma(\frac{1}{2})}\Delta^{\frac{n-1}{2}}\int\limits_{\Omega}\hat{f}(\omega,(\omega,x))\,d\omega$$

holds, where $\Delta^{\frac{n-1}{2}}$ is the Laplace operator of degree $(n-1)/2$.

Other similar formulas are contained in the papers by (Semyanistyi, 1960; 1961; 1966; Gel'fand *et al.*, 1980).

If $n=2$, $x_1=x$, $x_2=y$, $x^2+y^2 \leq 1$

$$\hat{f}(\varphi,p) = \int\limits_{x\cos\varphi+y\sin\varphi=p} f(x,y)\,dl, \quad 0 \leq \varphi \leq 2\pi, \quad -1 \leq p \leq 1,$$

then the different inversion formulas may be written on the basis of Theorem 4.2.

$$\begin{aligned} f(x,y) &= \frac{1}{(2\pi)^2}\Delta\int\limits_0^\pi d\varphi\int\limits_{-1}^1 \hat{f}(\varphi,p)\ln|p-rn|\,dp \\ &= \frac{1}{(2\pi)^2}\int\limits_0^\pi d\varphi\int\limits_{-1}^1 \frac{\partial^2\hat{f}(\varphi,p)}{\partial p^2}\ln|p-rn|\,dp \end{aligned}$$

$$= -\frac{1}{(2\pi)^2}\int_0^{\pi} d\varphi \int_{-1}^{1} \frac{\partial \hat{f}(\varphi, p)}{\partial p}\frac{1}{|p - rn|}\, dp$$

$$= \frac{1}{(2\pi)^2}\int_0^{\pi} d\varphi \int_{-1}^{1} \hat{f}(\varphi, p)\frac{1}{(p - rn)^2}\, dp$$

$$= \frac{1}{(4\pi)^2}\int_{-\infty}^{\infty}\int_{-\infty}^{\infty}\left[\int_{-1}^{1} \hat{f}\left(\arctan\frac{\omega_1}{\omega_2}, t\right) e^{it\sqrt{\omega_1^2+\omega_2^2}}\, dt\right] e^{i\omega_1 x + i\omega_2 y}\, d\omega_1\, d\omega_2,$$

$$rn = x\cos\varphi + y\sin\varphi.$$

Now, we consider the problem of integral geometry to find the function $f(x)$, $x \in R^3$ such that

$$\hat{f}(\gamma) = \int_{\gamma} f(x)\, dl,$$

where γ is a line passing through a smooth curve $\Gamma \subset R^3$. The function $f(\gamma)$ is supposed to be known for all lines such that $\gamma \cap \Gamma \neq \emptyset$.

Theorem 4.3. *(Kirillov, 1961) A function*

$$f(x) \in C^{\infty}(R^3), \quad \sup_{x\in R^3}(1 + |x|^m)\left|\frac{\partial^{\alpha} f}{\partial x_1^{\alpha_1}, \ldots, \partial x_n^{\alpha_n}}\right| < \infty, \quad \forall m \geq 0, \quad \alpha \geq 0$$

is restored by $\hat{f}(\gamma)$ and the inversion formulas hold if and only if $\{x : (x, \omega) = p\} \cap \Gamma \neq \emptyset$ for almost all (ω, p), $|\omega| = 1$, $p \in R$.

Tuy (1983) discovered the inversion formulas for the functions with compact support when the curve is bounded, provided that almost every hyperplane intersecting the support of the unknown function meets the curve Γ transversally in some point. Finch (1985) investigated the analogous inversion procedure for the class of the curves which did not satisfy the hypotheses of Tuy.

4.3 AN INVERSION FORMULA IN A PROBLEM OF INTEGRAL GEOMETRY ON THE SPHERE

Integral geometry problems involving the determination of functions given on a sphere from circular integrals assume that the integrated functions are even. In this section, we make an attempt to remove this natural and important restriction by changing the operation of integration of the desired function. Making use of an inversion formula constitutes the element of newness in this section.

Let n and p be points of Euclidean sphere $\omega = \{x \in R^3, \ |x| = 1\}$, $\omega(n) = \{p \in \omega, \ (n, p) \geq 0\}$ and let $n \in \omega$, $p \in \omega$ be the hemisphere; here, (n, p) is the inner product of the vectors $n \in \omega$ and $p \in \omega$. Let Δ be the Laplace-Beltrami operator on ω, and be $\Gamma(n) = \{p \in \omega, \ (n, p) = 0\}$ the great circle bounding the disk orthogonal to the vector n. Let a and b be constants, ds the element of length on $\Gamma(n)$, and $d\omega$ the element of area on ω.

Consider the integral geometry problem of finding the function $f(n) \in C^\infty(\omega)$, $n \in \omega$, if the function

$$\varphi(n) = 2a \int\limits_{\Gamma(n)} f(p)\,\mathrm{d}s + b \int\limits_{\omega(n)} (\Delta + 2) f(p)\,\mathrm{d}\omega_p \tag{4.5}$$

is known. Note that if $2a = 1$ and $b = 0$, then

$$\varphi(n) = \int\limits_{\Gamma(n)} f(p)\,\mathrm{d}s, \quad n \in \omega,$$

and in this case it is only possible to find the even part of $f(p)$ from $\varphi(n)$.

Note also that by virtue of the equality

$$\int\limits_{\omega(n)} (\Delta + 2) f\,\mathrm{d}\omega_p = \int\limits_{\Gamma(n)} \frac{\partial f}{\partial n}\,\mathrm{d}s + 2 \int\limits_{\omega(n)} f\,\mathrm{d}\omega_p,$$

relation (4.5) may be rewritten in the form

$$\varphi(n) = \int\limits_{\Gamma(n)} \left(2af + b\frac{\partial f}{\partial n}\right)\mathrm{d}s + 2b \int\limits_{\omega(n)} f\,\mathrm{d}\omega_p.$$

Theorem 4.4. *If a, $b \neq 0$, $a + b \neq 0$, $\varphi(n) \in C^\infty(\omega)$ and $\int\limits_{\omega} n\varphi(n)\,\mathrm{d}\omega = 0$, then there exists a solution $f(p)$, and*

$$f(n) = \frac{1}{4\pi} \int\limits_{\omega} \Phi\{1 - (n,p)\ln\left(1 - (n,p)\right)\}\,\mathrm{d}\omega_p,$$

where

$$\begin{aligned} \Phi(n) \;=\; & -\frac{1}{8\pi^2}\frac{\mathrm{d}}{\mathrm{d}t} \int\limits_{(n,p)^2 > t} \frac{(\Delta + 2)\varphi_1(p)(n,p)\,\mathrm{d}\omega_p}{\sqrt{(n,p)^2 - t}}\Bigg|_{t=0} \\ & -\frac{1}{4\pi^2 b} \int\limits_{\omega(n)} \frac{\Delta\varphi^-(p)}{(n,p)}\,\mathrm{d}\omega_p, \end{aligned}$$

$$\varphi_1(p) = \frac{1}{a}\varphi^+(p) - \frac{b}{4\pi a(a+b)} \int\limits_{\omega} \varphi(p)\,\mathrm{d}\omega_p,$$

$$\varphi^\pm = \frac{\varphi(p) \pm \varphi(-p)}{2}.$$

We shall sketch the proof. Using Green's formula, we obtain (Blaschke, 1956)

$$2\int\limits_{\Gamma(n)} f\,\mathrm{d}s = \int\limits_{\omega} |(n,p)|\,(\Delta + 2) f(p)\,\mathrm{d}\omega_p. \tag{4.6}$$

Therefore, in accordance with (4.5) and (4.6), equation (4.5) for $f(p)$ takes the form

$$\int_{\omega} \{ |(n,p)|\, a + b\chi((n,p)) \} (\Delta + 2) f(p)\, d\omega_p; \tag{4.7}$$

here, χ is the Heaviside function. Equation (4.7) is an integral equation of the first kind for $(\Delta + 2)f$, and it turns out that (Anikonov, 1995a)

$$\begin{aligned} (\Delta + 2) f(n) = \Phi(n) &= -\frac{1}{8\pi^2} \frac{d}{dt} \int_{(n,p)^2 > t} \frac{(\Delta + 2)\varphi_1(p)(n,p)\, d\omega_p}{\sqrt{(n,p)^2 - t}} \Bigg|_{t=0} \\ &\quad - \frac{1}{4\pi^2 b} \int_{\omega(n)} \frac{\Delta \varphi^-(p)}{(n,p)}\, d\omega_p \end{aligned} \tag{4.8}$$

for $a, b \neq 0$, $a + b \neq 0$, where

$$\varphi_1(p) = \frac{1}{a}\varphi^+(p) - \frac{b}{4\pi a(a+b)} \int_{\omega} \varphi(p)\, d\omega_p, \quad \varphi^{\pm} = \frac{\varphi(p) \pm \varphi(-p)}{2}.$$

Using the Weingarten formula (Blaschke, 1930), from (4.8) we obtain the required result:

$$f(n) = \frac{1}{4\pi} \int_{\omega} \Phi(p) \{ 1 - (n,p) \ln (1 - (n,p)) \}\, d\omega_p.$$

4.4 AN ESTIMATE IN A PROBLEM OF INTEGRAL GEOMETRY

Let D be a domain in three-dimensional Euclidean space R^3 that is homeomorphic to a ball and has smooth boundary Γ. Assume that with each point x of the closed domain $\bar{D}$ there is associated a sphere $\omega(x)$ about x with radius $t(x)$.

We consider a problem of integral geometry: find a function $f(x) \in C^{\infty}(R^3)$, $x \in \bar{D}$ given the function

$$\hat{f}(x) = \int_{\omega(x)} f(y)\, d\omega,$$

where $d\omega$ is the area element of the sphere $\omega(x)$, $x \in \bar{D}$.

Theorem 4.5. *Suppose that the function* $t(x) > 0$, $x \in D$, $t(x)|_{\Gamma} = 0$, *is continuous and almost everywhere continuously differentiable in the domain* $\bar{D}$, *and that*

$$\sum_{i=1}^{3} \left(\frac{\partial t}{\partial x_i} \right)^2 \geq 1$$

almost everywhere in $\bar{D}$. *Then*

$$\int_D f^2(x)\, dx \leq \frac{1}{(4\pi)^2} \int_D \left| \operatorname{grad} \frac{\hat{f}(x)}{t(x)} \right|^2 dx, \tag{4.9}$$

and this inequality becomes an equality in the case when

$$\sum_{i=1}^{3}\left(\frac{\partial t}{\partial x_i}\right)^2 = 1. \tag{4.10}$$

PROOF. In the domain $Q = \{(x,t) : x \in D, \quad 0 < t < t(x)\}$, we consider the function

$$u(x,t) = \int\limits_{\omega(x,t)} \frac{f(y)}{t}\, \mathrm{d}\omega,$$

where $\omega(x,t)$ is the sphere of radius t in R^3 about a point $x \in D$ and $\mathrm{d}\omega$ is the area element of $\omega(x,t)$.

Since $f(x) \in C^\infty(\mathrm{R}^3)$, the function $u(x,t)$ is infinitely differentiable, and we have the equalities (Bers *et al.*, 1974)

$$\frac{\partial^2 u}{\partial t^2} = \Delta u, \quad u\big|_{t=0} = 0, \quad \frac{\partial u}{\partial t}\bigg|_{t=0} = 4\pi f(x); \tag{4.11}$$

here, Δ is the Laplace operator in R^3. The following boundary condition for $u(x,t)$ is a consequence of the statement of the integral geometry problem:

$$u\big|_{t=t(x)} = \frac{\hat{f}(x)}{t(x)}, \quad x \in D. \tag{4.12}$$

Since $t(x) > 0$ for $x \in D$, $t(x) = 0$ for $x \in \Gamma$, and D is homeomorphic to a ball in R^3, the domain Q is homeomorphic to a ball in R^4 and has boundary consisting of $\bar{D}$ and the surface M given by

$$M = \{(x,t) : x \in \bar{D}, \quad t = t(x)\}.$$

Using an algebraic identity connected with the wave operator (Bers *et al.*, 1974), the equalities (4.11) and the Gauss-Ostrogradskii formula, we have

$$\begin{aligned}
&-2\int\limits_Q \frac{\partial u}{\partial t}\left(\frac{\partial^2 u}{\partial t^2} - \Delta u\right)\mathrm{d}x\,\mathrm{d}t \\
&= \int\limits_Q \left[2\sum_{i=1}^{3}\frac{\partial}{\partial x_i}\left(\frac{\partial u}{\partial t}\frac{\partial u}{\partial x_i}\right) - \frac{\partial}{\partial t}\left(\sum_{i=1}^{3}\left(\frac{\partial u}{\partial x_i}\right)^2 + \left(\frac{\partial u}{\partial t}\right)^2\right)\right]\mathrm{d}x\,\mathrm{d}t \\
&= -\int\limits_M \left[\sum_{i=1}^{3}\left(\frac{\partial u}{\partial x_i} + \frac{\partial u}{\partial t}\frac{\partial t}{\partial x_i}\right)^2 + \left(\sum_{i=1}^{3}\left(\frac{\partial t}{\partial x_i}\right)^2 - 1\right)\left(\frac{\partial u}{\partial t}\right)^2\right]\mathrm{d}x \\
&\quad + \int\limits_D \left(\frac{\partial u}{\partial t}\right)^2 \mathrm{d}x = 0
\end{aligned}$$

From this, by the relations

$$\left(\frac{\partial u}{\partial x_i} + \frac{\partial u}{\partial t}\frac{\partial t}{\partial x_i}\right)\bigg|_{t=t(x)} = \frac{\partial}{\partial x_i}\frac{\hat{f}(x)}{t(x)},$$

$$\left.\frac{\partial u}{\partial t}\right|_{t=0} = 4\pi f(x), \quad \sum_{i=1}^{3}\left(\frac{\partial t}{\partial x_i}\right)^2 - 1 \geq 0,$$

we get the inequality (4.9), with equality when (4.10) holds.

The theorem is proved.

4.5 A RADON TRANSFORMATION WITH VARIABLE ATTENUATION

Some problems of tomography lead to a Radon transformation with attenuation for functions with compact support:

$$\int_{x\cdot w=p} f(x)\,\mathrm{e}^{\gamma(x,\varphi)}\,\mathrm{d}l = g(w,p). \tag{4.13}$$

Here, $x = (x_1, x_2) \in Q \subset \mathrm{R}^2$, $p \in \mathrm{R}^2$, $w = (\cos\varphi, \sin\varphi)$, $0 \leq \varphi \leq 2\pi$, $f \in L_1(Q)$ is the source density, $x \cdot w$ denotes the scalar product, Q is a bounded convex domain in R^2, $\mathrm{d}l$ is the linear measure on the line $x \cdot w = p$, and $\gamma(x,\varphi)$ is a function which is defined by the attenuation coefficient $\mu(x,\varphi)$ through the formula

$$\gamma(x,\varphi) = \int_0^\infty \mu(x + tx^\perp, \varphi)\,\mathrm{d}t, \quad w^\perp = (-\sin\varphi, \cos\varphi),$$

where μ is a smooth function, 2π-periodic in φ, and $\mu(x,\varphi) = 0$ if $x \in Q$. In a number of cases, for instance, in medical emission tomography and plasma diagnosis problems, in addition to the function f, the function μ is also unknown. Results of investigations of similar problems can be found in Anikonov (1983); Natterer (1983); Anikonov (1985); Anikonov and Shneiberg (1991). We propose a method for the simultaneous determination of the two functions f and γ from the protection data $g(w,p)$ (Anikonov and Shneiberg, 1991). It turns out that the problem of determining the aforementioned functions is divided into two stages: the first stage is connected with the nonlinear problem of finding the function γ from the relations (4.14) and (4.16) given below, which are defined by the projection data, while the second stage is the linear problem of finding $f(x)$ when function $\gamma(x,\varphi)$ is known. Here, we are concerned only with the first stage. Suppose that all we know about the function γ is prior information that there are smooth functions $a(\varphi)$, $b(\varphi)$, and $H(p,\varphi)$ such that

$$\frac{\partial\gamma(x,\varphi)}{\partial\varphi} = a(\varphi)\left.\left(\frac{\partial H(p,\varphi)}{\partial p}x\cdot w^\perp + \frac{\partial H(p,\varphi)}{\partial\varphi}\right)\right|_{p=x\cdot w} + b(\varphi)H(x\cdot w,\varphi). \tag{4.14}$$

Define the operator $\mathbf{L}_H$ by the formula

$$(\mathbf{L}_H g)(\varphi,\sigma) = \int_{-\infty}^{\infty} g(w,p)\,\mathrm{e}^{-\sigma H(p,\varphi)}\,\mathrm{d}p = \iint_{\mathrm{R}^2} f(x)\,\mathrm{e}^{\gamma(x,\varphi)-\sigma H(x\cdot w,\varphi)}\,\mathrm{d}x. \tag{4.15}$$

Since the function $f(x)$ is assumed to have compact support, the function $(\mathbf{L}_H g)(\varphi,\sigma)$ is entire in the variable σ. The following result is readily deduced from (4.14) and (4.15).

Theorem 4.6. *If condition (4.14) is satisfied, the following relation holds:*

$$\left[\frac{\partial \mathbf{L}_H g}{\partial \varphi} + b(\varphi)\frac{\partial \mathbf{L}_H g}{\partial \sigma}\right]\bigg|_{\sigma=a(\varphi)} = 0. \tag{4.16}$$

We will show in our examples how relations (4.14) and (4.15) are used. We consider two ways of choosing the functions a, b, and H and, correspondingly, a certain class of function γ, defined by (4.15):

1. In relation (4.14) let $a(\varphi) = \mu(\varphi)$ and $b(\varphi) = \mu'(\varphi)$, where μ and H are arbitrary smooth functions. In this case we can assume $\gamma(x,\varphi) = \mu(\varphi)H(p,\varphi)$; here the projection data are found by means of the formula

$$g(w,p) = \int\limits_{x\cdot w=p} f(x)\,\mathrm{e}^{\mu(\varphi)H(p,\varphi)}\,\mathrm{d}l = \mathrm{e}^{\mu(\varphi)H(p,\varphi)}(\mathbf{R}f)(w,p), \tag{4.17}$$

where $\mathbf{R}f$ is the Radon transform of the function f. Relation (4.16) becomes differential equation for the unknown function μ:

$$\left[\frac{\partial \mathbf{L}_H g}{\partial \varphi} + \mu'(\varphi)\frac{\mathbf{L}_H g}{\partial \sigma}\right]\bigg|_{\sigma=\mu(\varphi)} = 0. \tag{4.18}$$

For a given function, $H(p,\varphi)$ is found from (4.18), and the function $\gamma(x,\varphi)$ is uniquely determined from the formula $\gamma = \mu H$. Substituting in (4.18) the expression of the operator $\mathbf{L}_H$ given by (4.15), we obtain the equality

$$\int\limits_{-\infty}^{\infty} g(w,p)\,\mathrm{e}^{-\mu(\varphi)H(p,\varphi)}\,\mathrm{d}p \equiv \text{const}\,. \tag{4.19}$$

In view of (4.17), the last relation is the Cavalieri condition for the Radon transform.

2. Now, in (4.14), we set $a(\varphi) = \mu'(\varphi)$, $b(\varphi) = -\mu(\varphi)$, and $H(p,\varphi) = p$. In this case, γ can be used in the form $\gamma(x,\varphi) = \mu(\varphi)x\cdot w^{\perp}$, so

$$g(w,p) = (\mathbf{R}_\mu f)(w,p) = \int\limits_{x\cdot w=p} f(x)\mathrm{e}^{\mu(\varphi)x\cdot w^{\perp}}\,\mathrm{d}l. \tag{4.20}$$

In our case, transformation (4.20) is called the exponential Radon transformation. This tranformation was studied in Solmon and Hazou (1988, 1989) and Finch and Hertle (1987). Here, relation (4.16) becomes a nonlinear differential equation for the sought function μ that is not solved with respect to the derivative μ' and can be written in the form

$$\int\limits_{-\infty}^{\infty}\left[\frac{\partial g(w,p)}{\partial \varphi} + \mu(\varphi)pg(w,p)\right]\mathrm{e}^{\mu'(\varphi)p}\,\mathrm{d}p = 0. \tag{4.21}$$

We note that for $\mu'(\varphi) \equiv 0$, equation (4.21) coincides with one of the relations obtained in Kuchment and L'vin (1990) for the exponential Radon transform with constant μ. The differential equation for $\mu'(\varphi) \equiv 0$, which includes a derivative with respect to φ, was considered in Bellini *et al.* (1979).

Let us now consider the solutions of equation (4.21) under the assumption that the function $f(x)$ is symmetric: $f(x) = f_0(|x|)$. In this case,

$$(\mathbf{L}_H g)(\varphi,\sigma) = \int_{-\infty}^{\infty} g(w,p)\,\mathrm{e}^{-\sigma p}\,\mathrm{d}p = \iint_{\mathrm{R}^2} f(x)\,\mathrm{e}^{\mu_0(\varphi)x\cdot w^\perp - \sigma x\cdot w}\,\mathrm{d}x = F(\sigma^2 + \mu_0^2(\varphi)), \quad (4.22)$$

where

$$F(\lambda) = \int_0^\infty r f_0(r) J_0(\mathrm{i}\, r\sqrt{\lambda})\,\mathrm{d}r,$$

J_0 is the zeroth-order Bessel function, and $\gamma(x,\varphi) = \mu_0(\varphi)x\cdot w^\perp$. From (4.22), it follows that, in our case, (4.21) admits two series of solutions:

$$\mu(\varphi) = \sqrt{\mu_0^2 + c}$$

where c is an arbitrary constant;

$$[\mu'(\varphi)]^2 + \mu_0^2(\varphi) = c_0, \quad c_0 \geq \max_{0\leq\varphi\leq 2\pi} \mu_0^2(\varphi), \quad F'(c_0) = 0.$$

One can show that any solution of series 1 is compatible with the projection data $g = \mathbf{R}_{\mu_0} f_0$; i.e., one can find a function f_1 such that $g = \mathbf{R}_\mu f_1$, whereas there are no solutions from series 2.

In conclusion, we note that the problem of determining $f(x)$ and $\gamma(x,\varphi)$ from $\mathbf{L}_H g$ using equation (4.15) may be regarded as one of the variants of the inverse problem of scattering theory.

4.6 SOME RELATIONS IN TOMOGRAPHY

Consider transport equations of the type

$$\frac{\partial w}{\partial x}\cos\varphi + \frac{\partial w}{\partial y}\sin\varphi = \mu(x,y)w + \lambda(x,y), \quad (4.23)$$

$(x,y) \in D \subset \mathrm{R}^2$, $0 \leq \varphi \leq 2\pi$, functions $\mu(x,y)$ and $\lambda(x,y)$ do not depend on variable φ. These are frequently met with in tomography, and the functions are often unknown.

From (4.23), we can derive an equation including function $\lambda(x,y)$:

$$\frac{\partial}{\partial\varphi}\left(\frac{\partial w}{\partial x}\cos\varphi + \frac{\partial w}{\partial y}\sin\varphi\right) = \mu(x,y)\frac{\partial w}{\partial\varphi}. \quad (4.24)$$

The equation (4.24) is an equation with non-constant coefficients which depends on variable φ. It is correct in the case $\mu = 0$. It seems that the following interesting theorem holds:

Theorem 4.7. *The function* $w(x,y,\varphi) \in C^2$,

$$w(x,y,\varphi) = \frac{u_0(x,y)}{2} + \sum_{k=1}^{\infty}\left(u_k(x,y)\cos k\varphi + v_k(x,y)\sin k\varphi\right)$$

satisfies equation:

$$\frac{\partial}{\partial\varphi}\left(\frac{\partial w}{\partial x}\cos\varphi+\frac{\partial w}{\partial y}\sin\varphi\right)=\mu(x,y)\frac{\partial w}{\partial\varphi}$$

if and only if functions $u(x,y,t)$, $v(x,y,t)$:

$$u(x,y,t)=u_0(x,y)+\sum_{k=1}^{\infty}\frac{u_k(x,y)t^k}{k!}$$

$$v(x,y,t)=\sum_{k=1}^{\infty}\frac{v_k(x,y)t^k}{k!}$$

satisfy the system of equations

$$\frac{\partial u}{\partial x}-\frac{\partial v}{\partial y}+\frac{\partial^2}{\partial t^2}\left(\frac{\partial u}{\partial x}+\frac{\partial v}{\partial y}\right)=2\mu(x,y)\frac{\partial u}{\partial t}$$

$$\frac{\partial v}{\partial x}+\frac{\partial u}{\partial y}+\frac{\partial^2}{\partial t^2}\left(\frac{\partial v}{\partial x}-\frac{\partial u}{\partial y}\right)=2\mu(x,y)\frac{\partial v}{\partial t}$$

and the formula holds that

$$\lambda(x,y)=\frac{1}{2}\left[\frac{\partial u_1}{\partial x}+\frac{\partial v_1}{\partial y}-\mu(x,y)u_0\right].$$

PROOF. If $w(x,y,\varphi)=\dfrac{u_0(x,y)}{2}+\sum\limits_{k=1}^{\infty}\left(u_k(x,y)\cos k\varphi+v_k(x,y)\sin k\varphi\right)$, then the functions $u_k(x,y)$, $v_k(x,y)$ satisfy the system:

$$\begin{cases}\dfrac{\partial u_k}{\partial x}+\dfrac{\partial u_{k+2}}{\partial x}-\dfrac{\partial v_k}{\partial y}+\dfrac{\partial v_{k+2}}{\partial y}=2\mu u_{k+1},\\[2ex] \dfrac{\partial v_k}{\partial x}+\dfrac{\partial v_{k+2}}{\partial x}+\dfrac{\partial u_k}{\partial y}-\dfrac{\partial u_{k+2}}{\partial y}=2\mu u_{k+1},\\[2ex] \lambda+\mu u_0=\dfrac{1}{2}\left[\dfrac{\partial u_1}{\partial x}+\dfrac{\partial v_1}{\partial y}\right], \qquad k=0,1,2,\ldots\end{cases}$$

From here we can see that functions $u(x,y,t)=u_0+\sum\limits_{k=1}^{\infty}\dfrac{u_k t^k}{k!}$, $v(x,y,t)=\sum\limits_{k=1}^{\infty}\dfrac{v_k t^k}{k!}$ satisfy the system

$$\begin{cases}\dfrac{\partial u}{\partial x}-\dfrac{\partial v}{\partial y}+\dfrac{\partial^2}{\partial t^2}\left(\dfrac{\partial u}{\partial x}+\dfrac{\partial v}{\partial y}\right)=2\mu(x,y)\dfrac{\partial u}{\partial t},\\[2ex] \dfrac{\partial v}{\partial u}+\dfrac{\partial u}{\partial y}+\dfrac{\partial^2}{\partial t^2}\left(\dfrac{\partial v}{\partial x}-\dfrac{\partial u}{\partial y}\right)=2\mu(x,y)\dfrac{\partial v}{\partial t},\\[2ex] \lambda(x,y)=\dfrac{1}{2}\left[\dfrac{\partial u_1}{\partial x}+\dfrac{\partial v_1}{\partial y}\right]-\mu u_0.\end{cases}$$

and the converse is correct too.

4.7 RELATIONS ASSOCIATED WITH INTEGRAL GEOMETRY

We discuss here only two results which were recently obtained when integrating the unknown function over segments of a straight line.

Let $\gamma(y,z)$ be a segment of a straight line in R^n, $n \geq 1$, which connects the points $y \in \mathrm{R}^n$ and $z \in \mathrm{R}^n$. We consider the function

$$w(y,z) = \int\limits_{\gamma(y,z)} \lambda(x)\,|\mathrm{d}x|\,,$$

where $\lambda(x)$, $x \in \mathrm{R}^n$, is a certain, say, continuous function.

In linear tomography problems, the function $\lambda(x)$ is sought in a domain e. g. $|x| \leq 1$, and the function $w(y,z)$ is assumed to be known for some pairs of points (y,z), e. g. for $|y| = 1$ and $|z| = 1$.

Both theoretically and practically, it is advisable to have formulas that specify the functions $\lambda(x)$ and $w(y,z)$ by a single function F. Thus, we should emphasize that for $n = 1$, according to the Newton-Leibnitz formula, this function exists. If

$$w(y,z) = \int\limits_{y}^{z} \lambda(x)\,\mathrm{d}x$$

$$w(y,z) = F(z) - F(y)$$

and $\lambda(x) = F'(x)$, where $F(x)$ is a continuously differentiable function. We assume analogous conditions in this work.

Theorem 4.8. *Let $F(\xi,p)$, $\xi \in B$, $p \in \mathrm{R}^1$, where B is a k-dimensional set in R^n, $k = 1,\dots,n$, be a complex-valued function continuous in ξ and continuously differentiable with respect to p. Then the functions*

$$\lambda(x) = \int\limits_{B} \frac{\partial F(\xi,(\xi,x))}{\partial p}\,\mathrm{d}s_\xi, \quad (\xi,x) = \sum_{j=1}^{n} \xi_j x_j,$$

$$w(y,z) = |y-z| \int\limits_{B} \frac{F(\xi,(y,\xi)) - F(\xi,(z,\xi))}{(\xi, y-z)}\,\mathrm{d}s_\xi$$

satisfy the relation

$$w(y,z) = \int\limits_{\gamma(y,z)} \lambda(x)\,|\mathrm{d}x|\,.$$

In particular, if

$$F(\xi,p) = \frac{\hat{\lambda}(\xi)\,\mathrm{e}^{\mathrm{i}p}}{\mathrm{i}}, \quad B = \mathrm{R}^n,$$

then

$$\lambda(x) = \int\limits_{\mathrm{R}^n} \mathrm{e}^{-\mathrm{i}(\xi,x)}\,\hat{\lambda}(\xi)\,\mathrm{d}\xi,$$

$$w(y,z) = |y-z| \int\limits_{\mathbf{R}^n} \hat{\lambda} \frac{e^{i(y,\xi)} - e^{i(z,\xi)}}{(\xi, y-z)} \, d\xi.$$

The theorem holds for the functionals whose extremals are straight lines.

Theorem 4.9. *Let $F(\xi,p)$, $\xi \in B$, $p \in \mathbf{R}^1$ (where B is a k-dimensional set in $\mathbf{R}^n$, $k = 1,\dots,n$) be a function continuous in ξ and continuously differentiable with respect to p. Then the functions:*

$$\lambda(x,x') = \int\limits_B \frac{\partial F(\xi,(\xi,x))}{\partial p} |(x',\xi)| \, ds_\xi,$$

$$w(y,z) = \int\limits_B \frac{F(\xi,(\xi,y)) - F(\xi,(z,\xi)) \, |(\xi, y-z)|}{(\xi, y-z)} \, ds_\xi$$

satisfy the relation:

$$w(y,z) = \int\limits_{\gamma(y,z)} \lambda(x, dx), \quad y \in \mathbf{R}^n, \quad z \in \mathbf{R}^n.$$

We can check the assertion of the above theorems directly.

4.8 UNIQUENESS THEOREMS FOR THE INVERSE PROBLEMS OF KINEMATIC AND INTEGRAL GEOMETRY

In this section, we prove uniqueness theorems for the solutions of inverse problems of linear and nonlinear integral geometry in the classes of piecewise analytic functions. Many formulas are used in proofs of the theorems.

The linearization of inverse kinematic problems leads to problems of integral geometry Anikonov (1979, 1980).

We consider the nonlinear multidimensional inverse problems in integral geometry and kinematics (Subsections 4.8.1 and 4.8.2). Those problems appear to be ill-posed. We prove that their solutions are unique in classes of piecewise analytic functions, with a few constraints on objects which are domain of integration. In Subsection 4.8.3, we prove the general uniqueness theorem for a planar problem that generalizes the problem of kinematics and integral geometry in the case of regular extremals connecting all points of the domain boundary. This is a serious restriction that allows us to obtain the general result.

4.8.1 Uniqueness theorems for the solution of a generalized problem of integral geometry in the class of piecewise analytic functions

1. Let E^{n+1} denote the $(n+1)$-dimensional Euclidean space of variables (x,y), $x \in E^n$, $-\infty < y < \infty$, and let E_h^{n+1} be the half-space $y > h$, $x \in E^n$. By $\{\omega\}$, we denote a collection of $(n+1)$-dimensional open sets $\omega(\xi,t)$, $\xi \in E^n$, $t > 0$, which cover the half-space $E_0^{n+1} = \{x,y : y > 0\}$ and have properties:

(1) for all $t > 0$, $h > 0$ and $t > h$ the intersection

$$\omega(\xi, t, h) = \omega \cap E_0^{n+1}$$

is an open $(n+1)$-dimensional set;

(2) $\omega(\xi, t, h)$ continuously depends on (ξ, t, h) and

$$\lim_{t \to h} \omega(\xi, t, h) = (\xi, h)$$

where (ξ, h) is a point in the hyperplane $y = h$.

The set of all hemispheres

$$\omega(\xi, t) = \{(x, y) : |x - \xi|^2 + y^2 < t, \quad y > 0\}, \quad \xi \in E^n, \quad t > 0$$

is an example of a collection of sets $\{\omega\}$. We assume that $B_k(\xi, t)$, $k = 0, 1, 2, \ldots, n+1$, are k-dimensional manifolds (for $k = 0$ $B_0(\xi, t)$ consists of a finite number of points) which satisfy the condition: for any k: $0 \le k \le n+1$, and $t > h$ the intersection

$$B_k(\xi, t, h) = B_k(\xi, t) \cap E_h^{n+1}$$

belongs to the set $\omega(\xi, t, h)$. We integrate over the manifolds $B_k(\xi, t)$. By $M_k(x, y, \xi, t, \lambda)$ we denote the real functions of the variables

$$x \in E^n, \quad \xi \in E^n, \quad y > 0, \quad t > 0, \quad -\infty < \lambda < \infty, \quad k = 0, 1, \ldots, n+1.$$

Let us examine the equation for a function $\lambda(x, y)$:

$$M\lambda = \sum_{k=0}^{n+1} \int_{B_k(\xi,t)} M_k(x, y, \xi, t, \lambda)\, d\sigma_k = u(\xi, t), \tag{4.25}$$

where $d\sigma_k$ is a volume element of the manifold $B_k(\xi, t)$; for $k = 0$ we define

$$\int_{B_0} M_0\, d\sigma_0 = \sum_{i,j} M_0(x^i, y^j, \xi, t, \lambda(x^i, y^j)),$$

summing over all points in $B_0(\xi, t)$). Note that if $M_k = 0$ for all $k \neq l$, $0 < l \le n+1$, and $M_l = \lambda$, then (4.25) has the form

$$M\lambda = \int_{B_l(\xi,t)} \lambda(x, y)\, d\sigma_l = u(\xi, t). \tag{4.26}$$

In this case, the problem of determining the function $\lambda(x, y)$ from $u(\xi, t)$, $\xi \in E^n$, $t > 0$ is called the problem of integral geometry.

Let $\{\lambda\}_\infty$ denote the class of functions $\lambda(x, y)$ which are piecewise analytic with respect to x in the domain $y \ge 0$; this class is defined as: $\lambda(x, y) \in \{\lambda\}_\infty$, if there exist infinitely differentiable functions $a_k^i(x)$ and numbers h_i, $h_0 \ge 0$, $h_{i+1} > h_i$ such that

$$\lambda(x, y) = \sum_{k=0}^{\infty} a_k^i(x)(y - h_i)^k, \quad h_i \le y < h_{i+1}, \quad x \in E^n.$$

Below, we use the general uniqueness theorem for a solution in the class of analytic functions for quasimonotonic operators (Anikonov, 1978). By definition, the operator

$$M: \quad M\lambda = \tau, \quad \lambda(x,y) \in \{\lambda\}, \tau \in \{\tau\}$$

is called the quasimonotonic operator with respect to the covering $\{\omega\}$ of the domain $y > 0$, $x \in E^n$ provided that the inequality $\lambda_1(x,y) > \lambda_2(x,y)$ for all $(x,y) \in \omega_0 \in \{\omega\}$ implies $M\lambda_1 \neq M\lambda_2$.

If the covering $\{\omega\}$ consists of open hemispheres $\omega(\xi,t)$ with centres in E^n:

$$\omega(\xi,t) = \{(x,y) : |x-\xi|^2 + y^2 < t, \quad y > 0\}, \quad \omega(\xi,t) \in \{\omega\},$$

the quasimonotony guarantees that the solution of the operator equation $M\lambda = \tau$ is unique in the class of infinitely differentiable functions $\lambda(x,y)$ analytic in y for $x \in E^n$, $y \geq 0$ (Anikonov, 1978).

Theorem 4.10. *Let the function $M_k(x,y,\xi,t,\lambda)$ be nondecreasing in the variable λ for all k, $k = 0,1,\ldots,n+1$, and at least one of these functions be strictly monotone increasing in λ. If $M\lambda_1 = M\lambda_2$, $\lambda_i(x,y) \in \{\lambda\}_\infty$, then $\lambda_1(x,y) = \lambda_2(x,y)$, $y \geq 0$.*

PROOF. One of the functions M_k, $k = 0,1,\ldots,n+1$, is a strictly monotone increasing function in λ, and $B_k(\xi,t,h) \subset \omega(\xi,t,h)$; therefore, for any $h \geq 0$

$$M_h\lambda = \sum_{k=0}^{n+1} \int\limits_{B_k(\xi,t)} M_k(x,y,\xi,t,\lambda)\,\mathrm{d}\sigma_k, \quad t > h,$$

is a quasimonotone operator with respect to the covering of E_h^{n+1} by the sets $\omega(\xi,t,h)$, (Anikonov, 1978).

We suppose that $\lambda_1(x,y)$ and $\lambda_2(x,y)$ are two solutions of the operator equation

$$M\lambda = u(\xi,t),$$

where the operator M is defined in (4.25). If $\lambda_1(x,y) \neq \lambda_2(x,y)$, by definition of the class $\{\lambda\}_\infty$ there exist $\tilde{h}_0 \geq 0$ and $\varepsilon > 0$ such that

(i) $\lambda_1(x,y) \equiv \lambda_2(x,y), \quad x \in E^n, \quad 0 \leq y < \tilde{h}_0,$

(ii) the functions $\lambda_i(x,y)$ are analytic with respect to y in the strip

$$\tilde{h}_0 \leq y \leq \tilde{h}_0 + \varepsilon, \quad x \in E^n$$

and $\lambda_1(x,y) \neq \lambda_2(x,y)$ for

$$\tilde{h}_0 \leq y \leq \tilde{h}_0 + \varepsilon, \quad x \in E^n.$$

Using (i), (ii) and the additive property of integration, we have

$$\sum_{k=0}^{n+1} \int\limits_{B_k(\xi,t,\tilde{h}_0)} M_k(x,y,\xi,t,\lambda_1(x,y))\,\mathrm{d}\sigma_k = \sum_{k=0}^{n+1} \int\limits_{B_k(\xi,t,\tilde{h}_0)} M_k(x,y,\xi,t,\lambda_2(x,y))\,\mathrm{d}\sigma_k,$$

where $B_k(\xi, t, \tilde{h}_0)$ again denotes the intersection of the manifold $B_k(\xi, t)$ and $E^{n+1}_{\tilde{h}_0}$. Since

$$M_{\tilde{h}_0}\lambda = \sum_{k=0}^{n+1} \int\limits_{B_k(\xi,t,\tilde{h}_0)} M_k\big(x, y, \xi, t, \lambda(x, y)\big)\, d\sigma_k$$

is quasimonotone operator and $\lambda_i(x, y)$ are analytic functions in y for $\tilde{h}_0 \leq y \leq \tilde{h}_0 + \varepsilon$, hence, according to Anikonov (1978, Theorem 1.1), $\lambda_1(x, y) = \lambda_2(x, y)$. However, this contradicts (ii). The proof is complete.

2. Of interest are uniqueness theorems for solutions in the classes of piecewise analytic functions in domains with curvilinear interfaces. Let us first describe such a class.

We assume that $D_0 \subset E^n$ is a fixed n-dimensional domain, homeomorphic to the ball $|x| < 0$. By $D_1 \supseteq D_2 \supseteq \ldots \supseteq D_k$ we denote a sequence of domains belonging and homeomorphic to D_0. Let $\{a\}_k$ be the set of all analytic functions $a_k(x)$, defined in a closed domain $\bar{D}_k$, $k = 0, 1, 2, \ldots$. We define in D_0 the classes $\{\lambda\}_k$ of functions λ_k recursively:

$$\{\lambda\}_0 = \{a\}_0, \quad \lambda_{k+1}(x) \in \{\lambda\}_{k+1}, \quad \lambda_{k+1} = \begin{cases} \lambda_k(x), & x \in \bar{D}_0 \setminus \bar{D}_{k+1} \\ a_{k+1}(x), & x \in \bar{D}_{k+1} \end{cases}$$

We put

$$\{\tilde{\lambda}\}_\infty = \bigcup_{k=0}^{\infty} \{\lambda\}_k.$$

Thus, in order to uniquely determine a function $\lambda \in \{\tilde{\lambda}\}_\infty$, we have to specify the domains $D_1, D_2, \ldots, D_k, \ldots$, and analytic functions $a_0(x), a_1(x), \ldots, a_k(x), \ldots, a_k(x) \in \{a\}_k$, $x \in \bar{D}_k$. We search for the solution of the problem of integral geometry in the class $\{\tilde{\lambda}\}_\infty$. We integrate over surfaces $\gamma(\xi, t)$ which are defined as follows.

We suppose that each point ξ belonging to the boundary S_0 of the domain D_0 is put into correspondence with a set of domains $B(\xi, t) \subset D_0$, $t > 0$, which have smooth boundaries and are homeomorphic to the ball $|x| < 1$, so that

1. $B(\xi, t)$ continuously depends on ξ and t; and if $t_1 > t_2$, then $\bar{B}(\xi, t_1) \supset \bar{B}(\xi, t_2)$, where $\bar{B}$ denotes the closure of B;
2. $\lim\limits_{t \to 0} B(\xi, t) = \xi$, $\quad \lim\limits_{t \to \infty} B(\xi, t) = D_0$,
3. the intersection $\bar{B}(\xi, t) \cap S_0$ is homeomorphic to an $(n-1)$-dimensional ball.

By $\gamma(\xi, t)$ we denote the boundary of $B(\xi, t)$ in D_0.

Let us consider the problem of determining the function $\lambda(x)$, $x \in D_0$, provided that the function

$$u(\xi, t) = \int\limits_{\gamma(\xi,t)} M\big(x, \xi, t, \lambda(x)\big)\, d\sigma = M\lambda,$$

is given in the domain $\xi \in S_0$, $t > 0$; here, M is a measurable function strictly monotone increasing with respect to λ, and $d\sigma$ is a surface element of γ. Before formulating the result, let us give a definition.

Definition 4.1. The domain $\tilde{D} \subset D_0$ with a boundary $\tilde{S}$ is quasicomplex with respect to domains $B(\xi, t)$, if nonempty intersections $B \cap \tilde{D}$ have the properties 1–3, namely, the sets $\tilde{B} = B(\xi, t) \cap \tilde{D}$ can be parametrized with parameters $\tilde{\xi} \subset \tilde{S}$ and $\tilde{t} > 0$ under the conditions 1–3.

Theorem 4.11. *We suppose that the conditions 1–3 are satisfied, and D_k, $k = 0, 1, 2, \dots$, are quasiconvex domains with piecewise analytic boundaries. If*

$$M\lambda_1 = M\lambda_2, \quad \lambda_i(x) \in \{\tilde{\lambda}\}_\infty, \quad i = 1, 2,$$

then

$$\lambda_1(x) = \lambda_2(x), \quad x \in \bar{D}_0.$$

PROOF. We first show that

$$M\lambda = \int\limits_{\gamma(\xi,t)} M\big(x, \xi, t, \lambda(x)\big)\, d\sigma$$

is a quasimonotone operator with respect to the covering of D_0 by open balls with the centres at points $\xi \in S_0$. Let $\omega_0 = \omega(\xi_0, t_0)$ be the intersection of the domain D_0 and the sphere $K(\xi_0, t_0)$ with the centre $\xi_0 \in S_0$ and radius $t_0 > 0$. We assume $\lambda_1(x) > \lambda_2(x)$, $x \in \omega_0$. According to 1 and 2, there exists a surface $\gamma_0 = \gamma(\xi_0, t_0)$ such that $\gamma_0 \subset \omega_0$. Hence, in view of the monotony of the function M, we have

$$\int\limits_{\gamma_0} M\big(x, \xi, t, \lambda_1(x)\big)\, d\sigma > \int\limits_{\gamma_0} M\big(x, \xi, t, \lambda_2(x)\big)\, d\sigma.$$

which shows that the operator M is quasimonotone. Furthermore, if $D_0 = D_1 = D_2 = \dots = D_k = \dots$, the sought-for solution $\lambda(x) \in \{\tilde{\lambda}\}_\infty$ is an analytic function in the domain $\bar{D}_0$ by the definition of the class $\{\tilde{\lambda}\}_\infty$, and the proof follows from the theorem of quasimonotone operators. Let us suppose that there exists a number $k_0 \geq 1$ such that $D_0 \neq D_{k_0}$, and if $\lambda = \{a_0(x), a_1(x), \dots, a_k(x), \dots\}$, then $a_0(x) \neq a_{k_0}(x)$. With no loss of generality, we put $k_0 = 1$; otherwise, we can renumerate the sequence $\{D\}_k$. Thus, we put $D_0 = D_1$,

$$a_0(x) \neq a_1(x). \tag{4.27}$$

We first prove that the function $a_0(x)$, $x \in D_0$, is determined uniquely, and then specify the domain D_1. We denote $P_0 = D_0 \setminus \bar{D}_1$. Since P_0 is an open set, there exists a neighbourhood

$$\omega_0 \subset P_0, \quad \omega_0 = \omega(\xi_0, t_0), \quad \xi_0 \in S_0, \quad t_0 > 0.$$

Let us consider the surfaces $\gamma(\xi, t)$ which belong to ω_0. For these surfaces, it is evident that

$$M\lambda = \int\limits_{\gamma(\xi,t)} M\big(x, \xi, t, a_0(x)\big)\, d\sigma.$$

This and the theorem of quasimonotone operators (Anikonov, 1978) imply that the analytic function $a_0(x)$, $x \in \bar{D}_0$ is determined uniquely. In the domain $\xi \in S_0$, $t > 0$ we introduce the function

$$u_0(\xi, t) = Ma_0 = \int\limits_{\gamma(\xi,t)} M\big(x, \xi, t, a_0(x)\big)\, d\sigma.$$

We define the set T_0 of points (ξ, t) as follows: a point (ξ_0, t_0) belongs to T_0 if

1. $u_0(\xi_0, t_0) = u(\xi_0, t_0)$,

2. For all ξ and t such that $\bar{B}(\xi, t) \subset \bar{B}(\xi_0, t_0)$; similarly $u_0(\xi, t) = u(\xi, t)$.

Let us put $\tilde{P}_0 = \cup B(\xi, t)$, $(\xi, t) \in T_0$ and show that $\tilde{P}_0 = D_0 \setminus D_1$; thereby, we uniquely determine the domain D_1. By the definition of quasimonotony, we have the inclusion

$$D_0 \setminus D_1 \subset \tilde{P}_0.$$

Therefore, we are to prove that $\tilde{P}_0 \cap D_1 = \emptyset$. We suppose that the converse is true, namely, there exists a set $B_0 = B(\xi_0. t_0)$, $(\xi_0, t_0) \in T_0$, such that

$$B_0 \cap D_1 = B_1^0 \neq \emptyset. \tag{4.28}$$

Let us consider the intersections

$$B_1 = B(\xi, t) \cap D_1, \quad B_1 = B(\xi_1, t_1), \quad \xi_1 \in S_1, \quad t_1 > 0$$

and denote by $\gamma_1(\xi_1, t_1)$ the surfaces corresponding to the sets $B_1(\xi_1, t_1)$. If $D_1 = D_2 = D_3 = \ldots = D_k = \ldots$, we have for any surface $\gamma_1(\xi_1, t_1) \subset B_0$

$$M_1 a_1 = \int_{\gamma_1} M\big(x, \xi, t, a_1(x)\big)\, d\sigma = \int_{\gamma_1} M\big(x, \xi, t, a_0(x)\big)\, d\sigma = M_1 a_0. \tag{4.29}$$

This and the theorem of quasimonotone operators (Anikonov, 1978) imply that $a_1(x) = a_0(x)$, $x \in D_1$. This equation contradicts (4.27). For example, if $D_1 \neq D_2$, there exists a set $\tilde{B}_0^1$ such that

$$\tilde{B}_0^1 = B(\xi_0, t_0) \cap D_1, \quad \tilde{B}_0^1 \subset B_0^1, \quad \tilde{B}_0^1 \cap D_1 = \emptyset$$

because the boundaries S_k of the domains D_k are piecewise analytic.

Equation (4.29) is valid for any surface $\gamma_1 \subset \tilde{B}_0^1$, and we again have a contradiction: $a_1(x) = a_0(x)$, $x \in D_1$. Thus, the domain D_1 and the function $a_0(x)$ are determined uniquely. We now introduce the function

$$u_1(\xi_1, t_1) = \int_{\gamma_1(\xi_1, t_1)} M\big(x, \xi, t, \lambda(x)\big)\, d\sigma, \quad \xi_1 \in S_1, \quad t_1 > 0.$$

Since $a_0(x)$ and D_1 are determined uniquely, the function $u_(\xi_1, t_1)$ is known. Repeating the above arguments, we find the function $a_1(x)$, $x \in D_1$, and the domain D_2 etc. The theorem is proved.

Instability of solutions in the above problems of integral geometry

Let $ds^2 = B^2(x, y)(dx^2 + dy^2)$ be a fixed analytic metric in a closed domain $\bar{D}$ with boundary L, and let $\gamma(\xi, \eta)$ be its geodesic that connects points ξ and η of an arc $l \leq L$; in particular, $\bar{D}$ may be the half-space $y \geq 0$.

Let us consider the problem of determining the function $\lambda(x,y)$, $(x,y) \in D$, provided that the function

$$w(\xi,\eta) = \int\limits_{\gamma(\xi,\eta)} \lambda(x,y)\sqrt{dx^2+dy^2}, \quad \xi \in l, \quad \eta \in l$$

is known. Here, we show that for $l \neq L$ this problem is ill-posed. Let $u(x,y)$ and $v(x,y)$ be arbitrary harmonic conjugate functions given in the domain D; $v_0(p)$ and $u_0(p)$ are their boundary values, $p \in l$.

By $\alpha(\xi,\eta)$ and $\beta(\xi,\eta)$ we denote the angles formed by the geodesic $\gamma(\xi,\eta)$ and the x-axis at the points ξ and η respectively.

We introduce the functions

$$\begin{gathered} w(\xi,\eta) = a_0(\eta)\left[u_0(\eta)\cos\beta + v_0(\eta)\sin\beta\right] - a_0(\xi)\left[u_0(\xi)\cos\alpha + v_0(\xi)\sin\alpha\right], \\ \lambda(x,y) = u(x,y)\frac{\partial B}{\partial x} + v(x,y)\frac{\partial B}{\partial y} + \frac{\partial u}{\partial x}B. \end{gathered} \tag{4.30}$$

We easily check that

$$w(\xi,\eta) = \int\limits_{\gamma(\xi,\eta)} \lambda(x,y)\sqrt{dx^2+dy^2}. \tag{4.31}$$

If we now specify $u_0(p)$ and $v_0(p)$, the function $w(\xi,\eta)$ is determined uniquely. However, in order to identify the solution we have to find the conjugate harmonic functions $u(x,y)$ and $v(x,y)$ in D from the boundary values $u_0(p)$ and $v_0(p)$, $p \in l$, using (4.29) and (4.30). This problem of analytic continuation is known to be ill-posed if $l \neq L$.

Note that if $u_0(p)$ and $v_0(p)$ are small, $w(\xi,\eta)$ is also small, according to the formula for $w(\xi,\eta)$; however, the solution $\lambda(x,y)$ may be large because of (4.29).

4.8.2 Uniqueness theorems for a multidimensional inverse kinematic problem

1. Let E^{n+1} denote the $(n+1)$-dimensional Euclidean space of variables (x,y), $x \in E^n$, $-\infty < y < \infty$, $n \geq 1$; let Ω denote the sphere in E^n:

$$\Omega = \{x : |x| = 1\},$$

its radius-vector is $\mathbf{n}$.

Problem 4.1. In the half-space $y \geq 0$ find an infinitely differentiable function $\lambda(x,y)$, $0 < \lambda < M < \infty$, such that

1. in the metric

$$ds^2 = \lambda^2(x,y)(dx^2+dy^2), \quad dx^2 = \sum_{i=1}^{n} dx_i^2,$$

there exists at least one piecewise smooth shortest curve $\gamma(a,b,\mathbf{n})$ belonging to the half-space $y \geq 0$ and connecting the points $a\mathbf{n}$ and $b\mathbf{n}$, where $-\alpha < a < \alpha$ and $-\alpha < b < \alpha$ for a constant α;

2. the function

$$\tau(a,b,\mathbf{n}) = \int\limits_{\gamma(a,b,\mathbf{n})} \lambda(x,y)\sqrt{dx^2+dy^2}$$

is given in the domain $|a| < \alpha$, $|b| < \alpha$, $\mathbf{n} \in \Omega$.

Note that the problem is local and the geodesics are locally shortest; therefore, we merely formulate the problem for the shortest curves.

Before presenting the results, we note that if the shortest $\gamma(a,b,\mathbf{n})$ in the metric

$$ds^2 = \lambda^2(x,y)(dx^2+dy^2)$$

is fully in the hyperplane $y = 0$ for all a, b, $\mathbf{n}$, then $\lambda(x,y)$ can be determined from $\tau(a,b,\mathbf{n})$ only for $y = 0$. Therefore, if we are to find the function $\lambda(x,y)$ in the half-space $y > 0$, we require that $\gamma(a,b,\mathbf{n})$ not be in this hyperplane. We use the function

$$\tau_0(a,b,\mathbf{n}) = \int\limits_{\gamma_0(a,b,\mathbf{n})} \lambda_0(x)\,|dx|,$$

where

$$\lambda_0(x) = \lambda(x,0) = \lim_{\substack{a \to b = t \\ t\mathbf{n} = x}} \frac{\tau(a,b,\mathbf{n})}{|b-a|},$$

$\gamma_0(a,b,\mathbf{n})$ connecting the points $a\mathbf{n}$ and $b\mathbf{n}$ is the shortest curve in the metric

$$ds_0^2 = \lambda_0(x)\,|dx|, \quad |x| < \alpha.$$

It is evident that the function $\tau_0(a,b,\mathbf{n})$ is determined by $\tau(a,b,\mathbf{n})$ only. The lemma is true.

Lemma 4.1. *If $\tau(a,b,\mathbf{n}) \neq \tau_0(a,b,\mathbf{n})$, the intersection of the shortest $\gamma(a,b,\mathbf{n})$ and the half-space $y > 0$ has a positive linear measure.*

PROOF. We suppose that the intersection of $\gamma(a,b,\mathbf{n})$ and the half-space $y > 0$ has a zero linear measure. By definition, $\gamma(a,b,\mathbf{n})$ is a piecewise smooth curve, and hence $\gamma(a,b,\mathbf{n})$ should be fully in the hyperplane. Accordingly, we have

$$\begin{aligned}\tau(a,b,\mathbf{n}) &= \int\limits_{\gamma(a,b,\mathbf{n})} \lambda(x,y)\sqrt{dx^2+dy^2} \\ &= \int\limits_{\gamma(a,b,\mathbf{n})} \lambda(x,0)\,|dx| = \int\limits_{\gamma_0(a,b,\mathbf{n})} \lambda_0(x)\,|dx| = \tau_0(a,b,\mathbf{n}).\end{aligned}$$

Therefore, $\tau(a,b,\mathbf{n}) = \tau_0(a,b,\mathbf{n})$, which contradicts the conditions of the lemma. Lemma 4.1 is proved.

REMARK 4.1. For the two-dimensional case, i. e. $n = 1$, we may take the condition

$$\tau(a,b) < \tau(a,b_1) + \tau(b_1,b), \quad a < b_1 < b$$

instead of $\tau(a,b,\mathbf{n}) \neq \tau_0(a,b,\mathbf{n})$. Under this condition, any internal point of $\gamma(a,b)$ cannot belong to the line $y = 0$. Let $\{\lambda\}$ denote the set of all functions $\lambda(x,y)$ infinitely differentiable for $x \in E^n$, $y \geq 0$, and analytic with respect to y for $0 \leq y \leq h_\lambda$, where $h_\lambda > 0$ is the positive constant.

Theorem 4.12. *Suppose that*

1. $\int\limits_{\gamma_1(a,b,\mathbf{n})} \lambda_1(x,y)\sqrt{dx^2+dy^2} = \int\limits_{\gamma_2(a,b,\mathbf{n})} \lambda_2(x,y)\sqrt{dx^2+dy^2} = \tau(a,b,\mathbf{n}),\ |a| < \alpha,$ $|b| < \alpha,\ \mathbf{n} \in \Omega;$

2. $\lambda_i(x,y) \in \{\lambda\},\ i = 1,2;$

3. $\tau(a,b,\mathbf{n}) \neq \tau_0(a,b,\mathbf{n})$ *for any point* $(a,b,\mathbf{n})$.

Then there exists h_0 *such that*

$$\lambda_1(x,y) = \lambda_2(x,y), \quad 0 \le y \le h_0, \quad |x| < \alpha.$$

PROOF. Let $\left\{\tilde{\lambda}\right\}$ be the set of all infinitely differentiable functions $\lambda(x,y)$, satisfying the problem conditions and 3. We introduce the operator

$$M\lambda = \int\limits_{\gamma(a,b,\mathbf{n})} \lambda(x,y)\sqrt{dx^2+dy^2} = \tau(a,b,\mathbf{n}).$$

Let $\{\omega\}$ denote the covering of the half-space $y > 0$ by open hemispheres $\omega(\xi,t)$:

$$\omega(\xi,t) = \left\{(x,y),\ y > 0,\ |x-\xi|^2 + y^2 < t^2\right\}, \quad \xi \in E^n, \quad t > 0, \quad |\xi| < \alpha.$$

Let us show that M is quasimonotone with respect to the covering $\{\omega\}$ of the half-space $y > 0$. Indeed, we suppose that $\lambda_1(x,y) > \lambda_2(x,y)$, $(x,y) \in \omega_0 \in \{\omega\}$. In the metrics

$$ds_i^2 = \lambda_i^2(x,y)(dx^2+dy^2),$$

we consider any two extremals $\gamma_i(a,b,\mathbf{n})$ fully contained in the closure $\bar{\omega}_0$. According to Lemma 4.1 and the condition 3, the intersection of each extremal $\gamma_i(a,b,\mathbf{n})$ and the half-space $y > 0$ has a positive measure. Therefore, in view of the inequality $\lambda_1(x,y) > \lambda_2(x,y)$, $(x,y) \in \omega_0$, we have

$$\int\limits_{\gamma_1(a,b,\mathbf{n})} \lambda_1(x,y)\sqrt{dx^2+dy^2} > \int\limits_{\gamma_1(a,b,\mathbf{n})} \lambda_2(x,y)\sqrt{dx^2+dy^2} \ge \int\limits_{\gamma_2(a,b,\mathbf{n})} \lambda_2(x,y)\sqrt{dx^2+dy^2}$$

which implies that the operator M is quasimonotone. Consequently, the inverse kinematic problem in the domain $0 \le y \le h_0$, $|x| < \alpha$, has a unique solution in the class $\{\lambda\}$; here, h_0 is a constant (Anikonov, 1978). The proof is complete.

REMARK 4.2. Instead of the functional

$$\int\limits_{\gamma(a,b,\mathbf{n})} \lambda(x,y)\sqrt{dx^2+dy^2}$$

we can consider the more general one, for example,

$$\int_{\gamma(a,b,\mathbf{n})} \lambda(x,y) M\left(x, y, \frac{dx}{dt}, \frac{dy}{dt}\right) \sqrt{dx^2 + dy^2},$$

where M is a fixed positive function and $\gamma(a,b,\mathbf{n})$ is an extremal of this functional. Under the same assumption, a similar uniqueness theorem is valid.

2. Let us suppose, in addition, that the class of functions which contains the sought-for solution satisfies the condition:

in any hyperplane $y = h$, $h \geq 0$, any two points p^h and q^h are connected by a unique geodesic $\gamma(p^h, q^h)$ in the metric

$$ds^2 = \lambda^2(x,y)\sqrt{dx^2 + dy^2},$$

the geodesic $\gamma(p^h, q^h)$ is in the half-space $y \geq h$; the intersection of $\gamma(p^h, q^h)$ with the open half-space $y > h$ has a strictly positive measure.

For a two-dimensional case, this condition is satisfied if $\lambda(x,y)$ is a strictly decreasing monotone function [with respect to y] and $\Delta \ln \lambda \geq 0$; here Δ is the Laplacian. Recall that $\{\lambda\}_\infty$ is a class of functions $\lambda(x,y)$ piecewise analytic in y and infinitely differentiable with respect to x; namely, for each $\lambda \in \{\lambda\}_\infty$ there exist infinitely differentiable functions a_k^i and a monotone increasing sequence h_i such that

$$\lambda(x,y) = \sum_{k=0}^{\infty} a_k^i (y - h_i)^k, \quad h_i \leq y < h_{i+1}, \quad x \in E^n.$$

Theorem 4.13. *If any points p and q in the hyperplane $y = 0$ satisfy the equality*

$$\int_{\gamma_1(p,q)} \lambda_1 \sqrt{dx^2 + dy^2} = \int_{\gamma_2(p,q)} \lambda_2 \sqrt{dx^2 + dy^2} = \tau(p,q)$$

and $\lambda_i \in \{\lambda\}_\infty$, then $\lambda_1(x,y) = \lambda_2(x,y)$, $x \in E^n$, $y \geq 0$.

PROOF. Let h be an arbitrary positive number, and p^h and q^h be points in the hyperplane $y = h$. Since the geodesics are locally shortest, the operator

$$M_h = \int_{\gamma(p^h,q^h)} \lambda(x,y)\sqrt{dx^2 + dy^2}, \tag{4.32}$$

is quasimonotone (see the proof of Theorem 4.12)

If $\lambda_1(x,y)$ and $\lambda_2(x,y)$ are two solutions and $\lambda_1 \not\equiv \lambda_2$, by the definition of the class $\{\lambda\}_\infty$ there exist two numbers $\tilde{h}_0 \geq 0$ and $\varepsilon > 0$ such that

(i) $\lambda_1(x,y) \equiv \lambda_2(x,y)$, $x \in E^n$, $0 \leq y \leq \tilde{h}_0$;

(ii) $\lambda_i(x,y)$, $i = 1, 2$, are analytic with respect to y in the strip $\tilde{h}_0 \leq y < \tilde{h}_0 + \varepsilon$, $x \in E_n$ and

$$\lambda_1(x,y) \neq \lambda_2(x,y)$$

in this strip.

Let us show that if $\tau_1(p,q) = \tau_2(p,q)$ and the condition (i) holds, then

$$\tau_1(p^{\tilde{h}_0}, q^{\tilde{h}_0}) = \tau_2(p^{\tilde{h}_0}, q^{\tilde{h}_0}), \tag{4.33}$$

where $p^{\tilde{h}_0}$ and $q^{\tilde{h}_0}$ are points in the hyperplane $y = \tilde{h}_0$. Since the operator of (4.32) is quasimonotone, then

$$\lambda_1(x,y) = \lambda_2(x,y), \quad x \in E^n, \quad h_0 \le y < h_0 + \varepsilon$$

which contradicts (ii) and, therefore, allows us to prove the theorem.

We are to prove (4.33). Let α and β be unit vectors tangent to the geodesic $\gamma(p,q)$ in the metric $\mathrm{d}s^2 = \lambda^2(\mathrm{d}x^2 + \mathrm{d}y^2)$ at the respective points p and q in the hyperplane $y = 0$. Let us show that they are determined by the function

$$\tau(p,q) = \int\limits_{\gamma(p,q)} \lambda(x,y)\sqrt{\mathrm{d}x^2 + \mathrm{d}y^2}. \tag{4.34}$$

For definiteness, we integrate in (4.34) from p to q. The familiar equations hold:

$$\lambda_0(p)\alpha = -\left(\frac{\partial\tau}{\partial p_1}, \frac{\partial\tau}{\partial p_2}, \ldots, \frac{\partial\tau}{\partial p_n}, -\sqrt{\lambda_0^2(p) + \sum_{i=1}^{n}\left(\frac{\partial\tau}{\partial p_i}\right)^2}\right), \tag{4.35}$$

$$\lambda_0(q)\beta = \left(\frac{\partial\tau}{\partial q_1}, \frac{\partial\tau}{\partial q_2}, \ldots, \frac{\partial\tau}{\partial q_n}, -\sqrt{\lambda_0^2(q) + \sum_{i=1}^{n}\left(\frac{\partial\tau}{\partial q_i}\right)^2}\right), \tag{4.36}$$

where

$$\lambda_0(x) = \lambda(x,0) = \lim_{p\to q=x}\left[\int\limits_{\gamma(p,q)} \frac{\lambda\sqrt{\mathrm{d}x^2+\mathrm{d}y^2}}{|p-q|}\right].$$

Equations (4.35) and (4.36) yield α and β. By the above condition, the geodesic $\gamma(p,q)$ connects the points p and q in the hyperplane $y = 0$; the vectors tangent to this geodesic at the points p and q are known. Therefore, the system of equations

$$\frac{\partial\lambda}{\partial x_i} = \frac{\mathrm{d}}{\mathrm{d}s}\left(\lambda\frac{\mathrm{d}x_i}{\mathrm{d}s}\right), \quad i = 1,2,\ldots,n, \quad \left|\frac{\mathrm{d}x}{\mathrm{d}s}\right|^2 + \left(\frac{\mathrm{d}y}{\mathrm{d}s}\right)^2 = 1,$$

$$\frac{\partial\lambda}{\partial y} = \frac{\mathrm{d}}{\mathrm{d}s}\left(\lambda\frac{\mathrm{d}y}{\mathrm{d}s}\right)$$

is provided with the Cauchy data at the points p and q. Hence, we conclude: if the function $\lambda(x,y)$ is known in the strip

$$x \in E^n, \quad 0 \le y \le \tilde{h}_0. \tag{4.37}$$

then the parts of $\gamma(p,q)$ belonging to this strip are uniquely determined. According to (i), $\lambda_1(x,y) = \lambda_2(x,y)$; consequently, in view of the above arguments, the parts of $\gamma_1(p,q)$ and $\gamma_2(p,q)$ inside the strip (4.37) are identical. Hence, using the additive property of integration, we obtain (4.33). The theorem is proved.

3. Let us consider a metric

$$ds^2 = \lambda^2(x,y)(dx^2 + dy^2)$$

and a function $\mu(x,y) > 0$ in the half-plane $y \geq 0$. We put

$$w(a,x,y) = \int\limits_{\gamma(a,x,y)} \mu(x,y)\sqrt{dx^2 + dy^2}, \tag{4.38}$$

where $\gamma(a,x,y)$ is a geodesic connecting the points $(a,0)$ and (x,y), $a \in E^1$ $x \in E^1$, $y \geq 0$.

Problem 4.2. Determine two functions $\lambda(x,y)$ and $\mu(x,y)$ in the half-space $y \geq 0$ provided that the functions

$$\begin{aligned} f(a,b) &= w(a,b,0), \\ \varphi(a,b) &= \left.\frac{\partial w}{\partial y}\right|_{y=0,x=b}, \quad -\infty < a < b < \infty \end{aligned}$$

are given. Hereafter, we assume that the class of functions $\{\lambda\}$ is such that the geodesics of metric $ds^2 = \lambda^2(dx^2 + dy^2)$ have the above property.

Theorem 4.14. *If*

1. $f_1(a,b) = f_2(a,b)$, $\varphi_1(a,b) = \varphi_2(a,b)$, $a \in E^1$, $b \in E^1$;
2. $\lambda_i \in \{\lambda\}_\infty$, $\mu_i \in \{\mu\}_\infty$, $\lambda_1(x,0) = \lambda_2(x,0)$

then $\lambda_1 = \lambda_2$, $\mu_1 = \mu_2$, $x \in E^n$, $y \geq 0$.

PROOF. For the points in $\gamma(a,x,y)$, we rewrite equation (4.38) as

$$\frac{dw}{ds} = \frac{\partial w}{\partial x}\frac{dx}{ds} + \frac{\partial w}{\partial y}\frac{dy}{ds} = \mu(x,y), \tag{4.39}$$

where s is the Euclidean length of γ. Note that

$$\lambda_i(x,0)\mu_i(x,0) = \lim_{b \to a = x} \frac{f_i(a,b)}{|b-a|},$$

where the functions $\lambda_i(x,y)$ and $\mu_i(x,y)$ satisfy the condition of this theorem. Consequently, from $\lambda_1(x,0) = \lambda_2(x,0)$ we obtain

$$\mu_1(x,0) = \mu_2(x,0).$$

Let $\lambda(x,y)$ be one of the functions $\lambda_1(x,y)$ and $\lambda_2(x,y)$. For the function

$$\tau(a,x,y) = \int\limits_{\gamma(a,x,y)} \lambda(x,y)\sqrt{dx^2 + dy^2}$$

we have

$$\frac{\partial \tau}{\partial x} = \lambda(x,y)\frac{dx}{ds}, \quad \frac{\partial \tau}{\partial y} = \lambda(x,y)\frac{dy}{ds}.$$

In view of these relations, we get from (4.39)

$$\frac{\partial w}{\partial x}\frac{\partial \tau}{\partial y} + \frac{\partial w}{\partial y}\frac{\partial \tau}{\partial y} = \lambda(x,y)\mu(x,y).$$

We put in this equation $y = 0$ and $x = b$ to obtain

$$\frac{\partial f}{\partial b}\frac{\partial \tau}{\partial b} + \varphi(a,b)\frac{\partial \tau}{\partial y} = \lambda(b,0)\mu(b,0) = g(b).$$

Note that

$$\frac{\partial \tau}{\partial y} = -\sqrt{\lambda^2(b,0) - \left(\frac{\partial \tau}{\partial b}\right)^2}.$$

Consequently,

$$\frac{\partial f}{\partial b}\frac{\partial \tau}{\partial b} - \varphi(a,b)\sqrt{\lambda_0^2(b) - \left(\frac{\partial \tau}{\partial b}\right)^2} = g(b). \tag{4.40}$$

Since f, φ, g, and $\lambda(b,0)$ are known and $\tau(a,a) = 0$, we find the unique function

$$\tau(a,x,y) = \int\limits_{\gamma(a,x,y)} \lambda(x,y)\sqrt{dx^2 + dy^2}.$$

This implies

$$\int\limits_{\gamma_1(a,b)} \lambda_1(x,y)\sqrt{dx^2+dy^2} = \int\limits_{\gamma_2(a,b)} \lambda_2(x,y)\sqrt{dx^2+dy^2}.$$

By Theorem 4.13, we have

$$\lambda_1(x,y) = \lambda_2(x,y), \quad x \in E^1, \quad y \geq 0.$$

Since the $\lambda(x,y)$ is unique, its geodesics $\gamma(a,b)$ are also determined uniquely. Consequently,

$$f(a,b) = \int\limits_{\gamma(a,b)} \mu(x,y)\sqrt{dx^2+dy^2}, \tag{4.41}$$

where the curves $\gamma(a,b)$ are already known. If the curves $\gamma(a,b)$ are given, the determination of $\mu(x,y)$ from $f(a,b)$ is the problem of integral geometry. We check that

$$M\mu = \int\limits_{\gamma(a,b)} \mu(x,y)\sqrt{dx^2+dy^2}$$

is a quasimonotone operator; therefore, by Theorem 4.11:

$$\mu_1(x,y) = \mu_2(x,y), \quad x \in E_1, \quad y \geq 0.$$

The proof is complete.

In Subsection 4.8.3 we prove a stronger theorem for a case of the closed contour of extremal cones.

4.8.3 Uniqueness theorem for an inverse problem

We consider a problem that generalizes kinematic inverse problems of seismology and problems of integral geometry, provided that the integrals are taken along analytic curves. How many solution functions can we determine in problems of this kind? If the data are specified for all pairs of points in a unit circumference, the number of identifiable unknown functions is twice the number of prescribed ones. In the disk $K = \{(x,y) : x^2 + y^2 \leq 1\}$, we consider the set $\{\gamma\}$ of all extremals of the functional

$$\int_\gamma [\lambda(x,y) + a(x,y)\cos\theta + b(x,y)\sin\theta]\sqrt{dx^2 + dy^2}. \tag{4.42}$$

Here,

$$\cos\theta = \frac{dx}{\sqrt{dx^2 + dy^2}}, \quad \sin\theta = \frac{dy}{\sqrt{dx^2 + dy^2}},$$

$\lambda > 0$, a and b are triply continuously differentiable functions. Hereafter we consider only those functionals (4.42) whose extremals have the property: any two points of the disk K can be connected by a unique curve γ of the set $\{\gamma\}$. Note that for $a = 0$ and $b = 0$, the set $\{\gamma\}$ coincides with the set of geodesics in the metric

$$ds^2 = \lambda^2(x,y)(dx^2 + dy^2).$$

Let the parameter z denote the arc of the circumference ∂K:

$$\partial K = \{(x,y) : x^2 + y^2 = 1\}, \quad 0 \leq z \leq 2\pi.$$

We identify any point in ∂K with the corresponding value of z, and consider ∂K to be an oriented curve.

Let $\gamma(z,t)$ denote the curve in $\{\gamma\}$ that connects the points z and t of the circumference ∂K, and let $\theta_0(z,t)$ denote the angle $\theta(x,y,z)|_{(x,y)=(\cos z, \sin z)}$, formed by this geodesic and the x-axis.

Let us examine whether we can uniquely determine $\lambda(x,y)$, $a(x,y)$, $b(x,y)$ (which define the functional) and continuously differentiable functions $\mu(x,y)$, $c(x,y)$, $d(x,y)$, provided that the function $\theta_0(z,t)$ and

$$\int_{\gamma(z,t)} [\mu(x,y) + c(x,y)\cos\theta + d(x,y)\sin\theta]\sqrt{dx^2 + dy^2}$$

are known in the domain $0 \leq z \leq 2\pi$, $0 \leq t \leq 2\pi$. In general, this problem is nonlinear because the (unknown) extremal $\gamma(z,t)$ depends on the sought-for functions $\lambda(x,y)$, $a(x,y)$, $b(x,y)$.

Note that for $\lambda = \mu$ and $a = b = c = d = 0$ this problem reduces to the kinematic inverse problem of seismology, provided that $w_0(z,t)$ is given for all pairs of points (t,z) in the circumference ∂K.

The function $\theta_0(z,t)$ is uniquely determined by $w_0(z,t)$. On the other hand, if the functions $\lambda(x,y)$, $a(x,y)$, $b(x,y)$ are fixed and $c = d = 0$, the above problem reduces to that of integral geometry because the curves $\{\gamma\}$ are the known extremals of (4.42). Under this assumption, we can determine $\theta_0(z,t)$ as well. Other cases, e. g. $\lambda = \mu$, $a = c$,

$b = d$, generalize both the kinematic inverse problem and the inverse problem of integral geometry in the case of integrating along the extremals $\gamma(z,t)$.

Theorem 4.15. *Let the functions $\lambda(x,y)$ $\mu(x,y)$, $a(x,y)$, $b(x,y)$, $c(x,y)$, $d(x,y)$ be given at the circumference ∂K. Then two functions*

$$\theta_0(z,t), \quad w_0(z,t), \quad 0 \le z \le 2\pi, \quad 0 \le t \le 2\pi$$

uniquely determine four functions

$$\lambda(x,y), \quad \mu(x,y), \quad \frac{\partial a}{\partial y} - \frac{\partial b}{\partial x}, \quad \frac{\partial c}{\partial y} - \frac{\partial b}{\partial x}$$

inside the disk K.

PROOF. Let λ, μ, a, b, c, d be a solution to the problem. We consider the functions

$$w(z,x,y) = \int\limits_{\gamma(z,x,y)} [\mu + c\cos\theta + d\sin\theta]\sqrt{\mathrm{d}x^2 + \mathrm{d}y^2}, \tag{4.43}$$

$$v(z,x,y) = \int\limits_{\gamma(z,x,y)} [\lambda + a\cos\theta + b\sin\theta]\sqrt{\mathrm{d}x^2 + \mathrm{d}y^2}, \tag{4.44}$$

Here $\gamma(z,x,y)$ stands for the extremals of (4.42) which connects the points $z \in \partial K$ and $(x,y) \in K$.

We have from (4.43)

$$\frac{\partial w}{\partial x}\cos\theta + \frac{\partial w}{\partial y}\sin\theta = \mu + \cos\theta \cdot c + \sin\theta \cdot d.$$

Let us rewrite this equation as

$$\left(\frac{\partial w}{\partial x} - c\right)\cos\theta + \left(\frac{\partial w}{\partial y} - d\right)\sin\theta = \mu. \tag{4.45}$$

Since $\gamma(z,x,y)$ is the extremal of (4.42), we have

$$\left(\frac{\partial v}{\partial x} - a\right) = \lambda\cos\theta, \quad \left(\frac{\partial v}{\partial y} - b\right) = \lambda\sin\theta. \tag{4.46}$$

This implies, in particular,

$$\left(\frac{\partial v}{\partial x} - a\right)^2 + \left(\frac{\partial v}{\partial y} - b\right)^2 = \lambda^2. \tag{4.47}$$

Let λ_0, μ_0, a_0, b_0, c_0, and d_0 be the respective values of the functions λ, μ, a, b, c, and d on ∂K. By the condition of the theorem, these are the known functions of points in ∂K.

Assuming

$$(x,y) = (\cos\theta, \sin\theta), \quad (x,y) \in \partial K$$

we obtain from (4.45) and (4.46) the boundary conditions for the functions $w(z,x,y)$ and $v(z,x,y)$ at $\partial K \times \partial K$:

$$\left(\frac{\partial w}{\partial x} - c_0\right)\cos\theta_0 + \left(\frac{\partial w}{\partial y} - d_0\right)\sin\theta_0 = \mu_0, \qquad \frac{\partial w}{\partial t} = \frac{\partial w}{\partial x}\cos t + \frac{\partial w}{\partial y}\sin t, \tag{4.48}$$

$$\frac{\partial v}{\partial x} = a_0 + \lambda_0\cos\theta_0, \quad \frac{\partial v}{\partial y} = b_0 + \lambda_0\sin\theta_0. \tag{4.49}$$

Using (4.49), we first prove that the functions

$$\lambda(x,y) \quad \text{and} \quad \frac{\partial a}{\partial y} - \frac{\partial b}{\partial x}$$

are uniquely determined. We suppose that there exist two solutions

$$\lambda_1(x,y), \quad \frac{\partial a_1}{\partial y} - \frac{\partial b_1}{\partial x}$$

and

$$\lambda_2(x,y), \quad \frac{\partial a_2}{\partial y} - \frac{\partial b_2}{\partial x}.$$

We denote

$$\tilde\lambda = \lambda_1 - \lambda_2, \quad \tilde a = a_1 - a_2, \quad \tilde b = b_1 - b_2, \quad \tilde\theta = \theta_1 - \theta_2, \quad \tilde v(z,x,y) = v_1 - v_2,$$

$$v_i = \int\limits_{\gamma_i(z,x,y)} [\lambda_i + a_i\cos\theta_i + b_i\sin\theta_i]\sqrt{dx^2 + dy^2},$$

Using (4.47), we obtain

$$\left(\frac{\partial \tilde v}{\partial x} - \tilde a\right)\left(\frac{\partial v_1}{\partial x} - a_1 + \frac{\partial v_2}{\partial x} - a_2\right) + \left(\frac{\partial \tilde v}{\partial y} - \tilde b\right)\left(\frac{\partial v_1}{\partial y} - b_1 + \frac{\partial v_2}{\partial y} - b_2\right) = \tilde\lambda.$$

We denote the left-hand side of this equation by $L\tilde v$. The function $\tilde\lambda$ is independent of z; therefore,

$$\frac{\partial}{\partial z}L\tilde v = 0.$$

Let us consider the expression

$$\frac{\lambda_1}{\lambda_2}\left[\left(\frac{\partial \tilde v}{\partial x} - a\right)\sin\theta_1 - \left(\frac{\partial v}{\partial y} - \tilde b\right)\cos\theta_1\right]\frac{\partial}{\partial z}L\tilde v$$
$$-L\tilde v\frac{\partial}{\partial z}\frac{\lambda_1}{\lambda_2}\left[\left(\frac{\partial \tilde v}{\partial x} - \tilde a\right)\sin\theta_1 - \left(\frac{\partial \tilde v}{\partial y} - \tilde b\right)\cos\theta_1\right]. \tag{4.50}$$

Since

$$\frac{\partial}{\partial z}L\tilde v = 0$$

the above expression is equal to

$$-\frac{\partial}{\partial z}\left[\frac{\lambda_2}{\lambda_1}\left[\left(\frac{\partial\tilde{v}}{\partial x}-\tilde{a}\right)\sin\theta_1-\left(\frac{\partial\tilde{v}}{\partial y}-\tilde{b}\right)\cos\theta_1\right]L\tilde{v}\right].$$

On the other hand, in view of the equations

$$\frac{\partial v_i}{\partial x}-a_i=\lambda_i\cos\theta_i,\quad \frac{\partial v_i}{\partial y}-b_i=\lambda_i\sin\theta_i,\quad i=1,2.$$

the calculation shows that (4.50) is equal to

$$-\left[\left(\frac{\partial\tilde{v}}{\partial x}-\tilde{a}\right)^2+\left(\frac{\partial\tilde{v}}{\partial y}-\tilde{b}\right)^2\right]\frac{\partial\theta_1}{\partial z}+\frac{\partial}{\partial y}\left(\frac{\partial\tilde{v}}{\partial x}\frac{\partial\tilde{v}}{\partial z}\right)-\frac{\partial}{\partial x}\left(\frac{\partial\tilde{v}}{\partial y}\frac{\partial\tilde{v}}{\partial z}\right)$$
$$-\frac{\partial}{\partial z}\left[\tilde{a}\frac{\partial\tilde{v}}{\partial y}-\tilde{b}\frac{\partial\tilde{v}}{\partial x}\right]+\left[\lambda_1\lambda_2\left(\sin^2\tilde{\theta}-\cos^2\tilde{\theta}\right)+\lambda_2^2\cos\tilde{\theta}\right]\frac{\partial\tilde{\theta}}{\partial z}.$$

We have thus obtained

$$\left[\left(\frac{\partial\tilde{v}}{\partial x}-\tilde{a}\right)^2+\left(\frac{\partial\tilde{v}}{\partial y}-\tilde{b}\right)^2\right]\frac{\partial\theta_1}{\partial z}=\left[\lambda_1\lambda_2\left(\sin^2\tilde{\theta}-\cos^2\tilde{\theta}\right)+\lambda_2^2\cos\tilde{\theta}\right]\frac{\partial\tilde{\theta}}{\partial z}$$
$$-\frac{\partial}{\partial z}\left\{\frac{\lambda_2}{\lambda_1}\left[\left(\frac{\partial\tilde{v}}{\partial x}-\tilde{a}\right)\sin\theta_1-\left(\frac{\partial\tilde{v}}{\partial y}-\tilde{b}\right)\cos\theta_1\right]L\tilde{v}-\tilde{a}\frac{\partial\tilde{v}}{\partial y}-\tilde{b}\frac{\partial\tilde{v}}{\partial x}\right\}\tag{4.51}$$
$$+\frac{\partial}{\partial y}\left(\frac{\partial\tilde{v}}{\partial x}\frac{\partial\tilde{v}}{\partial z}\right)-\frac{\partial}{\partial x}\left(\frac{\tilde{v}}{\partial y}\frac{\partial\tilde{v}}{\partial z}\right).$$

Since

$$v(0,x,y)=\tilde{v}(2\pi,x,y),\quad \theta(0,x,y)=\theta(2\pi,x,y),\quad i=1,2,$$
$$\left.\frac{\partial\tilde{v}}{\partial x}\right|_{\partial K\times\partial K}=0,\quad \left.\frac{\partial\tilde{v}}{\partial y}\right|_{\partial K\times\partial K}=0,$$

then (4.51) and the Stokes formula lead to

$$\iint\left[\left(\frac{\partial\tilde{v}}{\partial x}-a\right)^2+\left(\frac{\partial\tilde{v}}{\partial y}-b\right)^2\right]\frac{\partial\theta_1}{\partial z}\,\mathrm{d}x\,\mathrm{d}y\,\mathrm{d}z=0.\tag{4.52}$$

By this condition, two points of the disk K are connected by a unique extremal γ; therefore, $\frac{\partial\theta_1}{\partial z}\neq 0$. Consequently, (4.52) implies the equations

$$\frac{\partial\tilde{v}}{\partial x}=\tilde{a},\quad \frac{\partial\tilde{v}}{\partial y}=\tilde{b}$$

which, in turn, yield $\frac{\partial\tilde{a}}{\partial y}-\frac{\partial\tilde{b}}{\partial x}=0$. Besides, in view of (4.46), we have

$$\frac{\partial\tilde{v}}{\partial x}-\tilde{a}=\lambda_1\cos\theta_1-\lambda_2\cos\theta_2,\quad \frac{\partial\tilde{v}}{\partial y}-\tilde{b}=\lambda_1\sin\theta_1-\lambda_2\sin\theta_2.$$

Therefore $\lambda_1^2 = \lambda_2^2$, and since both λ_1 and λ_2 are positive, we have $\lambda_1(x,y) = \lambda_2(x,y)$, $(x,y) \in K$. We have demonstrated that the functions

$$\lambda(x,y) \quad \text{and} \quad \frac{\partial a}{\partial y} - \frac{\partial b}{\partial x}$$

are specified uniquely. Let us prove that the functions

$$\mu(x,y) \quad \text{and} \quad \frac{\partial c}{\partial y} - \frac{\partial d}{\partial x}$$

are also specified uniquely.

Let us first show that the curves γ are considered to be known. Indeed, the extremal $\gamma(z,x,y)$ is specified by the functions

$$\lambda(x,y) \quad \text{and} \quad \frac{\partial a}{\partial y} - \frac{\partial b}{\partial x}$$

only; these functions are determined uniquely and, hence, $\gamma(z,x,y)$ is unique. In order to check this we suppose that locally the extremal is described by $y = y(x)$ [or $x = x(y)$]. Then we can rewrite (4.42) as

$$\int\limits_{\gamma} \left[\lambda\sqrt{1+y'^2} + a + by'\right] \mathrm{d}x.$$

The Euler equation for this case is

$$\frac{\partial \lambda}{\partial y}\sqrt{1+y'^2} + \frac{\partial a}{\partial y} + \frac{\partial b}{\partial y}y' = \frac{\mathrm{d}}{\mathrm{d}x}\left(\lambda\frac{y'}{\sqrt{1+y'^2}}\right) + \frac{\partial b}{\partial x} + \frac{\partial b}{\partial y}y'.$$

This yields

$$\frac{\mathrm{d}}{\mathrm{d}x}\left(\lambda\frac{y'}{\sqrt{1+y'^2}}\right) = \frac{\partial \lambda}{\partial y}\sqrt{1+y'^2} + \frac{\partial a}{\partial y} - \frac{\partial b}{\partial x}.$$

which is what we were to prove.

Thus, we see that the problem of determining the functions

$$\mu(x,y) \quad \text{and} \quad \frac{\partial c}{\partial y} - \frac{\partial d}{\partial x}$$

is linear. We denote

$$\tilde{\mu} = \mu_1 - \mu_2, \quad \tilde{c} = c_1 - c_2, \quad \tilde{d} = d_1 - d_2, \quad \tilde{w} = w_1 - w_2$$

and obtain from (4.45) and (4.48)

$$M\tilde{w} = \left(\frac{\partial \tilde{w}}{\partial x} - \tilde{c}\right)\cos\theta + \left(\frac{\partial \tilde{w}}{\partial y} - \tilde{d}\right)\sin\theta = \tilde{\mu}(x,y),$$

$$\left.\frac{\partial \tilde{w}}{\partial x}\right|_{\partial K \times \partial K} = 0, \quad \left.\frac{\partial \tilde{w}}{\partial y}\right|_{\partial K \times \partial K} = 0, \quad \tilde{w}(0,x,y) = \tilde{w}(2\pi,x,y).$$

Since $\mu(x,y)$ is independent of z, then

$$\frac{\partial}{\partial z} M\tilde{w} = 0.$$

Similarly, we consider the expression

$$\left[\left(\frac{\partial \tilde{w}}{\partial x} - \tilde{c}\right)\sin\theta + \left(\frac{\partial \tilde{w}}{\partial y} - \tilde{d}\right)\cos\theta\right]\frac{\partial}{\partial z} M\tilde{w}$$
$$-M\tilde{w}\frac{\partial}{\partial z}\left[\left(\frac{\partial \tilde{w}}{\partial x} - \tilde{c}\right)\sin\theta + \left(\frac{\partial \tilde{w}}{\partial y} - \tilde{d}\right)\cos\theta\right].$$

It turns out to be equal to

$$-\left[\left(\frac{\partial \tilde{w}}{\partial x} - \tilde{c}\right)^2 + \left(\frac{\partial \tilde{w}}{\partial y} - \tilde{d}\right)^2\right]\frac{\partial \theta}{\partial z} + \left(\frac{\partial \tilde{w}}{\partial x} - \tilde{c}\right)\frac{\partial}{\partial z}\left(\frac{\partial \tilde{w}}{\partial y} - \tilde{d}\right)$$
$$- \left(\frac{\partial \tilde{w}}{\partial y} - \tilde{d}\right)\frac{\partial}{\partial z}\left(\frac{\partial \tilde{w}}{\partial x} - \tilde{c}\right).$$

The functions $\tilde{c}(x,y)$ and $\tilde{d}(x,y)$ are independent of z; therefore,

$$\left(\frac{\partial \tilde{w}}{\partial x} - \tilde{c}\right)\frac{\partial}{\partial z}\left(\frac{\partial \tilde{w}}{\partial y} - \tilde{d}\right) - \left(\frac{\partial \tilde{w}}{\partial y} - \tilde{d}\right)\frac{\partial}{\partial z}\left(\frac{\partial \tilde{w}}{\partial x} - \tilde{c}\right)$$
$$= \frac{\partial \tilde{w}}{\partial x}\frac{\partial^2 \tilde{w}}{\partial z \partial y} - \frac{\partial \tilde{w}}{\partial y}\frac{\partial^2 \tilde{w}}{\partial z \partial x} - \frac{\partial}{\partial z}\left[\tilde{c}\frac{\partial \tilde{w}}{\partial y} - \tilde{d}\frac{\partial \tilde{w}}{\partial x}\right]$$
$$= \frac{\partial}{\partial y}\left(\frac{\partial \tilde{w}}{\partial x}\frac{\partial \tilde{w}}{\partial z}\right) - \frac{\partial}{\partial x}\left(\frac{\partial \tilde{w}}{\partial y}\frac{\partial \tilde{w}}{\partial z}\right) - \frac{\partial}{\partial z}\left[\tilde{c}\frac{\partial \tilde{w}}{\partial y} - \tilde{d}\frac{\partial \tilde{w}}{\partial x}\right].$$

As above, we have

$$\frac{\partial \tilde{w}}{\partial x} = \tilde{c}, \quad \frac{\partial \tilde{w}}{\partial y} = \tilde{d}, \tag{4.53}$$

which yields

$$\frac{\partial \tilde{c}}{\partial y} - \frac{\partial \tilde{d}}{\partial x} = 0.$$

We now consider the function

$$\tilde{w}_1(z,x,y) = \int\limits_{\gamma(z,x,y)} \mu(x,y)\sqrt{\mathrm{d}x^2 + \mathrm{d}y^2}.$$

It satisfies the equation

$$\frac{\partial \tilde{w}_1}{\partial x}\cos\theta + \frac{\partial \tilde{w}_1}{\partial y}\sin\theta = \tilde{\mu}(x,y) \tag{4.54}$$

with zero boundary condition at $\partial K \times \partial K$:

$$\left.\frac{\partial \tilde{w}_1}{\partial x}\right|_{\partial K \times \partial K} = 0, \quad \left.\frac{\partial \tilde{w}_1}{\partial y}\right|_{\partial K \times \partial K} = 0.$$

Using the above result (see (4.53)), we therefore conclude

$$\frac{\partial \tilde{w}_1}{\partial x} = 0, \quad \frac{\partial \tilde{w}_1}{\partial y} = 0, \quad (x, y) \in K, \quad 0 \le z \le 2\pi.$$

We substitute these expressions in (4.54) and obtain $\mu(x, y) = 0$, $(x, y) \in K$. The theorem is proved.

REMARK 4.3. When proving solution uniqueness in the problem of integral geometry, we have never used the fact that $\{\gamma\}$ is a set of extremals. The functions $\mu(x, y)$, $\frac{\partial c}{\partial y} - \frac{\partial d}{\partial x}$ are unique, for example, for any set $\{\gamma\}$ of triply continuously differentiable curves with the properties:

1. Any point in the disk K and any direction at the point specify a single curve $\gamma \in \{\gamma\}$.
2. Any two points of K can be connected by a unique curve $\gamma \in \{\gamma\}$.

CHAPTER 5

The analytical representation of solutions of some inverse and ill-posed problems

5.1 ANALYTICAL REPRESENTATION SOLUTIONS OF INVERSE PROBLEMS FOR PARABOLIC AND KINETIC EQUATIONS

In this section, we consider the equation

$$a\frac{\partial w}{\partial t} = \frac{\partial^2 w}{\partial \xi \partial \eta} + \lambda(\xi, t)w. \tag{5.1}$$

Here, a is a complex constant; ξ, η are in general, complex variables belonging to some domains of the complex plane; $|t| < t_0$.

The sought functions in equation (5.1) are two complex-valued functions $w(\xi,\eta,t)$, $\lambda(\xi,t)$, $\dfrac{\partial \lambda}{\partial \eta} = 0$, this fact being the characteristic feature of the inverse problems for differential equations.

Note that if $\xi = \dfrac{x+\mathrm{i}\,y}{2}$, $\eta = \overline{\xi} = \dfrac{x-\mathrm{i}\,y}{2}$, then $\dfrac{\partial^2 w}{\partial \xi \partial \eta} = \dfrac{\partial^2 w}{\partial x^2} + \dfrac{\partial^2 w}{\partial y^2}$.

Therefore, when $a = 1$ and when $a = -\mathrm{i}$, $\mathrm{i}^2 = -1$, the parabolic equation and the Schrödinger equation, respectively, are obtained from equation (5.1):

$$\begin{aligned} \frac{\partial w}{\partial t} &= \frac{\partial^2 w}{\partial x^2} + \frac{\partial^2 w}{\partial y^2} + \lambda(\xi,t)w, \\ -\mathrm{i}\frac{\partial w}{\partial t} &= \frac{\partial^2 w}{\partial x^2} + \frac{\partial^2 w}{\partial y^2} + \lambda(\xi,t)w. \end{aligned}$$

In view of this remark, the nonlinear inverse problem of determination of two complex-valued functions $w(\xi,\eta,t)$, $\lambda(\xi,t)$, in equation (5.1) can be formulated for the parabolic equation and for the Schrödinger equation.

For the required functions $w(\xi,\eta,t)$, $\lambda(\xi,t)$ we obtain in this section the implicit formulas depending on arbitrary continuous functions of one or two variables with compact supports. In the author's opinion these formulas can be useful in investigating various special inverse problems for the differential equations of the aforementioned types, in particular, the problems of control.

Theorem 5.1. *Let $f(p,q)$ and $\varphi(s)$, $p \in \mathrm{R}^1$, $q \in \mathrm{R}^1$, $s \in \mathrm{R}^1$, be continuous complex-valued functions with compact supports, and let $I_0(z) = \sum_{k=0}^{\infty} \frac{\left(\frac{z}{2}\right)^{2k}}{(k!)^2}$ be the modified Bessel function. Then the functions $w(\xi,\eta,t)$ $\lambda(\xi,t)$, defined by the formulas*

$$w(\xi,\eta,t) = \int_{-\infty}^{\infty} \varphi(s) \exp\left[\int_{-\infty}^{\infty}\int_{-\infty}^{\infty} f(p,q) \times \sum_{k=0}^{\infty} \frac{(-\mathrm{i}\, q\xi)^k \left\{ e^{\mathrm{i}\, sp} - \left(1 + \frac{\mathrm{i}\, sp}{1!} + \ldots + \frac{(\mathrm{i}\, sp)^{k-1}}{(k-1)!}\right)\right\} e^{-\frac{\mathrm{i}q}{a}t}}{k! s^k}\, dp\, dq\right] e^{s\eta}\, ds,$$

$$\lambda(\xi,t) = a\frac{\partial}{\partial t} \int_{-\infty}^{\infty}\int_{-\infty}^{\infty} f(p,q) I_0(2\sqrt{\xi pq})\, e^{-\frac{\mathrm{i}q}{a}t}\, dp\, dq,$$

satisfy the equation

$$a\frac{\partial w}{\partial t} = \frac{\partial^2 w}{\partial \xi \partial \eta} + \lambda(\xi,t) w.$$

COROLLARY 5.1. *Under the conditions of the theorem, the functions $w(x,y,t)$, $\lambda(x + \mathrm{i}\, y,t)$ defined by the formulas*

$$w(x,y,t) = \int_{-\infty}^{\infty} \varphi(s) \exp\left[\int_{-\infty}^{\infty}\int_{-\infty}^{\infty} f(p,q) \times \sum_{k=0}^{\infty} \frac{\left(-iq\frac{x+\mathrm{i}\, y}{2}\right)^k \left\{ e^{\mathrm{i}\, sp} - \left(1 + \frac{\mathrm{i}\, sp}{1!} + \ldots + \frac{(\mathrm{i}\, sp)^{k-1}}{(k-1)!}\right)\right\} e^{-\frac{\mathrm{i}q}{a}t}}{k! s^k}\, dp\, dq\right] e^{s\left(\frac{x-\mathrm{i}y}{2}\right)}\, ds,$$

$$\lambda(x + \mathrm{i}\, y,t) = a\frac{\partial}{\partial t} \int_{-\infty}^{\infty}\int_{-\infty}^{\infty} f(p,q) I_0(\sqrt{2(x+\mathrm{i}\, y)pq})\, e^{-(\mathrm{i}\, q/a)t}\, dp\, dq,$$

satisfy the equation

$$\frac{\partial w}{\partial t} = \frac{\partial^2 w}{\partial x^2} + \frac{\partial^2 w}{\partial y^2} + \lambda(x+\mathrm{i}\, y,t) w.$$

In view of the formulas obtained in this section, concrete inverse problems can be reduced to finding two functions $\varphi(s)$ and $f(p,q)$.

In this section, we consider also the inverse problem for a kinetic equation: to find functions $w(x,p,t)$, $\lambda(x,t)$, $\frac{\partial \lambda}{\partial p}$ in the domain $|x - x_0| < a$, $|p| \le b$, $|t - t_0| < c$, such that

$$\frac{\partial w}{\partial t} + \frac{\partial w}{\partial x} p = \lambda(x,t) f(w,p), \tag{5.2}$$

$$w\big|_{x=x_0} = Q(p,t), \tag{5.3}$$

given the differentiable functions $Q(p,t)$, $f(y,p) \not\equiv 0$, $y \in \mathrm{R}$, $|p| \le b$, $|t - t_0| < c$, and constants a, b, c, x_0, t_0.

REMARK 5.1. The function $\lambda(x,t)$ in (5.2) can depend on w, i. e. $\lambda(x,t) = F(w)$, where F is some functional: e. g. $\lambda(x,t) = \tilde{\lambda}(x,t) \int_{-b}^{b} w^2(x,t,p)\,\mathrm{d}p$. Here, $\tilde{\lambda}(x,t)$ is some function independent of p. Such a situation is typical for inverse problems (Anikonov, 1993a, 1994a). Let the function $\Phi(y,p)$ and its inverse function $\Phi^{-1}(y,p)$ in the variable y be defined by the equalities

$$\frac{\mathrm{d}}{\mathrm{d}y}[\Phi^{-1}(y,p)] = \frac{1}{f(y,p)}, \quad \Phi(\Phi^{-1}(y,p),p) = y.$$

Example 5.1. Let $f(y,p) = 1$. In this case, one can take the function $\Phi(y,p) = y$ as the function $\Phi(y,p)$ in Remark 5.1, which corresponds to the problems of integral geometry.

Example 5.2. Let $f(y,p) = y$. In this case, the function $\Phi(y,p)$ in Remark 5.1 can be taken in the form $\Phi(y,p) = \mathrm{e}^y$; then $\Phi^{-1}(y,p) = \ln y$. This case is typical for the problems of tomography.

Example 5.3. Let $f(y,p) = ry\left(1 - \frac{y}{k}\right)$, where $r > 0$, $k > 0$ are constants. As the function $\Phi(y,p)$ in Remark 5.1, one can take the function $\Phi(y,p) = \frac{k}{1+\mathrm{e}^{-ry}}$; then $\Phi^{-1}(y,p) = -\frac{1}{r}\left(\ln\frac{k}{y} - 1\right)$. The example is typical for description of biological and other evolutionary processes under corresponding restriction (Svirezhev, 1987).

Theorem 5.2. *Let all functions in (5.2), (5.3) be analytic. Then the functions $w(x,p,t)$, $\lambda(x,t)$ defined by the relations*

$$\begin{aligned} w(x,p,t) &= \Phi\left(\sum_{m=0}^{\infty}\sum_{k=0}^{\infty}\frac{(-1)^m(x-x_0)^m p^k}{m!(m+k)!}\frac{\partial^m}{\partial t^m}\left[\frac{\partial^{m+k}\Phi^{-1}(Q(p,t),p)}{\partial p^{m+k}}\right]_{p=0}, p\right) \\ &= \Phi\left(\Phi^{-1}(Q(p,t),p) + \sum_{m=1}^{\infty}\frac{(-1)^m(x-x_0)^m}{m!(m-1)!}\int_0^1\left[\frac{\partial^{2m}\Phi^{-1}(Q(\xi,t),\xi)}{\partial t^m\partial\xi^m}\right]_{\xi=py}\mathrm{d}y, p\right), \end{aligned} \tag{5.4}$$

$$\lambda(x,t) = \sum_{k=0}^{\infty}\frac{(-1)^m(x-x_0)^m}{m!^2}\frac{\partial^{m+1}}{\partial t^{m+1}}\left[\frac{\partial^m\Phi^{-1}(Q(p,t),p)}{\partial p^m}\right]_{p=0} \tag{5.5}$$

are a formal solution to the inverse problem (5.2), (5.3).

Theorem 5.2 is proved by direct verification.

Theorem 5.3. *Let given functions $f(y,p)$, $Q(p,t)$ be analytic in the domain $y \in \mathrm{R}$, $|p| < a$, $|t - t_0| < c$ and let the analytic functions $\Phi(y,p)$, $\Phi^{-1}(y,p)$ be defined uniquely; besides, for all $m \ge 0$ and $l \ge 0$*

$$\left|\frac{1}{m!l!}\frac{\partial^m}{\partial t^m}\frac{\partial^l}{\partial p^l}[\Phi^{-1}(Q(p,t),p)]_{p=0}\right| \le M, \tag{5.6}$$

where M is a constant. Then formulas (5.4), (5.5) define the unique analytic solution to the inverse problem (5.2), (5.3) in the domain $|x-x_0| \le a_1 \le a$, $|p| \le b_1 \le b$, $a_1 < 1$, $b_1 < 1$.

The proof of Theorem 5.3 follows from (5.4), (5.5), substituting M instead of the derivatives (5.6), we obtain decreasing geometrical progressions which are summed.

Example 5.4. Let $Q(p,t) = \Phi\left(\dfrac{t-\sqrt{1+p^2-p^2t^2}}{1+p^2}\right)$, where $\left[\Phi^{-1}(y)\right]' = \dfrac{1}{f(y)}$. Then the functions

$$w(x,p,t) = \Phi\left(\frac{t+px-\sqrt{1+p^2-(x-pt)^2}}{1+p^2}\right),$$

$$\lambda(x,t) = 1$$

satisfy the equation

$$\frac{\partial w}{\partial t} + \frac{\partial w}{\partial x}p = \lambda(x,t)f(w)$$

and the boundary condition

$$w\big|_{x=0} = Q(p,t).$$

Theorem 5.4. *Let function $\tilde{Q}(p,t) = \Phi\big(Q(p,t)\big)$ be determined by formula*

$$\tilde{Q}(p,t) = \int_{-\infty}^{\infty}\int_{-\infty}^{\infty} g(\xi,\eta)\, e^{i\xi p + i\eta t}\, d\xi\, d\eta,$$

where the function $g(\xi,\eta)$, $(\xi,\eta) \in R^2$ is a distribution of the expanded series. Then the following formulas for the inverse problem (5.2), (5.3) hold

$$w(\xi,\eta,t) = \Phi\Bigg(\int_{-\infty}^{\infty}\int_{-\infty}^{\infty} g(\xi,\eta) \times \sum_{m=0}^{\infty} \frac{(-i\eta x)^m \left\{e^{ix\xi} - \left[1 + \dfrac{ix\xi}{1!} + \ldots + \dfrac{(ix\xi)^{m-1}}{(m-1)!}\right]\right\} e^{-i\eta t}}{m!\,p^k}\, d\eta\, d\xi\Bigg),$$

$$\lambda(x,t) = \frac{\partial}{\partial t}\int_{-\infty}^{\infty}\int_{-\infty}^{\infty} g(\xi,\eta) I_0(2\sqrt{x\xi\eta})\, e^{i\eta t}\, d\xi\, d\eta,$$

$$I_0(z) = \sum_{k=0}^{\infty}\left(\frac{z}{2}\right)^{2k}\frac{1}{(k!)^2}$$

Let $\alpha(s)$, $\beta(s)$, $s_1 \le s \le s_2$ be a continuous functions and $f(z,s) = u(x,y,s) + i\,v(x,y,s)$, $z = x + iy$ be continuous function analytic for variable $z \in D_1 \subset R^2$. We shall suppose that D and D_1 are domains such that: if $(x,y) \in D$, then $(\alpha(s)x, \beta(s)y) \in D_1$, $s_1 \le s \le s_2$.

Theorem 5.5. *The functions $\lambda(x,y)$, $w(x,y,p,q)$:*

$$w(x,y,p,q) = \exp\int_{s_1}^{s_2}\left[\frac{u(x\alpha(s),y\beta(s),s)p\alpha(s)+v(x\alpha(s),y\beta(s),s)q\beta(s)}{\alpha^2(s)p^2+\beta^2(s)q^2}\right]ds,$$

$$\lambda(x,y) = \frac{1}{2}\int_{s_1}^{s_2}\frac{\partial}{\partial x}u(x\alpha(s),y\beta(s),s)+\frac{\partial}{\partial y}v(x\alpha(s),y\beta(s),s)\,ds$$

satisfy in domain $p^2+q^2\neq 0$, $(x,y)\in D$ the kinetic equation

$$\frac{\partial w}{\partial x}p+\frac{\partial w}{\partial y}q=\lambda(x,y)w(x,y,p,q).$$

REMARK 5.2. If $p=\cos\varphi$, $q=\sin\varphi$ then we have the transport equation

$$\frac{\partial w}{\partial x}\cos\varphi+\frac{\partial w}{\partial y}\sin\varphi=\lambda(x,y)w(x,y,\varphi).$$

5.2 NONLINEAR EQUATIONS AND INVERSE PROBLEMS

Consider the nonlinear equation

$$\frac{\partial w}{\partial t}=\Delta w+\lambda(x,w),\quad x\in D\subset \mathrm{R}^n,\quad t\geq 0,\quad n\geq 1. \tag{5.7}$$

We suppose that the two functions, $w(x,t)$, $\lambda(x,z)$ are unknown in (5.7), $z\in\mathrm{R}^1$.

Theorem 5.6. *Let $F(y)$, $b(y)$, $y\in\mathrm{R}^1$ be arbitrary smooth functions, $F'(0)=1$, $F''(0)=0$, $F(0)=0$ and function $v(x)$ be a solution of the equation $\Delta v=b(-v)$. Then the functions $w(x,t)$, $\lambda(x,t)$:*

$$w(x,t)=F\big(v(x)+\varphi(t)\big),\qquad \text{where}\quad \frac{d\varphi}{dt}=b(\varphi(t)),$$

$$\lambda(x,z)=-F'(F^{-1}(z))\big(\Delta v-b(F^{-1}(z)-v(x))\big)-F''(F^{-1}(z))\,|\mathrm{grad}\,v|^2,\quad \lambda(x,0)=0,$$

$F^{-1}(z)$ is the inverse function for $F(y)$, satisfy the equation (5.7).

The theorem is proved by direct substitution of the functions $w(x,t)$, $\lambda(x,z)$ into the equation (5.7).

In this section, we study nonlinear inverse problems for nonlinear equations. First, we consider the nonlinear equation

$$\frac{\partial w}{\partial t}=\frac{\partial^2 w}{\partial\xi\partial\eta}+\lambda(\xi,\eta,w),$$

where the functions $w(\xi,\eta,w)$, $\lambda(\xi,\eta,w)$, are unknown, $\xi=\xi_1+\mathrm{i}\,\xi_2$, $\eta=\eta_1+\mathrm{i}\,\eta_2$, $z=z_1+\mathrm{i}\,z_2$, $\mathrm{i}^2=-1$. If $\xi=\dfrac{x+\mathrm{i}\,y}{2}$, $\eta=\dfrac{x-\mathrm{i}\,y}{2}$, then the formula

$$\frac{\partial^2 w}{\partial\xi\partial\eta}=\frac{\partial^2 w}{\partial x^2}+\frac{\partial^2 w}{\partial y^2}$$

holds. In this case, we have the nonlinear parabolic equation

$$\frac{\partial w}{\partial t} = \frac{\partial^2 w}{\partial x^2} + \frac{\partial^2 w}{\partial y^2} + \lambda(x, y, w).$$

Let $D \subset \mathrm{R}^n$ be the domain, $\mathbf{L}$ is a linear operator of the second order

$$\mathbf{L} = \sum_{i,j=1}^{n} a_{ij}(x) \frac{\partial^2}{\partial x_i \partial x_j} + \sum_{j=1}^{n} a_j(x) \frac{\partial}{\partial x_j}, \quad x \in D, \quad n \geq 1.$$

Here, coefficients $a_{ij}(x)$, $a_j(x)$ are continuous and complex-valued.

Theorem 5.7. *If the function $v(x)$ is an arbitrary solution of the equation*

$$\mathbf{L}v + kv = 0, \quad x \in D, \quad k \text{ is complex constant}$$

and $F(y)$, $y \in \mathrm{R}^1$ is a twice differentiable or entire function which has inverse function $F^{-1}(y)$. Then the functions $w(x, t)$, $\lambda(x, z)$, $x \in D \subset \mathrm{R}^n$, $a \leq t \leq b$, $z \in \mathrm{R}^1$:

$$w(x,t) = F\left(v(x) + \mathrm{e}^{kt}\right),$$

$$\lambda(x,z) = kF'(F^{-1}(z))F^{-1}(z) - F''(F^{-1}(z)) \sum_{i,j=1}^{n} a_{ij}(x) \frac{\partial v}{\partial x_i} \frac{\partial v}{\partial x_j},$$

$$F'(y) = \frac{\mathrm{d}F}{\mathrm{d}y}, \quad F''(y) = \frac{\mathrm{d}^2 F}{\mathrm{d}y^2}$$

satisfy the nonlinear equation

$$\frac{\partial w}{\partial t} = \mathbf{L}w + \lambda(x, w), \quad x \in D, a \leq t \leq b. \tag{5.8}$$

The proof of this theorem consists in direct substitution of functions $w(x, t)$, $\lambda(x, z)$ in (5.8).

Theorem 5.8. *Let $F(q)$ be an arbitrary analytical function, which has inverse function $F^{-1}(p)$ and $\varphi(z)$, $\psi(z)$ be arbitrary analytical functions, k, c are constants, $k \neq 0$. Then the functions $w(\xi, \eta, t)$, $\lambda(\xi, \eta, p)$:*

$$w(\xi,\eta,t) = F\left(\ln\left[\frac{2}{k}\frac{\varphi'(\xi)\psi'(\eta)}{(\varphi(\xi)+\psi(\eta))^2}(kt+c)\right]\right),$$

$$\lambda(\xi,\eta,p) = F'(F^{-1}(p))\left[\mathrm{e}^{-F^{-1}(p)} - 1\right]\frac{2\varphi'(\xi)\psi'(\eta)}{(\varphi(\xi)+\psi(\eta))^2}$$
$$-F''(F^{-1}(p))\frac{\partial}{\partial \xi}\ln\frac{2}{k}\frac{\varphi'(\xi)\psi'(\eta)}{(\varphi(\xi)+\psi(\eta))^2}\frac{\partial}{\partial \eta}\ln\frac{2}{k}\frac{\varphi'(\xi)\psi'(\eta)}{(\varphi(\xi)+\psi(\eta))^2}$$

satisfy equation

$$\frac{\partial w}{\partial t} = \frac{\partial^2 w}{\partial \xi \partial \eta} + \lambda(x, y, w).$$

PROOF. Let $\ln\left[\frac{2}{k}\frac{\varphi'(\xi)\psi'(\eta)}{(\varphi(\xi)+\psi(\eta))^2}(kt+c)\right] = a(\xi,\eta,t)$ we have

$$\frac{\partial w}{\partial t} = F'(a)\frac{\partial a}{\partial t}, \quad \frac{\partial^2 w}{\partial\xi\partial\eta} = F''(a)\frac{\partial a}{\partial\xi}\frac{\partial a}{\partial\eta} + F'(a)\frac{\partial^2 a}{\partial\xi\partial\eta},$$

$$\frac{\partial a}{\partial t} = \frac{k}{k+c} \quad \text{and} \quad \frac{\partial^2 a}{\partial\xi\partial\eta} = \frac{\partial^2}{\partial\xi\partial\eta}\ln\frac{2}{k}\frac{\varphi'(\xi)\psi'(\eta)}{(\varphi(\xi)+\psi(\eta))^2}.$$

The function $u(\xi,\eta) = \ln\frac{2}{k}\frac{\varphi'(\xi)\psi'(\eta)}{(\varphi(\xi)+\psi(\eta))^2}$ satisfies the Liouville equation $\frac{\partial^2 u}{\partial\xi\partial\eta} = k\,e^u$. Hence, $\frac{\partial^2 u}{\partial\xi\partial\eta} = 2\frac{\varphi'(\xi)\psi'(\eta)}{(\varphi(\xi)+\psi(\eta))^2}$ and the theorem follows from

$$F^{-1}(w) = \ln\frac{2}{k}\frac{\varphi'(\xi)\psi'(\eta)}{(\varphi(\xi)+\psi(\eta))^2}(kt+c),$$

and

$$e^{-F^{-1}}(w) = \frac{k}{2}\frac{(\varphi(\xi)+\psi(\eta))^2}{\varphi'(\xi)\psi'(\eta)(kt+c)}.$$

REMARK 5.3. If ξ and η are real variables, then functions $\varphi(\xi)$ and $\psi(\eta)$ can be only twice differentiable. If we want to solve a concrete inverse problem, we have to find functions F, φ, ψ.

Let R^{n+1} be the real $(n+1)$ Euclidean space of variables (x,t), $x \in R^n$, $t \in R^1$ and

$$D^\alpha = \frac{\partial^{\alpha_1}}{\partial x_1^{\alpha_1}}\frac{\partial^{\alpha_2}}{\partial x_2^{\alpha_2}}\cdots\frac{\partial^{\alpha_n}}{\partial x_n^{\alpha_n}}, \quad \sum_{i=1}^n \alpha_i = m.$$

We consider the next nonlinear differential equation

$$\sum_{|\alpha|=m} a_\alpha D^\alpha \mu(u(x,t)) = \frac{\partial^m u}{\partial t^m}. \tag{5.9}$$

Here a_α are constants, $u(x,t)$, $\mu(y)$, $y \in R^1$ are unknown functions. We put $\mu(u(x,t)) = w(x,t)$, $u(x,t) = \mu^{-1}(w(x,t)) = \lambda(w(x,t))$. Using this remark, we have

$$\sum_{|\alpha|=m} a_\alpha D^\alpha w = \frac{\partial^m \lambda(w)}{\partial t^m}. \tag{5.10}$$

Let $a(y) \in C^\infty(R^1)$, $\varphi_j(y) \in C^\infty(R^1)$ $j = 1,2,\dots,n$ be some functions and allow that $a^{-1}(z)$ exists. We put $\varphi^\alpha(y) = \varphi_1^{\alpha_1}(y)\dots\varphi_n^{\alpha_n}(y)$.

Theorem 5.9. *Let $w(x,t)$, $\lambda(y)$ be determined by relations:*

$$w(x,t) = a\left(t + \sum_{j=1}^n x_j\varphi_j(w(x,t))\right),$$

$$\lambda(y) = \int \sum a_\alpha \varphi^\alpha\left(a^{-1}(y)\right) dy.$$

Then, these functions satisfy equation (5.10) and functions

$$u(x,t) = \lambda\big(w(x,t)\big), \quad \mu = \lambda^{-1}(y)$$

satisfy equation (5.9).

PROOF. Let $z(x,t)$ be a root of equation $z = t + \sum_{j=1}^{n} x_j\varphi_j$. We put $w(x,t) = a\big(z(x,t)\big)$. At first, we show that

$$D^{\alpha}w = \frac{\partial^{m-1}}{\partial t^{m-1}}\left[\varphi^{\alpha}(z)\frac{\partial w}{\partial t}\right]. \tag{5.11}$$

We shall prove (5.11) by induction supposing that

$$D^{\beta}w = \frac{\partial^{p-1}}{\partial t^{p-1}}\left[\varphi^{\beta}(z)\frac{\partial w}{\partial t}\right], \quad |\beta| = p,$$

and we need to show that

$$D^{\beta+1}w = \frac{\partial^{p}}{\partial t^{p}}\left[\varphi^{\beta+1}(z)\frac{\partial w}{\partial t}\right].$$

From equality $z = t + \sum_{j=1}^{n} x_j\varphi_j$, we have relations

$$\frac{\partial z}{\partial t} = \frac{1}{1 - \sum_{j=1}^{n} x_j\varphi_j}, \quad \frac{\partial z}{\partial x_k} = \frac{\varphi_k}{1 - \sum_{j=1}^{n} x_j\varphi_j} = \varphi_k\frac{\partial z}{\partial t},$$
$$\frac{\partial^2 z}{\partial t\partial z_k} = \varphi_k'(z)\left(\frac{\partial z}{\partial t}\right)^2 + \varphi_k(z)\frac{\partial^2 z}{\partial t^2}. \tag{5.12}$$

Using induction we can write

$$D^{\beta+1}w = \frac{\partial}{\partial x_k}D^{\beta}w = \frac{\partial}{\partial x_k}\frac{\partial^{p-1}}{\partial t^{p-1}}\left[\varphi^{\beta}(z)\frac{\partial w}{\partial t}\right] = \frac{\partial^{p-1}}{\partial t^{p-1}}\frac{\partial}{\partial x_k}\left(\varphi^{\beta}(z)\frac{\partial w}{\partial t}\right).$$

We have to show that

$$\frac{\partial}{\partial x_k}\left(\varphi^{\beta}(z)\frac{\partial w}{\partial t}\right) = \frac{\partial}{\partial t}\left(\varphi^{\beta}(z)\varphi_k(z)\frac{\partial w}{\partial t}\right).$$

For this it is sufficient to prove equality

$$\frac{\partial^2 w}{\partial t\partial x_k} = \frac{\partial}{\partial t}\left(\varphi_k(z)\frac{\partial w}{\partial t}\right). \tag{5.13}$$

We have $w = a(z)$ and the equality (5.12). Hence

$$\frac{\partial^2 w}{\partial t\partial x_k} = a''(z)\varphi_k(z)\left(\frac{\partial z}{\partial t}\right)^2 + a'(z)\left[\varphi_k'(z)\left(\frac{\partial z}{\partial t}\right)^2 + \varphi_k(z)\frac{\partial^2 z}{\partial t^2}\right].$$

On the other hand

$$\begin{aligned}\frac{\partial}{\partial t}\left(\varphi_k(z)\frac{\partial w}{\partial t}\right) &= \frac{\partial}{\partial t}\left(\varphi_k(z)a'(z)\frac{\partial w}{\partial t}\right)\\ &= a''(z)\varphi_k(z)\left(\frac{\partial z}{\partial t}\right)^2 + a'(z)\left[\varphi_k'(z)\left(\frac{\partial z}{\partial t}\right)^2 + \varphi_k(z)\frac{\partial^2 z}{\partial t^2}\right] = \frac{\partial^2 w}{\partial t\partial x_k}.\end{aligned}$$

The equality (5.13) is proved.

From equalities $z = t + \sum_{j=1}^{n} x_j\varphi_j$, $w(x,t) = a\big(z(x,t)\big)$, we obtain

$$w(x,t) = a\left(t + \sum_{j=1}^{n} x_j\varphi_j a^{-1}\big(w(x,t)\big)\right).$$

Multiplying (5.11) by a_α and summarizing we have

$$\sum_{|\alpha|=m} a_\alpha D^\alpha w = \frac{\partial^{m-1}}{\partial t^{m-1}}\left[a_\alpha\varphi^\alpha(z)\big(a^{-1}(w)\big)\frac{\partial w}{\partial t}\right] = \frac{\partial^m}{\partial t^m}\lambda(w).$$

The theorem is proved.

If we want to solve concrete inverse problems we should determine functions a, φ_j.

We consider now inverse problems for a nonlinear equation: to find function $\mu(y)$, $y \in \mathrm{R}^1$ if the conditions

1. $\dfrac{\mathrm{d}^2 w}{\mathrm{d}t^2} = k\mu(w)$, $0 \le t \le 1$, $k \ge 0$,
   $w(0,k) = w(1,k) = 0$,
2. $\left.\dfrac{\mathrm{d}w}{\mathrm{d}t}\right|_{t=0} = w_0(k)$, $k \ge 0$,

hold, the smooth function $w_0(x) > 0$ is known and such that function $f(x) = \sqrt{1 - \dfrac{w_0^2(k^2)}{k^2}}$ has an inverse function and $f(0) = 1$.

Theorem 5.10. *Let $a(p)$ be the inverse function for $\sqrt{1 - \dfrac{w_0^2(k^2)}{k^2}}$. Then the following formulas hold*

$$\begin{cases} \mu(y) = \dfrac{1}{2}\lambda^2(y), \\ \lambda^{-1}(p) = \dfrac{1}{\pi}\displaystyle\int_p^1 \frac{qa(q)\,\mathrm{d}q}{\sqrt{a^2 - p^2}}. \end{cases}$$

We can show one interesting example for the parabolic equation

$$\rho(x,t)\frac{\partial w}{\partial t} = \Delta w, \quad x \in D \subset \mathrm{R}^3, \quad t \ge 0.$$

Let $g(y)$, $y \in \mathrm{R}^1$ be an arbitrary smooth function and bounded $|g(y)| \le M$, $\forall y \in \mathrm{R}^1$ and let the function $\varphi(x,t)$ satisfy the equation

$$|\mathrm{grad}_x \varphi|^2 \frac{\partial \varphi}{\partial t} = \Delta \varphi.$$

The following formulas for functions $w(x,t)$ and $\rho(x,t)$ hold

$$\begin{cases} w(x,t) &= \dfrac{1}{\pi} \displaystyle\int_{-\infty}^{\infty} g\big(\varphi(x,t) + 2\sqrt{t}\xi\big)\, \mathrm{e}^{-\xi^2}\, d\xi, \\ \rho(x,t) &= |\mathrm{grad}_x \varphi|^2. \end{cases}$$

We consider an essentially nonlinear inverse problem: given functions $a(x)$ and $b(x)$, find the function $H(z_1, \ldots, z_n)$, $z = (z_1, \ldots, z_n) \in \mathrm{R}^n$ and the function $w(x,t)$, $x \in \mathrm{R}^n$, $0 \le t \le T$, $T > 0$ such that

$$\frac{\partial w}{\partial t} + H\left(\frac{\partial w}{\partial t_1}, \frac{\partial w}{\partial t_2}, \ldots, \frac{\partial w}{\partial t_n}\right) = 0, \tag{5.14}$$

$$w\big|_{t=0} = a(x), \quad w\big|_{t=T} = b(x), \quad x \in \mathrm{R}^n. \tag{5.15}$$

Let $A(x) = \mathrm{grad}\, a$, $B(x) = \mathrm{grad}\, b$. Suppose that there exist mappings $A^{-1}(x)$, $B^{-1}(x)$, $x \in \mathrm{R}^n$, inverse to the mappings $A(x)$ and $B(x)$. It occurs that the following theorem is formally valid.

Theorem 5.11. *The following formulas hold*

$$\begin{aligned} \mathrm{grad}\, H &= \frac{B^{-1}(x) - A^{-1}(x)}{T}, \\ \mathrm{grad}\, w &= A\left(x - t\frac{B^{-1}(\mathrm{grad}\, w) - A^{-1}(\mathrm{grad}\, w)}{T}\right) \end{aligned} \tag{5.16}$$

REMARK 5.4. In the case of quadratic forms: $a = \sum_{i,j=1}^{n} a_{ij}x_i x_j$, $b = \sum_{i,j=1}^{n} b_{ij}x_i x_j$ the mappings $A = \mathrm{grad}\, a$ and $B = \mathrm{grad}\, b$ are linear. If B^{-1} is determined, we get from (5.16)

$$\left[\left(1 - \frac{t}{T}\right) E + \frac{t}{T} AB^{-1}\right] \mathrm{grad}\, w = Ax,$$

where E is a unit matrix. Besides, if the matrix AB^{-1} has no negative eigenvalues, then the following inversion formula obviously holds

$$\mathrm{grad}\, w = \left[\left(1 - \frac{t}{T}\right) E + \frac{t}{T} AB^{-1}\right]^{-1} Ax.$$

Therefore, in this case, the inverse problem (5.14), (5.15) has an explicit solution:

$$\begin{aligned} \mathrm{grad}\, H &= \frac{B^{-1}(z) - A^{-1}(z)}{T}, \\ \mathrm{grad}\, w &= \left[\left(1 - \frac{t}{T}\right) E + \frac{t}{T} AB^{-1}\right]^{-1} Ax. \end{aligned}$$

We give formulas for coefficients and solutions of parabolic and hyperbolic equations.

Theorem 5.12. *Let $v(x) \neq 0$, $x \in D$, $\varphi(t)$, $a \leq t \leq b$ $F(z)$, $z \in \mathrm{R}^1$ be an arbitrary twice differentiable function and the function $F^{-1}(y)$, $y \in \mathrm{R}^1$ is inverse for the $F(z)$. Then the functions $w(x,t)$, $\lambda(x,z)$:*

$$\begin{aligned} w(x,t) &= F\big(v(x)\varphi(t)\big), \\ \lambda(x,z) &= F'(F^{-1}(z))\left[b\left(\frac{F^{-1}(z)}{v(x)}\right)v(x) - \frac{F^{-1}(z)}{v(x)}\Delta v\right] \\ &\quad -F''(F^{-1}(z))\left(\frac{F^{-1}(z)}{v(x)}\right)^2 |\mathrm{grad}\, v|^2, \end{aligned}$$

where $\dfrac{\mathrm{d}\varphi}{\mathrm{d}t} = b\big(\varphi(t)\big)$, satisfy the equation

$$\frac{\partial w}{\partial t} = \Delta w + \lambda(x,w).$$

Proof. We have

$$\frac{\partial w}{\partial t} = F'(v\varphi)v(x)\varphi' = F'v(x)b(\varphi),$$

because $\dfrac{\mathrm{d}\varphi}{\mathrm{d}t} = b\big(\varphi(t)\big)$,

$$\Delta w = F'(v\varphi)\varphi(t)\Delta v + F''(v\varphi)\varphi^2\,|\mathrm{grad}\, v|^2.$$

Substitute $\dfrac{\partial w}{\partial t}$, Δw, w, $\lambda(x,w)$ in the equation

$$\frac{\partial w}{\partial t} = \Delta w + \lambda(x,w);$$

we can see that

$$\begin{aligned} \frac{\partial w}{\partial t} - \Delta w &= F'\big(v(x)\varphi(t)\big)\big[b(\varphi)v(x) - \varphi(t)\Delta v\big] - F''(v\varphi)\,|\mathrm{grad}\, v|^2, \\ \lambda(x,w) &= F'(F^{-1}(z))\left[b\left(\frac{F^{-1}(z)}{v(x)}\right)v(x) - \frac{F^{-1}(z)}{v(x)}\Delta v\right] \\ &\quad -F''(F^{-1}(z))\left(\frac{F^{-1}(z)}{v(x)}\right)^2 |\mathrm{grad}\, v|^2\Bigg|_{z=F(\varphi v)} \\ &= F'\big(v(x)\varphi(t)\big)\big[b(\varphi)v - \varphi\Delta v\big] - F''(v\varphi)\varphi^2\,|\mathrm{grad}\, v|^2, \end{aligned}$$

From here the theorem follows.

Remark 5.5. If $\lambda(x,0) = 0$, then we have the following equation for the function $v(x)$:

$$F''(F^{-1}(0))\left[b\left(\frac{F^{-1}(0)}{v(x)}\right)v(x) - \frac{F^{-1}(0)}{v(x)}\Delta v\right] - F''(F^{-1}(0))\left(\frac{F^{-1}(0)}{v(x)}\right)^2 |\mathrm{grad}\, v|^2 = 0.$$

Theorem 5.13. *Let $f(y)$, $g(y)$, $y \in R^1$ be arbitrary and functions which belong to $C^2(R^1)$, and $v(x)$ be harmonic functions, $x \in D \subset R^n$. Then the functions $w(x,t)$, $\rho(x)$:*

$$\begin{aligned} w(x,t) &= \int_0^1 g\left(\frac{2}{m+2} v^{\frac{m+2}{2}}(x)(2p-1)+t\right) p^{\beta-1}(1-p)^{\beta-1}\,dp \\ &+ v(x)\int_0^1 g\left(\frac{2}{m+2} v^{\frac{m+2}{2}}(x)(2p-1)-t\right) p^{-\beta}(1-p)^{-\beta}\,dp, \end{aligned}$$

$$\rho(x) = |\operatorname{grad} v|^2 v^m, \quad m>0, \quad \beta = \frac{m}{2(m+2)},$$

satisfy equation

$$\rho(x)\frac{\partial^2 w}{\partial t^2} = \Delta w, \quad a \le t \le b, \quad x \in D \subset R^n.$$

This theorem is proved by direct verification.

Let $F(y)$, $y \in R^1$, $v(x)$, $x \in D \subset R^n$, $\varphi(t)$, $a \le t \le b$ be arbitrary twice differentiable functions.

The function $b(\varphi)$, $\varphi \in R^1$ will be determine by the formula: $\dfrac{d\varphi}{dt} = b(\varphi(t))$.

Theorem 5.14. *The functions $w(x,t)$, $\lambda(x,z)$, $x \in D \subset R^n$, $a \le t \le b$, $z \in R^1$ determined by formulas*

$$\begin{aligned} w(x,t) &= F(v(x)+\varphi(t)), \\ \lambda(x,z) &= F'(F^{-1}(z))\left[b'(F^{-1}(z)-v(x))b(F^{-1}(z)-v(x)) - \Delta v\right] \\ &\quad -F''(F^{-1}(z))\left[|\operatorname{grad} v|^2 - b^2(F^{-1}(z)-v(x))\right] \end{aligned}$$

satisfy the equation

$$\frac{\partial^2 w}{\partial t^2} = \Delta w + \lambda(x,w).$$

REMARK 5.6. Let $z = z(x)$ be a function such that $\lambda(x,z(x)) = 0$, $x \in D \subset R^n$. If we know the function $z(x)$, then using the formula for $\lambda(x,t)$ in the theorem we can write a nonlinear equation on function $v(x)$, $x \in D$:

$$\begin{aligned} &F'(F^{-1}(z(x)))\left[b'(F^{-1}(z(x))-v(x))b(F^{-1}(z(x))-v(x)) - \Delta v\right] \\ &-F''(F^{-1}(z(x)))\left[|\operatorname{grad} v|^2 - b^2(F^{-1}(z(x))-v(x))\right] = 0. \end{aligned}$$

Now we consider systems of a nonlinear equation.

Let F, v, b be twice differentiable mapping and $z = F(y)$, $y \in R^m$, $z \in R^m$ and $y = F^{-1}$, $x \in D \subset R^n$, where D is a domain, $z = b(y)$, $y \in R^m$, $z \in R^m$, $m \ge 1$, $n \ge 1$

and $\mathbf{L} = \sum_{i,j=1}^{n} a_{ij} \frac{\partial^2}{\partial x_i \partial x_j} + \sum_{j=1}^{n} a_j \frac{\partial}{\partial x_j}$ be a differential operator with continuous coefficients a_{ij}, a_j, $x \in D$. Let vector function $\varphi(t)$ be the solution of the system:

$$\frac{\mathrm{d}\varphi}{\mathrm{d}t} = b(\varphi) = b(\varphi_1, \varphi_2, \ldots, \varphi_m)$$

Theorem 5.15. *The functions $w(x,t)$, $\lambda(x,z)$, $x \in D \subset \mathrm{R}^n$, $z \in \mathrm{R}^m$, $\alpha < t < \beta$ which are determined by formulas:*

$$\begin{aligned} w(x,t) &= F\big(\varphi(t) + v(x)\big), \quad \frac{\mathrm{d}\varphi}{\mathrm{d}t} = b(\varphi) \\ \lambda(x,z) &= \sum_{k=1}^{m} \frac{\partial F\big(F^{-1}(z)\big)}{\partial y_k} \big[b_k\big(F^{-1}(z) - v(x)\big) - \mathbf{L}v_k\big] \\ &- \sum_{k,l=1}^{n} \sum_{i,j=1}^{n} a_{ij}(x) \frac{\partial^2 F\big(F^{-1}(z)\big)}{\partial y_k \partial y_l} \frac{\partial v_k}{\partial x_j} \frac{\partial v_l}{\partial x_i} \end{aligned}$$

satisfy a nonlinear system of the parabolic type equation

$$\frac{\partial w}{\partial t} = \mathbf{L}w + \lambda(x,w).$$

Theorem 5.16. *The functions $w(x,t)$, $\lambda(x,z)$, $x \in D \subset \mathrm{R}^n$, $z \in \mathrm{R}^m$, $\alpha < t < \beta$ which are determined by formulas:*

$$\begin{aligned} w(x,t) &= F\big(\varphi(t) + v(x)\big), \quad \frac{\mathrm{d}\varphi}{\mathrm{d}t} = b(\varphi) \\ \lambda(x,z) &= \sum_{k=1}^{m} \frac{\partial F\big(F^{-1}(z)\big)}{\partial y_k} \left[\frac{\partial b_k\big(F^{-1}(z) - v(x)\big)}{\partial \varphi_l} b_k\big(F^{-1}(z) - v(x)\big) - \mathbf{L}v_k\right] \\ &- \sum_{k,l=1}^{n} \sum_{i,j=1}^{n} a_{ij}(x) \frac{\partial^2 F\big(F^{-1}(z)\big)}{\partial y_k \partial y_l} \frac{\partial v_k}{\partial x_j} \frac{\partial v_l}{\partial x_i} \end{aligned}$$

satisfy a nonlinear system of the wave type equation

$$\frac{\partial^2 w}{\partial t^2} = \mathbf{L}w + \lambda(x,w).$$

5.3 FORMULAS FOR SOLUTION AND COEFFICIENTS OF DIFFERENTIAL EQUATIONS OF THE SECOND ORDER

Let $\mathbf{L} = \sum_{i,j=1}^{n} a_{ij}(x,t) \frac{\partial^2}{\partial x_i \partial x_j} + \sum_{j=1}^{n} a_j(x,t) \frac{\partial}{\partial x_j}$ be a differential operator of the second order with continuous coefficients $a_{ij}(x,t)$, $a_j(x,t)$, $x \in D \subset \mathrm{R}^n$, $\alpha \le t \le \beta$; D is a domain in Euclidean space R^n, $n \ge 1$; α, β are constants.

Lemma 5.1. *Let complex-valued twice differentiable functions $\varphi(x,t)$, $\tau(x,t)$, $\varphi \neq 0$, $\tau_t = \dfrac{\partial \tau}{\partial t} \neq 0$, $x \in D$, $\alpha < t < \beta$ satisfy the differential relation*

$$\left[\left(\frac{\sum_{i,j=1}^{n} a_{ij}\frac{\partial \tau}{\partial x_i}\frac{\partial \tau}{\partial x_j}}{\tau_t^2}\right)\frac{\partial^2 \tau}{\partial t^2} - \mathbf{L}\tau\right]\varphi + \left(\frac{2\sum_{i,j=1}^{n} a_{ij}\frac{\partial \tau}{\partial x_i}\frac{\partial \tau}{\partial x_j}}{\tau_t^2}\right)\frac{\partial \varphi}{\partial t}\frac{\partial \tau}{\partial t} = 2\sum_{i,j=1}^{n} a_{ij}\frac{\partial \varphi}{\partial x_i}\frac{\partial \tau}{\partial x_j} \tag{5.17}$$

and let function $f(y)$, $y \in \mathrm{R}^1$ be arbitrary twice differential or analytical. Then the function $w(x,t) = \varphi(x,t) f(\tau(x,t))$ satisfies equation

$$\left(\frac{\sum_{i,j=1}^{n} a_{ij}\frac{\partial \tau}{\partial x_i}\frac{\partial \tau}{\partial x_j}}{\tau_t^2}\right)\frac{\partial^2 w}{\partial t^2} = \mathbf{L}w + \left[\left(\frac{\sum_{i,j=1}^{n} a_{ij}\frac{\partial \tau}{\partial x_i}\frac{\partial \tau}{\partial x_j}}{\tau_t^2}\right)\frac{\partial^2 \varphi}{\partial t^2} - \mathbf{L}\varphi\right]\frac{w}{\varphi}. \tag{5.18}$$

Lemma 5.2. *Let complex-valued twice differentiable functions $\varphi(x,t)$, $\tau(x,t)$, $x \in D$, $\alpha \leq t < \beta$, $\varphi \neq 0$ satisfy the differential relation:*

$$\left[\left(\sum_{i,j=1}^{n} a_{ij}\frac{\partial \tau}{\partial x_i}\frac{\partial \tau}{\partial x_j}\right)\frac{\partial \tau}{\partial t}\right]\varphi = 2\sum_{i,j=1}^{n} a_{ij}\frac{\partial \varphi}{\partial x_i}\frac{\partial \tau}{\partial x_j} \tag{5.19}$$

and the function $F(t,y)$, $\alpha < t < \beta$, $y \in \mathrm{R}^1$ be arbitrary twice differentiable or analytic for variable solution of the equation

$$a\frac{\partial F}{\partial t} = \frac{\partial^2 F}{\partial y^2}, \qquad a = a_1 + \mathrm{i}\, a_2, \quad \mathrm{i}^2 = -1.$$

Then the function $w(x,t) = \varphi(x,t) F(t, \tau(x,t))$ satisfies equation

$$a\left(\sum_{i,j=1}^{n} a_{ij}\frac{\partial \tau}{\partial x_i}\frac{\partial \tau}{\partial x_j}\right)\frac{\partial w}{\partial t} = \mathbf{L}w + \left[a\left(\sum_{i,j=1}^{n} a_{ij}\frac{\partial \tau}{\partial x_i}\frac{\partial \tau}{\partial x_j}\right)\frac{\partial \varphi}{\partial t} - \mathbf{L}\varphi\right]\frac{w}{\varphi} \tag{5.20}$$

Proof of Lemmas 5.1, 5.2.

Let conditions of Lemma 5.1 hold. At first, we remark that if the function $\tau(x,t)$ is essentially complex-valued, then we suggest the function $f(y)$, $y \in \mathrm{R}^1$ is analytic. Differentiate twice with respect to t the function $w(x,t) = \varphi(x,t) f(\tau(x,t))$; this gives

$$\frac{\partial^2 w}{\partial t^2} = \frac{\partial^2 \varphi}{\partial t^2} f + \left(\frac{\partial \varphi}{\partial t}\frac{\partial \tau}{\partial t} + \varphi\frac{\partial^2 \tau}{\partial t^2}\right) f' + \varphi\left(\frac{\partial \tau}{\partial t}\right)^2 f''$$

and similarly we obtain equalities

$$\begin{aligned}\frac{\partial w}{\partial x_i} &= \frac{\partial \varphi}{\partial x_i} f + \varphi \frac{\partial \tau}{\partial x_i} f', \\ \frac{\partial^2 w}{\partial x_i \partial x_j} &= \frac{\partial^2 \varphi}{\partial x_i \partial x_j} f + \left(\frac{\partial \varphi}{\partial x_i}\frac{\partial \tau}{\partial x_j} + \frac{\partial \varphi}{\partial x_j}\frac{\partial \tau}{\partial x_i} + \varphi \frac{\partial^2 w}{\partial x_i \partial x_j} \right) f' + \varphi \frac{\partial \tau}{\partial x_i}\frac{\partial \tau}{\partial x_j} f''.\end{aligned}$$

So we have

$$\mathbf{L}w = \mathbf{L}\varphi f + \left[\varphi \mathbf{L}\tau + 2\sum_{i,j=1}^{n} a_{ij} \frac{\partial \varphi}{\partial x_i}\frac{\partial \tau}{\partial x_j} \right] f' + \left(\sum_{i,j=1}^{n} a_{ij} \frac{\partial \tau}{\partial x_i}\frac{\partial \tau}{\partial x_j} \right) \varphi f'',$$

$$\begin{aligned}\mathbf{L}w + \left[\frac{\sum_{i,j=1}^{n} a_{ij} \frac{\partial \tau}{\partial x_i}\frac{\partial \tau}{\partial x_j}}{\tau_t^2} \frac{\partial^2 \varphi}{\partial t^2} - \mathbf{L}\varphi \right] \frac{w}{\varphi} &= \left(\frac{\sum_{i,j=1}^{n} a_{ij} \frac{\partial \tau}{\partial x_i}\frac{\partial \tau}{\partial x_j}}{\tau_t^2} \right) \frac{\partial^2 \varphi}{\partial t^2} f \\ &+ \left[\varphi \mathbf{L}\tau + 2\sum_{i,j=1}^{n} a_{ij} \frac{\partial \varphi}{\partial x_i}\frac{\partial \tau}{\partial x_j} \right] f' + \left(\sum_{i,j=1}^{n} a_{ij} \frac{\partial \tau}{\partial x_i}\frac{\partial \tau}{\partial x_j} \right) \varphi f'',\end{aligned}$$

The functions $\tau(x,t)$ and $\varphi(x,t)$ satisfy equation (5.17). Hence, the following equality holds

$$\begin{aligned}&\left(\frac{\sum_{i,j=1}^{n} a_{ij} \frac{\partial \tau}{\partial x_i}\frac{\partial \tau}{\partial x_j}}{\tau_t^2} \right) \frac{\partial^2 w}{\partial t^2} - \mathbf{L}w - \left[\left(\frac{\sum_{i,j=1}^{n} a_{ij} \frac{\partial \tau}{\partial x_i}\frac{\partial \tau}{\partial x_j}}{\tau_t^2} \right) \frac{\partial^2 \varphi}{\partial t^2} - \mathbf{L}\varphi \right] \frac{w}{\varphi} \\ &= \left\{ \left[\left(\frac{\sum_{i,j=1}^{n} a_{ij} \frac{\partial \tau}{\partial x_i}\frac{\partial \tau}{\partial x_j}}{\tau_t^2} \right) \frac{\partial^2 \tau}{\partial t^2} - \mathbf{L}\tau \right] \varphi + \left(\frac{2\sum_{i,j=1}^{n} a_{ij} \frac{\partial \tau}{\partial x_i}\frac{\partial \tau}{\partial x_j}}{\tau_t^2} \right) \frac{\partial \varphi}{\partial t}\frac{\partial \tau}{\partial t} \right. \\ &\left. - \sum_{i,j=1}^{n} a_{ij} \left(\frac{\partial \varphi}{\partial x_i}\frac{\partial \tau}{\partial x_j} + \frac{\partial \varphi}{\partial x_j}\frac{\partial \tau}{\partial x_i} \right) \right\} f' = 0\end{aligned}$$

Lemma 5.1 is proved.

The proof of Lemma 5.2 is like that for Lemma 5.1

Lemma 5.3. *Let conditions of Lemmas 5.1 and 5.2 hold. If* $\tau(x,t) = \dfrac{\psi(x,t)}{\varphi(x,t)}$, $\varphi \neq 0$, *then relations (5.17), (5.19) can be represented as follows:*

$$\left[\left(\frac{\sum_{i,j=1}^{n} a_{ij} \frac{\partial}{\partial x_i}\frac{\psi}{\varphi}\frac{\partial}{\partial x_j}\frac{\psi}{\varphi})}{\left(\frac{\partial}{\partial t}\frac{\psi}{\varphi}\right)^2} \right) \frac{\partial^2 \psi}{\partial t^2} - \mathbf{L}\psi \right] \varphi - \left[\left(\frac{\sum_{i,j=1}^{n} a_{ij} \frac{\partial}{\partial x_i}\frac{\psi}{\varphi}\frac{\partial}{\partial x_j}\frac{\psi}{\varphi}}{\left(\frac{\partial}{\partial t}\frac{\psi}{\varphi}\right)^2} \right) \frac{\partial^2 \varphi}{\partial t^2} - \mathbf{L}\varphi \right] \psi = 0, \tag{5.21}$$

$$\left[\left(a\sum_{i,j=1}^{n} a_{ij}\frac{\partial}{\partial x_i}\frac{\psi}{\varphi}\frac{\partial}{\partial x_j}\frac{\psi}{\varphi}\right)\frac{\partial\psi}{\partial t}-\mathbf{L}\psi\right]\varphi-\left[\left(a\sum_{i,j=1}^{n} a_{ij}\frac{\partial}{\partial x_i}\frac{\psi}{\varphi}\frac{\partial}{\partial x_j}\frac{\psi}{\varphi}\right)\frac{\partial\varphi}{\partial t}-\mathbf{L}\varphi\right]\psi=0. \tag{5.22}$$

The proof is derived from Lemmas 5.1 and 5.2.

Lemma 5.4. *Let the function $\lambda(x,t)$ be known and the functions $\psi(x,t)$, $\varphi(x,t)\neq 0$ be solutions of a system of equations*

$$\left(\frac{\sum_{i,j=1}^{n} a_{ij}\frac{\partial}{\partial x_i}\left(\frac{\psi}{\varphi}\right)\frac{\partial}{\partial x_j}\left(\frac{\psi}{\varphi}\right)}{\left(\frac{\partial}{\partial t}\left(\frac{\psi}{\varphi}\right)\right)^2}\right)\frac{\partial^2\psi}{\partial t^2}=\mathbf{L}\psi+\lambda\psi,$$

$$\left(\frac{\sum_{i,j=1}^{n} a_{ij}\frac{\partial}{\partial x_i}\left(\frac{\psi}{\varphi}\right)\frac{\partial}{\partial x_j}\left(\frac{\psi}{\varphi}\right)}{\left(\frac{\partial}{\partial t}\left(\frac{\psi}{\varphi}\right)\right)^2}\right)\frac{\partial^2\varphi}{\partial t^2}=\mathbf{L}\varphi+\lambda\varphi.$$

Then the functions

$$\rho(x,t)=\frac{\sum_{i,j=1}^{n} a_{ij}\frac{\partial}{\partial x_i}\left(\frac{\psi}{\varphi}\right)\frac{\partial}{\partial x_j}\left(\frac{\psi}{\varphi}\right)}{\left(\frac{\partial}{\partial t}\left(\frac{\psi}{\varphi}\right)\right)^2},\quad w(x,t)=\varphi(x,t)f\left(\frac{\psi(x,t)}{\varphi(x,t)}\right),\quad f(y)\in C^2(R^1)$$

satisfy the equation

$$\rho(x,t)\frac{\partial^2 w}{\partial t^2}=\mathbf{L}w+\lambda w.$$

Lemma 5.5. *Let the function $\lambda(x,t)$ be known and the functions $\psi(x,t)$, $\varphi(x,t)\neq 0$ be solutions of a system of equations*

$$\left(\frac{\sum_{i,j=1}^{n} a_{ij}\frac{\partial}{\partial x_i}\left(\frac{\psi}{\varphi}\right)\frac{\partial}{\partial x_j}\left(\frac{\psi}{\varphi}\right)}{\left(\frac{\partial}{\partial t}\left(\frac{\psi}{\varphi}\right)\right)^2}\right)\frac{\partial\psi}{\partial t}=\mathbf{L}\psi+\lambda\psi,$$

$$\left(\frac{\sum_{i,j=1}^{n} a_{ij}\frac{\partial}{\partial x_i}\left(\frac{\psi}{\varphi}\right)\frac{\partial}{\partial x_j}\left(\frac{\psi}{\varphi}\right)}{\left(\frac{\partial}{\partial t}\left(\frac{\psi}{\varphi}\right)\right)^2}\right)\frac{\partial\varphi}{\partial t}=\mathbf{L}\varphi+\lambda\varphi.$$

Then the functions

$$\rho(x,t)=\sum_{i,j=1}^{n} a_{ij}\frac{\partial}{\partial x_i}\left(\frac{\psi}{\varphi}\right)\frac{\partial}{\partial x_j}\left(\frac{\psi}{\varphi}\right),\quad w(x,t)=\varphi(x,t)F\left(t,\frac{\psi(x,t)}{\varphi(x,t)}\right),$$

where $F(t,y)$ is the solution of the equation $a\dfrac{\partial F}{\partial t} = \dfrac{\partial^2 F}{\partial y^2}$, $a = a_1 + \mathrm{i}\,a_2$, *satisfy the equation*

$$\rho(x,t)\frac{\partial w}{\partial t} = \mathbf{L}w + \lambda w.$$

Lemma 5.6. *Let* $\mathbf{L} = \Delta = \sum_{j=1}^{n} \dfrac{\partial^2}{\partial x_j^2}$, *then relation (5.17) holds if* $\varphi = 1$ *and the function* $\tau(x,t)$ *is determined by one of the next equalities*

1. $\tau(x,t) = \tau_0\left(x_1 + \mathrm{i}\,x_3\sqrt{1+c^2} + \dfrac{\mathrm{i}\,ct}{a}, x_2 + cx_3 + \dfrac{t\sqrt{1+c^2}}{a}\right)$, $\tau_0(y,z)$ *is analytic functions;*

2. *the function* $\tau(x,t)$ *is a root of the equation*

$$\sum_{j=1}^{n} a_j(\tau) + b(\tau)t + d(\tau) = 0,$$

$$\sum_{j=1}^{n} a_j^2(z) = b^2(z)a^2,$$

$a_j(z)$, $b(z)$, $d(z)$ *are arbitrary twice differentiable or analytic functions.*

The proof of Lemma 5.6 can be made on the basis of Lemma 5.1 and with results which are contained in Anikonov and Vishnevskii (1994).

Now we consider a differential equation with coefficients which do not depend on variable t:

$$\rho(x)\frac{\partial^2 w}{\partial t^2} = \sum_{i,j=1}^{n} a_{ij}\frac{\partial^2 w}{\partial x_i \partial x_j} + \sum_{j=1}^{n} a_j \frac{\partial w}{\partial x_j} + \lambda w = \mathbf{L}w + \lambda(x)w.$$

Theorem 5.17. *Let* $f(y)$, $g(y)$, $y \in \mathrm{R}^1$ *be arbitrary twice differentiable functions, the function* $\tau(x)$ *be a solution of the equation*

$$\sum_{i,j=1}^{n} a_{ij}\frac{\partial \tau}{\partial x_i}\frac{\partial \tau}{\partial x_j} = \rho(x)$$

and the function $\varphi(x) \neq 0$ *is a solution of the transport equation:*

$$\varphi\left[\sum_{i=1}^{n} a_i \frac{\partial \tau}{\partial x_i} + \sum_{i,j=1}^{n} a_{ij}\frac{\partial^2 \tau}{\partial x_i \partial x_j}\right] + 2\sum_{i,j=1}^{n} a_{ij}\left(\frac{\partial \varphi}{\partial x_i}\frac{\partial \tau}{\partial x_j}\right) = 0.$$

Then functions $w(x,t)$, $\lambda(x)$, $\dfrac{\partial \lambda}{\partial t} = 0$ *which are determined by formulas*

$$w(x,t) = \varphi(x)\big[f(\tau(x)+t) + g(\tau(x)-t)\big], \quad \lambda(x) = -\frac{\mathbf{L}\varphi}{\varphi}$$

satisfy the equation

$$\rho(x)\frac{\partial^2 w}{\partial t^2} = \mathbf{L}w + \lambda(x)w.$$

Theorem 5.18. *Let $F(t,y)$, $y \in \mathrm{R}^1$, $\alpha < t < \beta$ be an arbitrary solution of the equation*

$$a\frac{\partial F}{\partial t} = \frac{\partial^2 F}{\partial y^2}, \quad a = a_1 + \mathrm{i}\, a_2$$

and the functions $\tau(x)$, $\varphi(x) \neq 0$ satisfy the equation of the ray method:

$$\sum_{i,j=1}^{n} a_{ij}\frac{\partial \tau}{\partial x_i}\frac{\partial \tau}{\partial x_j} = \rho(x),$$

$$\varphi\left[\sum_{i=1}^{n} a_i\frac{\partial \tau}{\partial x_i} + \sum_{i,j=1}^{n} a_{ij}\frac{\partial^2 \tau}{\partial x_i \partial x_j}\right] + 2\sum_{i,j=1}^{n} a_{ij}\left(\frac{\partial \varphi}{\partial x_i}\frac{\partial \tau}{\partial x_j}\right) = 0.$$

Then functions $w(x,t)$, $\lambda(x)$ which are determined by formulas

$$w(x,t) = \varphi(x)F(t,\tau(x)), \quad \lambda(x) = -\frac{\mathbf{L}\varphi}{\varphi},$$

satisfy the equation

$$a\rho(x)\frac{\partial w}{\partial t} = \mathbf{L}w + \lambda(x)w.$$

Theorem 5.19. *Let $f(y)$, $g(y)$, $y \in \mathrm{R}^1$ be arbitrary twice differentiable functions and the functions $\psi(x)$, $\varphi(x)$, $x \in D \subset \mathrm{R}^n$, $\varphi \neq 0$ be solutions of the equations*

$$\mathbf{L}\psi + \lambda(x)\psi = 0, \quad \mathbf{L}\varphi + \lambda(x)\varphi = 0$$

Then functions

$$\begin{aligned} w(x,t) &= \varphi(x)\left[f\left(\frac{\psi(x)}{\varphi(x)} + t\right) + g\left(\frac{\psi(x)}{\varphi(x)} - t\right)\right], \\ \rho(x) &= \sum_{i,j=1}^{n} a_{ij}\frac{\partial}{\partial x_i}\left(\frac{\psi}{\varphi}\right)\frac{\partial}{\partial x_j}\left(\frac{\psi}{\varphi}\right) \end{aligned}$$

satisfy the equation

$$\rho(x)\frac{\partial^2 w}{\partial t^2} = \mathbf{L}w + \lambda(x)w.$$

Theorem 5.20. *Let functions $\psi(x)$, $\varphi(x) \neq 0$ satisfy equations*

$$\mathbf{L}\psi + \lambda(x)\psi = 0, \quad \mathbf{L}\varphi + \lambda(x)\varphi = 0$$

and the function $F(t,y)$ be the solution of the equation

$$a\frac{\partial F}{\partial t} = \frac{\partial^2 F}{\partial y^2}, \quad a = a_1 + \mathrm{i}\, a_2.$$

Then the functions

$$\begin{aligned} w(x,t) &= \varphi(x) F\left(t, \frac{\psi(x)}{\varphi(x)}\right), \\ \rho(x) &= \sum_{i,j=1}^{n} a_{ij} \frac{\partial}{\partial x_i}\left(\frac{\psi}{\varphi}\right) \frac{\partial}{\partial x_j}\left(\frac{\psi}{\varphi}\right) \end{aligned}$$

satisfy the equation

$$a\rho(x)\frac{\partial w}{\partial t} = \mathbf{L}w + \lambda(x) w$$

The proofs of Theorems 5.17–5.20 follow from Lemmas 5.4, 5.5.

5.3.1 The differential inequality in inverse problems

Let functions $\psi(x)$, $\varphi(x) > 0$ be arbitrary solutions of the equation

$$\mathbf{L}\psi = \mathbf{L}\varphi = 0, \text{ where } \mathbf{L} = \sum_{i,j=1}^{n} a_{ij} \frac{\partial^2}{\partial x_i \partial x_j} + \sum_{i=1}^{n} a_i(x) \frac{\partial}{\partial x_i}, \quad x \in D \subset \mathrm{R}^n$$

and D be a domain in R^n; $a_{ij}(x)$, $a_i(x)$ are continuous functions such that

$$\sum_{i,j=1}^{n} a_{ij}\xi_i\xi_j > 0, \quad \forall \xi \in \mathrm{R}^n, \quad |\xi| \neq 0.$$

Suppose that the function $F(t,y)$, $\alpha \le t \le \beta$, $y \in \mathrm{R}^1$ is a solution of a differential inequality

$$\frac{\partial^2 F}{\partial t^2} \le \frac{\partial^2 F}{\partial y^2}.$$

Theorem 5.21. *The functions*

$$\begin{aligned} w(x,t) &= \varphi(x) F\left(t, \frac{\psi(x)}{\varphi(x)}\right), \\ \rho(x) &= \sum_{i,j=1}^{n} a_{ij} \frac{\partial}{\partial x_i}\left(\frac{\psi}{\varphi}\right) \frac{\partial}{\partial x_j}\left(\frac{\psi}{\varphi}\right) \end{aligned}$$

satisfy the differential inequality

$$\rho(x)\frac{\partial^2 w}{\partial t^2} \le \mathbf{L}w.$$

Suppose that function $F(t,y)$, $\alpha \le t \le \beta$, $y \in \mathrm{R}^1$, is the solution of a differential inequality of the parabolic type:

$$\frac{\partial F}{\partial t} \le \frac{\partial^2 F}{\partial y^2}$$

and like the above functions $u(x)$, $v(x) > 0$ are solutions of equation $\mathbf{L}u = \mathbf{L}v = 0$, $\sum_{i,j=1}^{n} a_{ij}\xi_i\xi_j > 0$.

Theorem 5.22. *The functions*

$$w(x,t) = \varphi(x)F\left(t, \frac{\psi(x)}{\varphi(x)}\right),$$

$$\rho(x) = \sum_{i,j}^{n} a_{ij}\frac{\partial}{\partial x_i}\left(\frac{\psi}{\varphi}\right)\frac{\partial}{\partial x_j}\left(\frac{\psi}{\varphi}\right)$$

satisfy the differential inequality

$$\rho(x)\frac{\partial w}{\partial t} \le \mathbf{L}w.$$

5.4 FORMULAS FOR SOLUTION OF INVERSE PROBLEMS FOR EVOLUTION EQUATIONS IN CONVOLUTION

We consider the equation

$$a(y)\frac{\partial w}{\partial t} = \int_{\mathrm{R}^n} b(x-x')w(x',y,t)\,\mathrm{d}x' + \int_{\mathrm{R}^n} c(x-x',t)\frac{\partial w(x',y,t)}{\partial y}\,\mathrm{d}x', \tag{5.23}$$

$x \in \mathrm{R}^n$, $y \in \mathrm{R}^1$, $\alpha < t < \beta$, the function $a(y)$ is continuous, $b(x,t)$, $c(x,t)$ are distributions.

Let functions $A(y)$, $\Phi(x,t)$ be determined by the relations

$$\frac{\mathrm{d}A}{\mathrm{d}y} = a(y), \quad \frac{\partial\Phi}{\partial t} = c(x,t)$$

and let $\hat{\Phi}(\omega,t)$ be the Fourier transform:

$$\hat{\Phi} = \int_{\mathrm{R}^n} \Phi(x,t)\,\mathrm{e}^{-\mathrm{i}\omega x}\,\mathrm{d}x.$$

We shall find two functions $w(x,t)$ and $b(x,t)$ which satisfy equation (5.23).

Theorem 5.23. *Let $f(y,\omega)$, $g(y,\omega)$ be arbitrary smooth functions, $y \in \mathrm{R}^1$, $\omega \in \mathrm{R}^n$. Then the functions*

$$w(x,t) = \int_{\mathrm{R}^n} f(A(y) + \hat{\Phi}(\omega,t),\omega)\,\mathrm{e}^{\int_{y_0}^{y} g(A(y)-A(\eta)+\hat{\Phi}(\omega,t),\omega)\,\mathrm{d}\eta}\,\mathrm{e}^{\mathrm{i}\omega x}\,\mathrm{d}\omega, \tag{5.24}$$

$$b(x,t) = -\frac{1}{(2\pi)^n}\int_{\mathrm{R}^n} g(\hat{\Phi}(\omega,t),\omega)\frac{\partial\Phi}{\partial t}\,\mathrm{e}^{\mathrm{i}\omega x}\,\mathrm{d}\omega \tag{5.25}$$

formally satisfy equation (5.23).

Theorem 5.24. *If function $f(y,\omega)$ has a compact support for variable ω and functions $A(y)$, $\hat{\Phi}(y,t)$, $g(y,\omega)$ are finite differential functions,*

$$\left| g(\hat{\Phi}(\omega,t),\omega)\frac{\partial \Phi}{\partial t}\right| \le c(1+|\omega|^m), \quad c>0, \quad m>0.$$

Then the formulas (5.24), (5.25) are correct.

5.5 THE OPERATOR FUNCTION METHOD AND INVERSE PROBLEMS

In this section we use Fantappié formulas (Fantappié, 1933) for studying inverse problems. We consider only two inverse problems for parabolic and wave equations. Other results can be seen in Anikonov (1978) and Shishko (1975).

First, we consider the inverse problem of finding the functions $w(x,y)$ and $f(x)$ such that

1. $$\frac{\partial^p w}{\partial x^p} + \sum_{k=1}^{p-1} a_k(y)\frac{\partial^k w}{\partial x^k} + b(y)\frac{\partial w}{\partial y} = f(x)h(y) + q(y),$$
$$\left.\frac{\partial^k w}{\partial x^k}\right|_{x=0} = 0, \quad k=0,1,2,\dots,p-1, \quad p\ge 2,$$
$$0 \le x \le 1, \quad -\infty < y < \infty;$$

2. $w|_{y=0} = \tau(x)$.

Here, all functions $a_k(y)$, $b(y)$, $h(y)$, $q(y)$, $\tau(y)$ are known and analytic. Let $\psi_1(u)$ be a function, inverse for $\psi(y) = \int\limits_0^y \frac{d\xi}{b(\xi)}$.

Put

$$F(x) = \frac{\partial^p \tau}{\partial x^p} - q(0) + \frac{1}{2}q'(0)b(0)x^2\delta_p^2 - \operatorname{Res}\lambda^{-3}\int\limits_0^x e^{\frac{x-\xi}{\lambda}}\int\limits_0^\xi q(\psi_1(\lambda^{p-1}(t-\xi)))\,dt\,d\xi;$$

the residium is calculated in point $\lambda = 0$, δ_p^2 is Kroneker symbol.

Theorem 5.25. *The inverse problem 1, 2 has unique analytic solution $f(x)$ and if $h(y)=1$, $a_k=0$, $k=0,1,2,\dots,p-1$, then the formula $f(x)=F(x)$ holds.*

PROOF. According to the Fantappié formulas, the direct problem 1 has solution $w(x,y)$ which is determined by the formula

$$w(x,y) = \frac{1}{2\pi i}\int\limits_C \lambda^{p-3}\int\limits_0^x e^{\frac{x-\xi}{\lambda}}\int\limits_0^\xi \exp\left[\sum_{k=0}^{p-1}\lambda^{p-k-1}\int\limits_\xi^t a_k(\omega)\,d\theta\right] g(t,u)\,dt\,d\xi\,d\lambda,$$

$$u=\psi_1\big(\lambda^{p-1}(t-\xi)+\psi(y)\big),\quad \omega=\psi_1\big(\lambda^{p-1}(\theta-\xi)+\psi(y)\big),$$

C is a closed smooth curve which contains point $\lambda=0$, $g(x,y)=f(x)h(y)+q(y)$ If we put $y=0$ in this formula we obtain relation

$$\tau(x)=\frac{1}{2\pi\,\mathrm{i}}\int_C \lambda^{p-3}\int_0^x e^{\frac{x-\xi}{\lambda}}\int_0^\xi \exp\left[\sum_{k=0}^{p-1}\lambda^{p-k-1}\int_0^t a_k(\omega)\,\mathrm{d}\theta\right] g(t,u)\,\mathrm{d}t\,\mathrm{d}\xi\,\mathrm{d}\lambda, \tag{5.26}$$

Here $u=\psi_1\big(\lambda^{p-1}(t-\xi)\big)$, $\omega=\psi_1\big(\lambda^{p-1}(\theta-\xi)\big)$. Differentiating (5.26) p times we obtain the integral equation on unknown function $f(x)$:

$$f(x)=\varphi(x)+\int_0^x K_3(x,t)f(t)\,\mathrm{d}t+\frac{1}{2\pi\,\mathrm{i}}\int_C\int_0^x\int_0^\xi N(\lambda,t,\xi,x)f(t)\,\mathrm{d}t\,\mathrm{d}\xi\,\mathrm{d}\lambda. \tag{5.27}$$

Here,

$$\begin{aligned}\varphi(x)&=\tau^{(p)}(x)-q(0)-q(0)\int_0^x K_1(x,t)\,\mathrm{d}t-b(0)q'(0)\int_0^x K_2(x,t)\,\mathrm{d}t\\&\quad-\frac{1}{2\pi\,\mathrm{i}}\int_C\lambda^{-3}\int_0^x e^{\frac{x-\xi}{\lambda}}\int_0^\xi\exp\left[\sum_{k=0}^{p-1}\lambda^{p-k-1}\int_\xi^t a_k(\omega)\,\mathrm{d}\theta\right]q(u)\,\mathrm{d}t\,\mathrm{d}\xi\,\mathrm{d}\lambda,\\ K_3(x,t)&=-h(0)K_1(x,t)-h'(0)K_2(x,t),\\ N(\lambda,t,\xi,x)&=\lambda^{-3}\exp\left(\frac{x-\xi}{\lambda}\sum_{k=0}^{p-1}\lambda^{p-k-1}\int_\xi^t a_k(\omega)\,\mathrm{d}\theta\right).\end{aligned}$$

Let $C=\{\lambda:|\lambda|=\rho\}$. From (5.27), a linear integral equation on $f(x)$ follows

$$f(x)=\varphi(x)+\int_0^x K(x,t)f(t)\,\mathrm{d}t, \tag{5.28}$$

where

$$K(x,t)=\frac{1}{2\pi\,\mathrm{i}}\int_{|\lambda|=\rho}\int_t^x N(\lambda,t,\xi,x)\,\mathrm{d}\xi\,\mathrm{d}\lambda+K_3(x,t)$$

The functions $\varphi(x)$, $K(x,t)$ are continuous. Hence, equation (5.28) has unique solution $f(x)$. If $a_k\equiv 0$, $h=1$, then $K(x,t)=0$ and $f(x)=\varphi(x)=F(x)$.

Theorem 5.25 is proved.

Now we shall study the other inverse problem for the wave equation: to find functions $w(x,y)$ and $b(y)$ such that

3. $\dfrac{\partial^2 w}{\partial x\partial y}+b(y)\dfrac{\partial w}{\partial x}+c(y)w=g(x,y)$,
   $w(0,y)=\varphi(y)$, $w(x,0)=\psi(x)$, $0\le x\le 1$, $0\le y\le 1$,
   $c(y)$, $g(x,y)$, $\varphi(y)$, $\psi(x)$ are known smooth functions;

4. the function $\tau(y) = w(v(y), y)$, $0 \le y \le 1$ is known. Here $v(y) > 0$ is a smooth known function, for example, $v(y) = 1$.

Put

$$\begin{aligned} u(y) &= \exp \int_0^y b(p)\,dp, \quad r(t,y) = 2\left((v(y)-t)\int_0^y c(p)\,dp\right)^{1/2}, \\ F(y) &= \frac{1}{\tau(y)-\varphi(y)}\Bigg[\psi(v(y)) - \varphi(0)I_0(r(0,y)) \\ &\quad - \int_0^y c(\xi)\,d\xi \int_0^{v(y)} \psi(t)(I_0)(r(t,y)) + I_2(r(t,y))\,dt\Bigg], \end{aligned}$$

where $I_0(z)$, $I_2(z)$ are Bessel functions.

Theorem 5.26. *If* $|\tau(y) - \varphi(y)| \ge \delta > 0$, $0 \le y \le 1$, *then the inverse problem has one solution. If* $c(y) = g(x,y) = 0$ *or* $g(x,y) = c(y)\varphi(y)$, $c(y) > 0$, *then the following formula holds*

$$u(y) = \exp \int_0^y b(p)\,dp = F(y).$$

PROOF. Using the approach of Fantappié it is possible to obtain solution of the problem 3. above by the formula

$$\begin{aligned} w(x,y) &= e^{-B(0,y)}\psi(x) + \frac{1}{2\pi i}\int_C \frac{1}{\lambda^2} e^{\tilde F(0,y,\lambda)} \int_0^x e^{\frac{x-t}{\lambda}} \psi(t)\,dt\,d\lambda \\ &\quad + \frac{1}{2\pi i}\int_C \frac{1}{\lambda} e^{\frac{x}{\lambda}} \int_0^y (\psi'(\xi) + b(\xi)\varphi(\xi))\, e^{\tilde F(\xi,y,\lambda)}\,d\xi\,d\lambda \qquad (5.29) \\ &\quad + \frac{1}{2\pi i}\int_C \frac{1}{\lambda}\int_0^x e^{\frac{x-t}{\lambda}} \int_0^y g(t,\xi)\, e^{\tilde F(\xi,y,\lambda)}\,d\xi\,dt\,d\lambda, \end{aligned}$$

$$B(\xi,y) = \int_\xi^y b(s)\,ds, \quad \tilde C(\xi,y) = \int_\xi^y c(p)\,dp,$$

$$\tilde F(\xi,y,\lambda) = -B(\xi,y) - \lambda C(\xi,y)$$

Using (5.29) we can write

$$\begin{aligned} w(x,y) &= \frac{1}{u(y)}\Bigg[\psi(x) - \tilde C(0,y)\int_0^x \sum_{k=0}^{\infty} \frac{(-1)^k (x-t)^k C^k(0,y)}{k!(k+1)!}\psi(t)\,dt \\ &\quad + \int_0^y (\varphi' + b\varphi)u(\xi)\sum_{k=0}^{\infty}\frac{(-1)^k (x)^k C^k(\xi,y)}{(k!)^2}\,d\xi \\ &\quad + \int_0^x\int_0^y g(t,\xi)u(\xi)\sum_{k=0}^{\infty}\frac{(-1)^k (x-t)^k C^k(\xi,y)}{(k!)^2}\,d\xi\,dt\Bigg] \end{aligned}$$

Hence, it follows that

$$w(x,y) = \frac{1}{u(y)}\Bigg[\psi(x) - \tilde{C}(0,y)\int_0^x \psi(t)\frac{I_1(\bar{r}_0)}{\bar{r}_0}\,\mathrm{d}t + \varphi(y)u(y) - \varphi(0)I_0(\bar{r}_2)$$

$$- 2x\int_0^y \varphi(\xi)C(\xi)u(\xi)\frac{I_1(\bar{r}_1)}{\bar{r}_1}\,\mathrm{d}\xi + \int_0^x\int_0^y g(t,\xi)u(\xi)I_0(r)\,\mathrm{d}\xi\,\mathrm{d}t\Bigg] \tag{5.30}$$

$$r = 2\sqrt{(x-t)\tilde{C}(\xi,y)}, \quad r_0 = r\big|_{\xi=0}, \quad r_1 = r\big|_{t=0}, \quad r_2 = r\big|_{t=0\,\xi=0}.$$

If we put in (5.30) $x = v(y)$, using the information in 4. above we obtain the Volterra equation

$$u(y) = F(y) + \int_0^y K(y,\xi)u(\xi)\,\mathrm{d}\xi, \tag{5.31}$$

for function $u(y)$ with continuos kernel $K(y,\xi)$. If $C(y) = g = 0$ or $g = c\varphi$, $K = 0$ and we have the formula $u(y) = F(y)$.

Theorem 5.26 is proved.

CHAPTER 6

Mathematical models in the problems of ethnogeny and evolution of populations

6.1 EQUATIONS ENCOUNTERED IN ECOLOGY

Problems of ethnogeny and evolution of populations should be related to the subject of ecology in its general sense. According to the encyclopaedic views, ecology is a biological science studying the organization and functioning of over-organizational systems involving various equations. The main problem of ecology is to study in detail and with the aid of quantitative methods the foundations of the structure and functioning of natural and man-made systems, in particular, socio-economic ones. In view of this it is conceived, and it should be specially accentuated, that problems of environment preservation are only a special case of problems of ecology. Such a general statement of ecological problems naturally implies the use of a wide class of various equations in qualitative and quantitative description of evolution processes (Davydov, 1986; Forrester, 1974; Kugaenko, 1991; Murray, 1977; Odum, 1983; Poluektov *et al.*, 1980; Svirezhev, 1967; Svirezhev and Lagofet, 1978; Volterra, 1931; Watt, 1968). First and foremost, equations of reaction-diffusion type,

$$\frac{\partial w}{\partial t} = \mathbf{A}w + F(x,t,w) + Q(x,t) \tag{6.1}$$

should be classified among the important types of equations encountered in ecology. Here vector $w(x,t) = (w_1, \ldots, w_m)$ characterizes an ecological, $\mathbf{A}$ is a propagational operator of this, $F(x,t,w)$ characterizes reactions and interactions, and $Q(x,t)$ describes sources. The following important special cases of equation (6.1) should be mentioned:

$\dfrac{\partial w}{\partial t} = k\dfrac{\partial^2 w}{\partial x^2} + qw - \alpha w^3$ — the Kolmogorov–Piskunov–Petrovskii equation.

$\dfrac{\partial w}{\partial t} = r\left(1 - \dfrac{w}{k}\right) w$ — the logistic equation widely employed in the theory of populations.

Here (cf. Izrael', 1976) the detriment to the m-th population due to anthropogenic action can be written as

$$A_m(t) = \int\limits_{\mathbf{R}^n} \sum_{i,j} \sigma_{ij}(\mathbf{x},t)\varepsilon_{ijm} N_m(\mathbf{x},t)\,\mathrm{d}\mathbf{x},$$

where σ_{ij} is the concentration of the i-th component in the j-th medium, ε_{ijm} is the toxicity of this for the m-th population, $N_m(\mathbf{x}, t)$ is the quantity of the m-th population, and $x \in D \subset \mathrm{R}^n$ is a spatial variable.

$$\frac{\partial w}{\partial t} = k\frac{\partial^2 w}{\partial x^2} + F(x,t,w),\ w = (w_1, w_2),\ F = (F_1, F_2) \text{ — Turing system of equations.}$$

$$\left.\begin{aligned} \frac{\partial w_1}{\partial t} &= \alpha w_1 - \beta w_1 w_2, \\ \frac{\partial w_2}{\partial t} &= K\beta w_1 w_2 - \mu w_2, \end{aligned}\right. \text{ — 'predator – prey' system of equations.}$$

$$\frac{\partial w}{\partial t} = \left(1 - (1 + C_2)\,|w|^2\right) w + (1 + \mathrm{i}\, C_2)\,\Delta w \text{ — Kuramoto–Tzuzuki equation.}$$

At present, evolution equations with discrete time

$$w_{t+h} = \mathbf{A} w_t + F(x, t, w_t) + Q(x, t),$$

where h is some number, gain wide popularity in ecology. The following equation with discrete time, which is well known in the theory of populations, can serve as an example:

$$w_{t+1} = a w_t (1 - w_t).$$

Sometimes the previous history should be taken into account. In such a case, there can arise equations like

$$\frac{\partial w}{\partial t} = r \left(1 - \frac{\int\limits_{-\infty}^{t} H(\tau) w(t - \tau)\, \mathrm{d}\tau}{K} \right) w.$$

Processes in biophysics are described by Korteweg-de Vries, Schrödinger, and Sine–Gordon equations:

$$\frac{\partial w}{\partial t} + \sigma w \frac{\partial w}{\partial x} + \frac{\partial^3 w}{\partial x^3} = 0,$$

$$-\mathrm{i}\,\frac{\partial w}{\partial t} = \Delta w + \Phi(x, t) w,$$

$$\frac{\partial^2 w}{\partial t^2} = \Delta w + \sin w.$$

When probabilistic processes are studied, differential equations of Kolmogorov type arise for transient and generating functions. Propagation of epidemics is modelled by a generating function $w(x, y, t)$ satisfying the equation

$$\frac{\partial w}{\partial t} = \lambda(x^2 - xy)\frac{\partial^2 w}{\partial x \partial y} + \mu(1 - x)\frac{\partial w}{\partial x}.$$

Here λ, μ are constants.

In mathematical simulation of environmental processes, the diffusion equation of the type

$$\frac{\partial w}{\partial t}+\alpha\frac{\partial w}{\partial x}+\beta\frac{\partial w}{\partial y}+\gamma\frac{\partial w}{\partial z}+\sigma w=\frac{\partial}{\partial z}\lambda\frac{\partial w}{\partial z}+\mu\Delta w+Q(x,y,z,t).$$

(cf. Marchuk, 1989) is quite popular.

Kinetic equations undoubtedly play a special role in ecology. Kinetic equations characterize the continuity of the motion of a substance and are fundamental equations of natural science. They are widely used for quantitative and qualitative description of physical, chemical, biological and other processes on a microscopic level. In view of the vital importance of kinetic equations in the theory of the motion of the matter under the action of forces, in particular in the theory of irreversible processes, they are often called 'master equations'.

As a rule, kinetic equations are integro-differential equations, which are often nonlinear in one or several distribution functions $F(x,p,t)$ of dynamic states of the matter depending on the spatial point x, momentum p, and time t. When strong interactions are present, quantum addends arise in kinetic equations. By way of examples, let us cite Liouville, Boltzmann, Vlasov, Fokker–Planck equations, Bogolyubov chain of equations, various equations of radiation transfer, and quantum equations for density matrix and for Wigner function. Kinetic equations and their applications are considered in detail in Alekseev (1983); Bogolyubov (1981); Case and Zweifel (1967); Ferziger and Kaper (1972); Haken (1987); Klimontovich (1982); Kompaneets (1977); Lifshitz and Pitaevskii (1979); Prigogine (1980); Tikhonov *et al.* (1987) and in the literature cited there.

Let us cite a number of examples of kinetic equations. Below we assume that the distribution function $F(x,p,t)$ is differentiable sufficiently many times and decreases sufficiently fast in the variable p; $x\in D$, $p\in \mathrm{R}^n$, where D is a region with a smooth boundary ∂D. The product $F(x,p,t)\Delta x\Delta p$ characterizes the number of particles in an element of phase space at the moment t. Of fundamental importance are Liouville equations characterizing the continuity of the motion of a substance. They have a representation of the kind

$$\frac{\partial F}{\partial t}+\{F,H\}=0,$$

where $F(x,p,t)$ is a distribution function of particles of any type; $x\in D$; $p\in\mathrm{R}^n$; $t\in\mathrm{R}$; $H=H(x,p,t)$ is the Hamilton function, and $\{F,H\}=\sum_{j=1}^{n}\left(\frac{\partial F}{\partial x^j}\frac{\partial H}{\partial p_j}-\frac{\partial F}{\partial p_j}\frac{\partial H}{\partial x^j}\right)$ is the Poisson bracket.

The Hamilton function is often of the form

$$H(x,p,t)=\frac{1}{2}|p|^2-\Phi(x,t),$$

which corresponds to the Liouville equation

$$\frac{\partial F}{\partial t}+\sum_{j=1}^{n}\left(\frac{\partial F}{\partial x^j}p_j+\frac{\partial F}{\partial p_j}\frac{\partial \Phi}{\partial x^j}\right)=0.$$

The theory of rare gases is based on Boltzmann equations which are usually written as

$$\frac{\partial F}{\partial t}+\sum_{j=1}^{n}\frac{\partial F}{\partial x^j}p_j=\mathrm{St}\ F.$$

Here St F is the collision integral characterizing the change of the distribution $F(x,p,t)$ with a pair of collisions of particles taken into account. In the case of the Boltzmann equation, St F has the following representation

$$\mathrm{St}\ F = \int\limits_{\mathbf{R}^n}\int\limits_{\mathbf{R}^n}\int\limits_{\mathbf{R}^n} w(p,p_1;p',p_1')\left[F(x,p,t)F(x,p_1,t) - F(x,p',t)F(x,p_1',t)\right] \mathrm{d}p_1\,\mathrm{d}p'\,\mathrm{d}p_1',$$

where $w(p,p_1;p',p_1')$ is a function characterizing the probability of collision of two particles with momenta p, p_1 with their transition into the state p', p_1', this function being such that $w(p,p_1;p',p_1') = w(p',p_1';p,p_1)$. The latter equality defines the detailed equilibrium: the process of collision is balanced by the reverse process.

Equations of Boltzmann type are widely used for qualitative and quantitative description of processes in gas dynamics.

In the presence of strong interactions, quantum kinetic equations arise. Quantum kinetic equations are integro-differential equations of the kind

$$\frac{\partial F}{\partial t} + \{F,H\} + \sigma F = \frac{\mathrm{i}}{(2\pi)^n h}\int\limits_{\mathbf{R}^{2n}} \left[\Phi\left(x - \frac{1}{2}hy, t\right) - \Phi\left(x + \frac{1}{2}hy, t\right)\right] \mathrm{e}^{\mathrm{i}\,y(p-p')}\, F(x,p',t)\,\mathrm{d}p'\,\mathrm{d}y + \mathrm{St}\ F + q(x,p,t), \tag{6.2}$$

where $F(x,p,t)$ is a quantum distribution function, H is the Hamilton function, Φ is the average potential, $\sigma(x,p,t)$ is the absorption, h is the Planck constant, St F is the collision integral, and $q(x,p,t)$ is the function characterizing sources. The presence of the integral

$$\int\limits_{\mathbf{R}^{2n}} [\Phi]\,\mathrm{e}^{\mathrm{i}\,y(p-p')}\, F\,\mathrm{d}p'\,\mathrm{d}y$$

determines the character of the equation. In the case $\sigma = 0$, St $F = 0$, $q = 0$, $H = |p|^2/2 - \Phi$, the quantum distribution function $F(x,p,t)$ is related to the wave function $\Psi(x,t)$ by the equality

$$F(x,p,t) = \frac{1}{(2\pi)^n}\int\limits_{\mathbf{R}^n} \Psi^*\left(x - \frac{1}{2}hy, t\right)\Psi\left(x + \frac{1}{2}hy, t\right)\mathrm{e}^{\mathrm{i}\,yp}\,\mathrm{d}y.$$

The wave function $\Psi(x,t)$ is the solution to the Schrödinger equation

$$\mathrm{i}\frac{\partial\Psi}{\partial t} = -\Delta\Psi + \Phi\Psi.$$

When expanded in terms of h, the quantum kinetic equation can be formally represented as an equation of infinite order in the variable p,

$$\frac{\partial F}{\partial t} + \{F,H\} + \sigma F = \sum_{\alpha} C_\alpha h^{2\alpha}\mathbf{D}_x^{2\alpha}\Phi\mathbf{D}_p^{2\alpha+1}F + \mathrm{St}\ F + q,$$

where C_α are constants; and $\mathbf{D}_x$, $\mathbf{D}_p$ are operators of differentiating with respect to x and p.

Various classical approximations are obtained from the latter equation. For example, taking into account only the term with $\alpha = 0$ we have

$$\frac{\partial F}{\partial t} + \{F, H\} + \sigma F = \sum_{j=1}^{n} \frac{\partial \Phi}{\partial x^j} \frac{\partial F}{\partial p_j} \operatorname{St} F + q.$$

If $H = |p|^2/2$, $\partial\sigma/\partial p_j = 0$, and $\operatorname{St} F = 0$, then the formalism of the Fourier transform of equation (6.2) with respect to p leads to the second order equation

$$\frac{\partial w}{\partial t} - \mathrm{i} \sum_{j=1}^{n} \frac{\partial^2 w}{\partial x^j \partial y^j} + \sigma w = \left[\Phi\left(x - \frac{1}{2}hy, t\right) - \Phi\left(x + \frac{1}{2}hy, t\right)\right] w + \hat{q},$$

$$\hat{q}(x, y, t) = \frac{1}{(2\pi)^n} \int_{\mathrm{R}^n} q(x, p, t)\, \mathrm{e}^{-\mathrm{i}yp}\, \mathrm{d}p$$

for the function

$$w(x, y, t) = \frac{1}{(2\pi)^n} \int_{\mathrm{R}^n} F(x, p, t)\, \mathrm{e}^{-\mathrm{i}yp}\, \mathrm{d}p.$$

As concerns linear equations of the theory of transfer, these are balance equations

$$\frac{\partial F}{\partial t} + \sum_{j=1}^{n} v^j \frac{\partial F}{\partial x^j} + |v|\,\sigma(x, v, t) F = \int_{\mathrm{R}^n} w(x, v, v', t) F(x, v', t)\, \mathrm{d}v' + q(x, v, t). \qquad (6.3)$$

Here $F(x, v, t)\Delta x \Delta v$ characterizes the number of particles, e.g., neutrons, in the volume $\Delta x \Delta v$ at the moment t; $w(x, v, v', t)$ is the scattering index; $q(x, v, t)$ is a function characterizing sources; and $1/\sigma(x, v, t)$ mean free path.

Linear equations (6.3) are obtained from nonlinear equations of Boltzmann type. The velocity in (6.3) is often assumed to be constant in absolute value. In this case, equations (6.3) after proper modifications are written as

$$\frac{\partial F}{\partial t} + |v|\,\Omega \operatorname{grad}_x F + \sigma F = \int_{\omega} J(x, \Omega, \Omega', t) F(x, \Omega, t)\, \mathrm{d}\Omega + q(x, \Omega, t), \qquad (6.4)$$

where $|\Omega| = 1$, and ω is a sphere of unit radius. Equation (6.4) is called a nonstationary linear equation of the theory of transfer (Case and Zweifel, 1967).

Electromagnetic and other processes are described by the Vlasov system of equations. This system of kinetic equations characterizes the interaction of charged particles in an electromagnetic field and is written as

$$\begin{cases} \dfrac{\partial F_e}{\partial t} + v\dfrac{\partial F_e}{\partial x} - e\left(E + \dfrac{1}{c}[v, B]\right)\dfrac{\partial F_e}{\partial v} = 0; \\ \dfrac{\partial F_i}{\partial t} + v\dfrac{\partial F_i}{\partial x} - ze\left(E + \dfrac{1}{c}[v, B]\right)\dfrac{\partial F_i}{\partial v} = 0; \\ \text{curl } E = -\dfrac{1}{c}\dfrac{\partial B}{\partial t}; \quad \text{curl } B = \dfrac{1}{c}\dfrac{\partial E}{\partial t} + \dfrac{4\pi}{c}J; \\ \text{div } B = 0; \quad \text{div } E = 4\pi\rho; \\ \rho = e\displaystyle\int\limits_{\mathbf{R}^3}(zF_i - F_e)\,\mathrm{d}v; \quad J = e\int\limits_{\mathbf{R}^3}(z_iF_i - F_e)v\,\mathrm{d}v. \end{cases} \tag{6.5}$$

Here F_e and F_i are the distribution functions of electrons and ions; ρ and J are the charge and current densities; E and B are the intensities of the electric and magnetic field; e is the electron charge; and z is the ion charge. The system of equations (6.5) is a nonlinear system of integro-differential equations with respect to F_e, F_i, E, B.

Fokker–Planck equations are usually used to describe slow processes of transfer and diffusion of a small admixture of a heavy gas in a light one. Fokker–Planck equations are linear integro-differential equations

$$\frac{\partial F}{\partial t} + \{F, H\} = \int\limits_{\mathbf{R}^n}\Big[w(p+q, q)f(x, p+q, t) - w(p, q)F(x, q, t)\Big]\,\mathrm{d}q. \tag{6.6}$$

In this equation, $w(p, q)$ is the probability of the change of the momentum of a heavy particle. Expanding the expression under the integral sign in (6.6) into a Taylor series one arrives at the Fokker–Planck differential kinetic equations

$$\frac{\partial F}{\partial t} + \{F, H\} = \sum_{j=1}^{n}\frac{\partial}{\partial p_j}A_jF + \sum_{j,k=1}^{n}\frac{\partial}{\partial p_j}\frac{\partial}{\partial p_k}B_{kj}F = 0,$$

$$A_j(p) = \int\limits_{\mathbf{R}^n} q_jw(p, q)\,\mathrm{d}q, \quad B_{kj}(p) = \int\limits_{\mathbf{R}^n} q_kq_jw(p, q)\,\mathrm{d}q.$$

Bogolyubov equations are a linked chain of kinetic equations with respect to k-particle distribution functions

$$\begin{aligned} &F_k(x_1, x_2, \ldots, x_k, p_1, p_2, \ldots, p_k, t) \\ &= V^k\int F_N(x_1, x_2, \ldots, x_N, p_1, p_2, \ldots, p_N, t)\,\mathrm{d}x_{k+1}\ldots\mathrm{d}x_N\,\mathrm{d}p_{k+1}\ldots\mathrm{d}p_N, \end{aligned}$$

$$k = 1, 2, \ldots, N,$$

where F_N is the solution to the Liouville equation $\dfrac{\partial F_N}{\partial t} + \{F_N, H_N\} = 0$; V is the volume of the region of the phase space in question.

The Bogolyubov chain of equations is written in the following form

$$\begin{cases} \dfrac{\partial F_k}{\partial t} + \{F_k, H_k\} = \displaystyle\int_{\mathrm{R}^n} \{\Phi_k, F_{k+1}\}\, \mathrm{d}x_{k+1}\, \mathrm{d}p_{k+1}, \\ k = 1, 2, \ldots, N, \end{cases}$$

where H_k is the Hamilton function; Φ_k is the potential.

Further we will use the kinetic equation of the kind

$$P * w + \sum_{j=1}^{m} \left(\frac{\partial w}{\partial x_j} * \frac{\partial H}{\partial p_j} - \frac{\partial w}{\partial p_j} * \frac{\partial H}{\partial x_j} \right) = 0.$$

Here $w(x,p,t)$ is the distribution function, $x \in \mathrm{R}^m$, $p \in \mathrm{R}^m$, $y \in \mathrm{R}^{n+1}$, $y = (t, y')$, $t \in \mathrm{R}^1$, $y' \in \mathrm{R}^n$, and $*$ denotes the convolution with respect to y, i.e.,

$$P * w = \int_{\mathrm{R}^{n+1}} P(x,p,y-\xi) w(x,p,\xi)\, \mathrm{d}\xi,$$

Functions $P(x,p,y)$, $H(x,p,y)$ characterize the absorption and energy.

6.2 FORMULAS AND EQUATIONS IN INVERSE PROBLEMS FOR GENERAL EVOLUTION EQUATIONS

Often, the entities sought in problems of ecology are not only solutions to evolutionary equations but also evolutionary operators themselves, in particular, some coefficients. In this section, two inverse problems for nonlinear general evolutionary equations are considered.

The general system of nonlinear differential evolutionary equations of reaction-diffusion type is considered,

$$\frac{\partial w}{\partial t} = \mathbf{A}w + \lambda(w), \quad w = (w_1, \ldots, w_n), \quad x \in D \subset \mathrm{R}^n, \quad t \geq 0, \tag{6.7}$$

where $\mathbf{A}$ is a known and generally speaking a nonlinear operator, which is independent of t; $\lambda = \big(\lambda_1(z), \ldots, \lambda_n(z)\big)$, where $z \in \mathrm{R}^n$, and $w(x)$ are two sought vector functions. We assume that the following initial information is given

$$w|_{t=0} = w_0(x), \quad \left.\frac{\partial w}{\partial t}\right|_{t=0} = w_1(x), \quad x \in D. \tag{6.8}$$

The inverse problem is considered to simultaneously determine the vector functions $w(x)$ and $\lambda(z)$ if (6.7) and (6.8) are satisfied.

Theorem 6.1. *Let the mapping $z = w_0(x)$ of the region D in R^n have the inverse one $x = w_0^{-1}(z)$. The the following formula holds for the sought vector function $\lambda(z)$,*

$$\lambda(z) = w_1\big(w_0^{-1}(z)\big) - \mathbf{A}w_0\big(w_0^{-1}(z)\big);$$

therefore, evolution of the initial state (6.8) is determined only by this state itself, independently of $\lambda(z)$,

$$\begin{cases} \dfrac{\partial w}{\partial t} = \mathbf{A}w + w_1\big(w_0^{-1}(w)\big) - \mathbf{A}w_0\big(w_0^{-1}(w)\big), \\ w\big|_{t=o} = w_0(x). \end{cases} \tag{6.9}$$

PROOF. Taking $t = 0$ in (6.7) and employing data (6.8) we have the equality

$$w_1(x) = \mathbf{A}w_0(x) + \lambda\big(w_0(x)\big). \tag{6.10}$$

Denote $w_0(x)$ by z, i.e., $z = w_0(x)$. Since by condition of the theorem the mapping $z = w_0(x)$ of the region D in R^n has an inverse one $\big(x = w_0^{-1}(z)\big)$, then we substitute the inverse mapping $w_0^{-1}(z)$ for x in (6.10) to arrive at the formula $\lambda(z) = w_1\big(w_0^{-1}(z)\big) - \mathbf{A}w_0\big(w_0^{-1}(z)\big)$ for function $\lambda(z)$. Then, replacing the function $\lambda(z)$ in (6.7) by this latter expression, we arrive at the evolutionary problem (6.9).

Thus, the inverse problem (6.7), (6.8) or the problem of evolution of $w(x,t)$ from the state $w_0(x)$, $w_1(x)$ is reduced to a topologo-differential problem, related to the mapping $z = w_0(x)$ of the region D in R^n.

Consider now the problem of simultaneous determination of two functions $w(x,t)$, $\lambda(x)$, $\dfrac{\partial \lambda}{\partial t} = 0$, if

$$\frac{\partial w}{\partial t} = \mathbf{A}w + \lambda(x)\mathbf{B}w, \tag{6.11}$$

$$w\big|_{t=o} = w_0(x), \quad w\big|_{t=T} = w_T(x), \tag{6.12}$$

where $\mathbf{A}$ is in general a nonlinear operator, and $\mathbf{B}$ is some operator which also is not necessarily linear.

Theorem 6.2. *The function* $w(x,t)$ *is the solution to the Cauchy problem for the following evolutionary equation of non-local type*

$$\frac{\partial w}{\partial t} = \mathbf{A}w + \frac{(w_T - w_0) - \displaystyle\int_0^T \mathbf{A}w\,dt}{\displaystyle\int_0^T \mathbf{B}w\,dt}\,\mathbf{B}w, \tag{6.13}$$

$$w\big|_{t=o} = w_0(x). \tag{6.14}$$

And in this case

$$\lambda(x) = \frac{(w_T(x) - w_0(x)) - \displaystyle\int_0^T \mathbf{A}w\,dt}{\displaystyle\int_0^T \mathbf{B}w\,dt}\,\mathbf{B}w.$$

By virtue of this theorem, the inverse problem (6.11), (6.12) is reduced to investigating the Cauchy problem (6.13), (6.14) for an evolution integro-differential equation.

Some results obtained with the aid of such an approach to the study of inverse problems for differential equations of the second order can be found in Annina (1987).

6.3 INVERSE PROBLEMS FOR EVOLUTION EQUATIONS OF THE THEORY OF POPULATIONS

One of the most important problems related to the preservation of specific diversity and quantity of populations in nature reserves, game reserves, and national parks is the forecasting of possible changes of these indicators due to anthropogenic, climatic, geological, etc., action on systems concerned. Quantity forecasting for animal communities is an important problem in such branches of agriculture as fishing industry, bee-keeping, pasturable cattle-breeding, etc., (Begon *et al.*, 1986). Underestimation of such a forecasting can lead either to unwarranted expenditures or, when measures aimed at nature preservation are undertaken, to an undesirable shift of inter-specific balance which in the long run can result in the death of the ecosystem and damage to human health and to economic and cultural activities.

One of the methods making it possible to solve such problems is the mathematical description of the processes involved. Equations of evolution type often arise in such a description of models.

Consider the equation

$$\begin{aligned} \frac{\partial w}{\partial t} &+ \frac{\partial w}{\partial x}v - \int\limits_{\mathrm{R}^{n+1}} \frac{\partial \Phi(x, y-\xi, t-\tau)}{\partial x}\frac{\partial w(x,\xi,v,\tau)}{\partial v}\,\mathrm{d}\tau\,\mathrm{d}\xi \\ &+ \int\limits_{\mathrm{R}^{n+1}} \lambda(x, y-\xi, t-\tau) w \,\mathrm{d}\tau\,\mathrm{d}\xi + \mathbf{A}_y u = 0, \end{aligned} \tag{6.15}$$

where $w(x,y,v,t)$ is a quantitative characteristic of some population; variables have the following sense: x is the physiological age, $y \in \mathrm{R}^n$ is a spatial variable, v is the rate of change of the physiological age, t is time; $\Phi(x,y,t)$ is the potential; $\lambda(x,y,t)$ is the absorption coefficient; and $\mathbf{A}_y$ is an operator depending only on y.

Such an equation has been studied in Anikonov (1991b) for the case when there is no dependence on the spatial variable y, and formulas have been obtained there connecting the functions $w(x,v,t)$, $\Phi(x,t)$, and $\lambda(x,t)$ with the aid of two arbitrary functions. It occurs that similar formulas can be obtained for equation (6.15) too.

Theorem 6.3. *Let*

$$\hat{\Phi}(x,\xi,\omega) = \int\limits_{\mathrm{R}^{n+1}} \Phi(x,y,t)\,\mathrm{e}^{-\mathrm{i}\,t\omega - \mathrm{i}<\xi,y>}\,\mathrm{d}t\,\mathrm{d}y,$$

and let $f(z,\xi,\omega)$ and $g(z,\xi,\omega)$ be differentiable or analytic functions such that functions $w(x,y,v,t) \in C^1$, and $\lambda(x,y,t)$ are defined by formulas

$$w(x,y,v,t) = \int_{R^{n+1}} f\left(\frac{1}{2}v^2 + \hat{\Phi}(x,\xi,\omega),\xi,\omega\right)$$

$$\times \exp\left(\int_{-v}^{v} g\left(\frac{1}{2}v^2 + \hat{\Phi}(x,\xi,\omega) - \frac{1}{2}\eta^2,\xi,\omega\right) d\eta\right) e^{i\omega t + i<\xi,y>} d\omega\, d\xi, \tag{6.16}$$

$$\lambda(x,y,t) = \frac{1}{(2\pi)^{n+1}} \int_{R^{n+1}} \left(2g\left(\hat{\Phi}(x,\xi,\omega),\xi,\omega\right)\frac{\partial\hat{\Phi}}{\partial x} - i\omega - A(\xi)\right) e^{i\omega t + i<\xi,y>} d\omega\, d\xi, \tag{6.17}$$

where $A(\xi)$ is the symbol of the operator $\mathbf{A}_y$. Then functions $w(x,y,v,t)$ and $\lambda(x,v,t)$ formally satisfy equation (6.15).

PROOF. Applying the inverse Fourier transform with respect to y and t to equation (6.15) we get

$$i\omega\hat{w} + v\frac{\partial\hat{w}}{\partial x} - \frac{\partial\hat{\Phi}}{\partial x}\frac{\partial\hat{w}}{\partial v} + \hat{\lambda}\hat{w} + A(\xi)\hat{u} = 0, \tag{6.18}$$

where

$$\hat{w} = \int_{R^{n+1}} w(x,y,v,t)\, e^{-i\omega t - i<\xi,y>} dt\, dy.$$

Hence, it follows that

$$\hat{w} = f\left(\frac{1}{2}v^2 + \hat{\Phi}(x,\xi,\omega),\xi,\omega\right) e^{\int_{-v}^{v} g\left(\frac{1}{2}v^2 + \hat{\Phi}(x,\xi,\omega) - \frac{1}{2}\eta^2,\xi,\omega\right) d\eta}.$$

Substituting $\hat{w}$ in (6.18) we obtain

$$\hat{\lambda}(x,\xi,\omega) = 2g\left(\hat{\Phi}(x,\xi,\omega),\xi,\omega\right)\frac{\partial\hat{\Phi}}{\partial x} - i\omega - A(\xi).$$

Returning to variables y, t we obtain formulas (6.16) and (6.17). The theorem is proved.

In all examples to follow, $\Phi = \Phi_0(x)\delta(t)\delta(y)$.

Example 6.1. Processes periodic in time rank high in the study of populations.

Let

$$\Phi_0(x) = x, \quad f(z,\xi,\omega) = f_0(z,\xi)\sum_k C_k\delta(\omega - s_k), \quad s_k = \text{const}, \quad C_k = \text{const}.$$

Then, following the considerations in Gel'fand and Shilov (1959) we conclude that the solution $w(x,y,v,t)$ to equation (6.15) will be

$$w(x,y,v,t) = \sum_k C_k e^{i s_k t} \int_{R^n} f_0\left(\frac{1}{2}v^2 + x,\xi\right) G(x,\xi,v,s_k)\, e^{i<\xi,y>} d\xi,$$

where

$$G(x,\xi,v,s) = \exp\left(\int_{-v}^{v} g\left(\frac{1}{2}v^2 + x - \frac{1}{2}\eta^2, \xi, s\right) d\eta\right).$$

In the present case, $\lambda(x,y,t)$ has the general form (6.17).

Example 6.2. Let $f(z,\xi,\omega) = f_0(z,\xi)(\operatorname{sgn}\omega - \delta(\omega)C)$, $C = \text{const}$, and let g be independent of ω and ξ. According to Gel'fand and Shilov (1959)

$$w(x,y,v,t) = H(x,y,v)(2\,\mathrm{i}\,t^{-1} - C),$$

where

$$H(x,y,v) = \int_{\mathrm{R}^n} f_0\left(\frac{1}{2}v^2 + \Phi_0, \xi\right) \exp\left(\int_{-v}^{v} g\left(\frac{1}{2}v^2 + \Phi_0 - \frac{1}{2}\eta^2\right) d\eta\right) \mathrm{e}^{\mathrm{i}<\xi,y>}\, d\xi.$$

Denoting $H_1(x,y,v) = \operatorname{Re}\big(H(x,y,v)\big)$, $H_2(x,y,v) = \operatorname{Im}\big(H(x,y,v)\big)$, one can rewrite the solution $w(x,y,v,t)$ as follows

$$w(x,y,v,t) = \Big[-H_1(x,y,v)C - 2H_2(x,y,v)t^{-1}\Big] + \mathrm{i}\Big[CH_2(x,y,v) + 2H_1(x,y,t)t^{-1}\Big].$$

The absorption function λ is as follows

$$\lambda(x,y,t) = 2g\big(\Phi_0(x)\big)\frac{\partial\Phi_0(x)}{\partial x}\delta(t)\delta(y) - \delta'(t)\delta(y) - \frac{\delta(t)}{(2\pi)^n}\int_{\mathrm{R}^n} A(\xi)\,\mathrm{e}^{\mathrm{i}<\xi,y>}\, d\xi.$$

Considering the solution at times $t \geq t_0 > 0$ we conclude that when t increases both real and imaginary parts of the solution at fixed x, y, and v tend to some limiting values.

Example 6.3. Let $f(x,\xi,\omega) = \delta(\xi)$, $g(z,\xi,\omega) = -g_1(z)\,|\omega|$, $g_1 > 0$. According to Oberhettinger (1973) the solution $w(x,y,v,t)$ is

$$w(x,y,v,t) = \frac{2G(x,v)}{G^2(x,v) + t^2},$$

where

$$G(x,v) = \int_{-v}^{v} g_1\left(\frac{1}{2}v^2 + \Phi_0(x) - \frac{1}{2}\eta^2\right) d\eta.$$

$$\lambda(x,y,t) = -\frac{g_1\big(\Phi_0(x)\big)}{t^2}\frac{\partial\Phi_0}{\partial x}\delta(y) - \delta'(t)\delta(y) - \frac{\delta(t)}{(2\pi)^n}\int_{\mathrm{R}^n} A(\xi)\,\mathrm{e}^{\mathrm{i}<\xi,y>}\, d\xi.$$

Now consider the problem of determining the functions f and g based on the information $w_0(y,v,t) = w|_{x=x_0}$. Denote

$$\hat{w}_0(\xi,v,\omega) = \int_{\mathrm{R}^{n+1}} w_0(y,v,t)\,\mathrm{e}^{-\mathrm{i}\,t\omega - \mathrm{i}<\xi,y>}\, dt\, dy.$$

As in Anikonov (1991b), the functions f and g can be found from $\hat{w}_0(\xi,v,\omega)$.

Theorem 6.4. *Let the function $w(x,y,v,t)$ be defined by formula (6.17), and let $\hat{\Phi}(x_0,\xi,\omega) = 0$. Then, for all x such that $\hat{\Phi}(x,\xi,\omega) \geq \hat{\Phi}(x_0,\xi,\omega)$ the following formulas hold*

$$|f(z,\xi,\omega)| = \frac{1}{(2\pi)^{n+1}}\left[\hat{w}_0(\xi,\sqrt{2z},\omega)\hat{w}_0(\xi,-\sqrt{2z},\omega)\right]^{1/2}, \qquad (6.19)$$

$$g(z,\xi,\omega) = \frac{1}{2\sqrt{2}\pi}\int_0^z \frac{\ln\left[\dfrac{\hat{w}_0(\xi,\sqrt{2z},\omega)}{\hat{w}_0(\xi,-\sqrt{2z},\omega)}\right]}{\sqrt{z-\eta}}\,\mathrm{d}\eta. \qquad (6.20)$$

PROOF. From formula (6.17) and condition $\hat{\Phi}(x_0,\xi,\omega) = 0$, we conclude that

$$\hat{w}_0(\xi,v,\omega) = (2\pi)^{n+1} f\left(\frac{1}{2}v^2,\xi,\omega\right)\exp\left(\int_{-v}^{v} g\left(\frac{1}{2}v^2 - \frac{1}{2}\eta^2,\xi,\omega\right)\mathrm{d}\eta\right),$$

whence we obtain

$$\left|f\left(\frac{1}{2}v^2,\xi,\omega\right)\right| = \frac{1}{(2\pi)^{n+1}}\left[\hat{w}_0(\xi,v,\omega)\hat{w}_0(\xi,-v,\omega)\right]^{1/2},$$

or

$$|f(z,\xi,\omega)| = \frac{1}{(2\pi)^{n+1}}\left[\hat{w}_0(\xi,\sqrt{2z},\omega)\hat{w}_0(\xi,\sqrt{2z},\omega)\right]^{1/2}.$$

We have the following relationship for g:

$$\int_0^v g\left(\frac{1}{2}v^2 - \frac{1}{2}\eta^2,\xi,\omega\right)\mathrm{d}\eta = \frac{1}{4}\ln\left[\frac{\hat{w}_0(\xi,v,\omega)}{\hat{w}_0(\xi,-v,\omega)}\right].$$

Introducing a new variable $\zeta = \frac{1}{2}v^2 - \frac{1}{2}\eta^2$ and denoting $z = \frac{1}{2}v^2$, we obtain

$$\int_0^z \frac{g(\zeta,\xi,\omega)}{\sqrt{z-\zeta}}\,\mathrm{d}\zeta = \frac{1}{2\sqrt{2}}\ln\left[\frac{\hat{w}_0(\xi,\sqrt{2z},\omega)}{\hat{w}_0(\xi,-\sqrt{2z},\omega)}\right].$$

Having solved this integral equation we arrive at formula (6.20). The theorem is proved.

6.4 ON MATHEMATICAL MODELLING OF ETHNIC PROCESSES

According to Gumilev (1989), ethnos is 'a collective body of people that has formed up naturally on the basis of original stereotype of behavior and which exists as an energetic system (structure), contrasting itself, as a consequence of a feeling of complementarity, to other similar collective bodies'.

Complementarity is 'a feeling of subliminal mutual sympathy of individuals that determines the division into friends and aliens'.

The basic concepts and conclusions of the theory developed by L. N. Gumilev are related to 'passionarity' which is treated as a dynamical characteristic of behavior of individuals of an ethnic system.

In particular, according to Gumilev, 'a passionary impulse of behavior of individuals is a behavioristic impulse directed opposite to the instinct of personal and specific self-preservation'; 'passionaries are those individuals whose passionary impulse of behavior exceeds the impulse due to the instinct of self-preservation'; 'a passionary field is the field due to the presence of biochemical energy'; 'and an ethnic field is the field of behavior and attractivity of individuals of an ethnic system, arising on the basis of a passionary field'.

Here, we propose the formalization of, first, the concepts of an ethnic field and a passionary field and, second, their dynamic connection by means of equations of motion. Following Gumilev we treat ethnos as a statistical ensemble in space-time with coordinates $\mathbf{y} = (\mathbf{y}', t)$, $\mathbf{y}' \in \mathrm{R}^n$, $t \in \mathrm{R}^1$, $n \geq 1$, which possesses passionarity (x, p), $x \in \mathrm{R}^m$, $p \in \mathrm{R}^m$, where x characterizes the potential capability of an individual for resolute actions, p is a passionary impulse, and R^l is a real Euclidean space. Let $w(x, p, \mathbf{y})$ denote the density of distribution of individuals of the given ensemble in the space R^{2m+n+1} of variables $(x, p, \mathbf{y})$, and let $H(x, p, \mathbf{y})$ denote the biochemical energy determining the passionary field. Let $P(x, p, \mathbf{y})$ characterize the appearance, disappearance, and motion of individuals in (physical) space. Here we propose the equations of 'motion' in the phase space of variables $(x, p, \mathbf{y})$ in the form

$$P * w + \sum_{j=1}^{m} \left(\frac{\partial w}{\partial x_j} * \frac{\partial H}{\partial p_j} - \frac{\partial w}{\partial p_j} * \frac{\partial H}{\partial x_j} \right) = 0, \tag{6.21}$$

where $*$ denotes convolution in the space-time variable $\mathbf{y} = (\mathbf{y}', t)$,

$$P * w = \int_{\mathrm{R}^{n+1}} P(x, p, \mathbf{y} - \mathbf{q}) w(x, p, \mathbf{q}) \, d\mathbf{q}.$$

Note that functions P, H can depend functionally on w.

Consider some examples.

Let $H = (x^2 + p^2)\delta(t)\delta(y)$. When $P = a(x,p)\delta'(t)\delta(y) + b(x,p)\delta(t)\delta'(y)$, $m = 1$, $\mathbf{y} = (y, t)$, equation (6.21) will transform into

$$a(x,p)\frac{\partial w}{\partial t} + b(x,p)\frac{\partial w}{\partial y} + 2p\frac{\partial w}{\partial x} - 2x\frac{\partial w}{\partial p} = 0.$$

In the case of $H = [R(p) + Q(x)]\delta(t)\delta(y)$, $P = a_1(x,p)\delta''(t)\delta(y) + a_2(x,p)\delta(t)\delta''(y) + a_3(x,p)\delta'(t)\delta(y) + a_4(x,p)\delta(t)\delta'(y) + a_5(x,p)\delta(t)\delta(y)$, $m = 1$, $\mathbf{y} = (y,t)$, equation (6.21) assumes the following form

$$a_1\frac{\partial^2 w}{\partial t^2} + a_2\frac{\partial^2 w}{\partial y^2} + a_3\frac{\partial w}{\partial t} + a_4\frac{\partial w}{\partial y} + a_5 w + R'(p)\frac{\partial w}{\partial x} - Q'(x)\frac{\partial w}{\partial p} = 0.$$

Now if $P = a(x,p)\delta(t)\,e^{-|y|} + b(x,p)\delta(t)\delta'(y)$, $m = 1$, $\mathbf{y} = (y,t)$, then equation (6.21) is as follows

$$a(x,p)\int_{-\infty}^{\infty} w(x,p,t,\xi)\,e^{-|y-\xi|}\,d\xi + b(x,p)\frac{\partial w}{\partial y} + 2p\frac{\partial w}{\partial x} - 2x\frac{\partial w}{\partial p} = 0.$$

When $H = 1$, and $P = \mu\delta(t)\delta(y) - k(x,p,t,y)$, $m = 1$, $\mathbf{y} = (y,t)$, we obtain a homogeneous integral equation with a difference kernel

$$\mu w(x,p,t,y) - \int_{-\infty}^{\infty}\int_{-\infty}^{\infty} k(x,p,t-\tau,y-\xi) w(x,p,\tau,\xi)\, d\xi\, d\tau = 0.$$

Let $P = f(w) + P_1$, где $f(s)$, $s \in \mathrm{R}$, is some function, and $P_1 = P_1(\mathbf{x},\mathbf{p},\mathbf{y})$. In this case, we have a nonlinear equation for w,

$$f(w) * w + P_1 * w + \sum_{j=1}^{m} \left(\frac{\partial w}{\partial x_j} * \frac{\partial H}{\partial p_j} - \frac{\partial w}{\partial p_j} * \frac{\partial H}{\partial x_j} \right) = 0.$$

In particular, with proper choice of f and P_1 we can obtain an analogue of a logistic equation,

$$\frac{\partial w}{\partial t} - r\left(1 - \frac{w}{k}\right) + \sum_{j=1}^{m} \left(\frac{\partial w}{\partial x_j} * \frac{\partial H}{\partial p_j} - \frac{\partial w}{\partial p_j} * \frac{\partial H}{\partial x_j} \right) = 0.$$

Knowing w we can introduce the function of entropy, $N(t)$, and the function of work, $A(t)$, as

$$N(t) = -\int_{\mathrm{R}^{2m+n}} w \ln w \, d\mathbf{x}\, d\mathbf{p}\, d\mathbf{y}', \quad A(t) = \frac{\displaystyle\int_{\mathrm{R}^{n+m}}\int_{|\mathbf{p}|\geq a} w\, d\mathbf{x}\, d\mathbf{p}\, d\mathbf{y}'}{\displaystyle\int_{\mathrm{R}^{2m+n}} w\, d\mathbf{x}\, d\mathbf{p}\, d\mathbf{y}'},$$

where $a > 0$ is some number.

The mathematical model proposed in the present paper consists of three functions $P(x,p,\mathbf{y})$, $H(x,p,\mathbf{y})$, and $w(x,p,\mathbf{y})$, which in general are unknown *a priori* and are subject to equation (6.21).

The results of the present research, which are presented below, consist, first, in a number of formulas for functions P, H, w, these formulas containing five arbitrary functions of a lesser number of variables; second, in the description of correctness classes for these formulas; and, last, in formulas for solutions to some inverse problems.

Theorem 6.5. *Let $f(z,\boldsymbol{\xi})$, $g(z,\boldsymbol{\xi})$, $h(z,\boldsymbol{\xi})$, $R(z,\boldsymbol{\xi})$, $Q(z,\boldsymbol{\xi})$, $z \in \mathrm{R}^1$, $\boldsymbol{\xi} \in \mathrm{R}^{n+1}$, be some differentiable or analytical complex-valued functions such that the following formulas define the functions $w(x,p,\mathbf{y})$, $P(x,p,\mathbf{y})$, and $H(x,p,\mathbf{y})$ correctly,*

$$w(x,p,\mathbf{y}) = \int_{\mathrm{R}^{n+1}} \Big\{ f(R(p,\boldsymbol{\xi}) + Q(x,\boldsymbol{\xi}),\boldsymbol{\xi}) \times e^{\int_0^p g\left(R(p,\boldsymbol{\xi}) - R(\eta,\boldsymbol{\xi}) + Q(x,\boldsymbol{\xi}),\boldsymbol{\xi}\right) d\eta + \int_{x_0}^{x} h\left(Q(x,\boldsymbol{\xi}) - Q(\eta,\boldsymbol{\xi}) + R(p,\boldsymbol{\xi}),\boldsymbol{\xi}\right) d\eta} \Big\} e^{i\mathbf{y}\boldsymbol{\xi}}\, d\boldsymbol{\xi}, \tag{6.22}$$

$$P(x,p,\mathbf{y}) = \int_{\mathrm{R}^{n+1}} \left[g(Q(x,\boldsymbol{\xi}),\boldsymbol{\xi}) \frac{\partial Q}{\partial x} - h(R(p,\boldsymbol{\xi}),\boldsymbol{\xi}) \frac{\partial R}{\partial p} \right] e^{i\mathbf{y}\boldsymbol{\xi}}\, d\boldsymbol{\xi}, \tag{6.23}$$

$$H(x,p,\mathbf{y}) = \int\limits_{\mathbf{R}^{n+1}} [R(p,\boldsymbol{\xi}) + Q(x,\boldsymbol{\xi})]\mathrm{e}^{\mathrm{i}\mathbf{y}\boldsymbol{\xi}}\,\mathrm{d}\boldsymbol{\xi}. \tag{6.24}$$

Then functions $w(x,p,\mathbf{y})$, $P(x,p,\mathbf{y})$, $H(x,p,\mathbf{y})$ *are a formal solution to equation (6.21)* $m = 1$.

PROOF. Let us apply the inverse Fourier transform in the variable $\mathbf{y}$ to equation (6.21),

$$\hat{P}(x,p,\boldsymbol{\xi}) = \int\limits_{\mathbf{R}^{n+1}} P(x,p,\mathbf{y})\mathrm{e}^{-\mathrm{i}\mathbf{y}\boldsymbol{\xi}}\,\mathrm{d}\mathbf{y}.$$

Then equation (6.21) transforms into

$$\hat{P}\hat{w} + \frac{\partial \hat{w}}{\partial x}\frac{\partial \hat{H}}{\partial p} - \frac{\partial \hat{w}}{\partial p}\frac{\partial \hat{H}}{\partial x} = 0. \tag{6.25}$$

Denote

$$S(x,p,\boldsymbol{\xi}) = \mathrm{e}^{\left(\int_0^p g(R(p,\boldsymbol{\xi})-R(\eta,\boldsymbol{\xi})+Q(x,\boldsymbol{\xi}),\boldsymbol{\xi})\,\mathrm{d}\eta + \int_{x_0}^x h(Q(x,\boldsymbol{\xi})-Q(\eta,\boldsymbol{\xi})+R(p,\boldsymbol{\xi}),\boldsymbol{\xi})\,\mathrm{d}\eta\right)},$$

$$f_1'(z,\boldsymbol{\xi}) = \frac{\partial f}{\partial z}.$$

then

$$\frac{\partial \hat{H}}{\partial p} = \frac{\partial R}{\partial p}, \quad \frac{\partial \hat{H}}{\partial x} = \frac{\partial Q}{\partial x},$$

$$\begin{aligned}\frac{\partial \hat{w}}{\partial x} &= f_1'(R(p,\boldsymbol{\xi}) + Q(x,\boldsymbol{\xi}),\boldsymbol{\xi})S(x,p,\boldsymbol{\xi})\frac{\partial Q}{\partial x} + f(R(p,\boldsymbol{\xi}) + Q(x,\boldsymbol{\xi}),\boldsymbol{\xi})S(x,p,\boldsymbol{\xi}) \\ &\times \left[\frac{\partial Q}{\partial x}\int\limits_0^p g_1'(R(p,\boldsymbol{\xi}) - R(\eta,\boldsymbol{\xi}) + Q(x,\boldsymbol{\xi}),\boldsymbol{\xi})\,\mathrm{d}\eta + h(R(p,\boldsymbol{\xi}),\boldsymbol{\xi})\right. \\ &\left. + \frac{\partial Q}{\partial x}\int\limits_{x_0}^x h_1'(Q(x,\boldsymbol{\xi}) - Q(\eta,\boldsymbol{\xi}) + R(p,\boldsymbol{\xi}),\boldsymbol{\xi})\,\mathrm{d}\eta\right],\end{aligned}$$

$$\begin{aligned}\frac{\partial \hat{w}}{\partial p} &= f_1'(R(p,\boldsymbol{\xi}) + Q(x,\boldsymbol{\xi}),\boldsymbol{\xi})S(x,p,\boldsymbol{\xi})\frac{\partial R}{\partial p} + f(R(p,\boldsymbol{\xi}) + Q(x,\boldsymbol{\xi}),\boldsymbol{\xi})S(x,p,\boldsymbol{\xi}) \\ &\times \left[g(Q(x,\boldsymbol{\xi}),\boldsymbol{\xi}) + \frac{\partial R}{\partial p}\int\limits_0^p g_1'(R(p,\boldsymbol{\xi}) - R(\eta,\boldsymbol{\xi}) + Q(x,\boldsymbol{\xi}),\boldsymbol{\xi})\,\mathrm{d}\eta\right. \\ &\left. + \frac{\partial R}{\partial p}\int\limits_{x_0}^x h_1'(Q(x,\boldsymbol{\xi}) - Q(\eta,\boldsymbol{\xi}) + R(p,\boldsymbol{\xi}),\boldsymbol{\xi})\,\mathrm{d}\eta\right].\end{aligned}$$

Substituting $\frac{\partial \hat{w}}{\partial x}$, $\frac{\partial \hat{w}}{\partial p}$, $\frac{\partial \hat{H}}{\partial x}$, $\frac{\partial \hat{H}}{\partial p}$, $\hat{P}$, $\hat{H}$ in (6.25) we satisfy ourselves that the statement of the theorem is true. The theorem is proved.

By virtue of Theorem 6.5 the problem of simulation of ethnic processes in the frames of a predetermined model is related to assignment or search for five functions f, g, h, R, Q.

Let functions $g(z,\boldsymbol{\xi})$, $h(z,\boldsymbol{\xi})$ be polynomials in z and $\boldsymbol{\xi}$, $\boldsymbol{\xi} \in \mathrm{R}^{n+1}$, $z \in \mathrm{R}^1$, and let functions $Q(x,\boldsymbol{\xi})$, $R(p,\boldsymbol{\xi})$ be polynomials in $\boldsymbol{\xi}$ with coefficients depending on x and p respectively.

Then, according to formulas (6.23), (6.24) the equation

$$P * w + \frac{\partial w}{\partial x} * \frac{\partial H}{\partial p} - \frac{\partial w}{\partial p} * \frac{\partial H}{\partial x} = 0$$

is a differential equation of the form

$$\sum_{|\alpha|\le m} \left\{ [A_\alpha(x) + B_\alpha(p)]\, D_y^\alpha w + N_\alpha(p) D_y^\alpha \frac{\partial w}{\partial x} - M_\alpha(x) D_y^\alpha \frac{\partial w}{\partial p} \right\} = 0,$$

where

$$\sum_{|\alpha|\le m} A_\alpha(x) D_y^\alpha = g\Big(Q(x,-\mathrm{i}\,D), -\mathrm{i}\,D\Big) \frac{\partial Q}{\partial x}(x,-\mathrm{i}\,D),$$

$$\sum_{|\alpha|\le m} B_\alpha(p) D_y^\alpha = h\Big(R(p,-\mathrm{i}\,D), -\mathrm{i}\,D\Big) \frac{\partial R}{\partial p}(x,-\mathrm{i}\,D),$$

$$\sum_{|\alpha|\le m} N_\alpha(p) D_y^\alpha = \frac{\partial R}{\partial p}(p,-\mathrm{i}\,D), \quad \sum_{|\alpha|\le m} M_\alpha(p) D_y^\alpha = \frac{\partial Q}{\partial x}(x,-\mathrm{i}\,D),$$

$D_y^\alpha = \dfrac{\partial^{\alpha_1}}{\partial y_1} \cdots \dfrac{\partial^{\alpha_{n+1}}}{\partial y_{n+1}}$, and m is the maximal power of polynomials appearing here.

Let us present some examples of concrete equations and their solutions obtained by formulas (6.22)–(6.24).

Example 6.4. Let $n = 1$, $\boldsymbol{\xi} = (\tau, \xi)$, $\mathbf{y} = (t, y)$,

$$Q(x,\boldsymbol{\xi}) = x^2, \quad R(p,\boldsymbol{\xi}) = p^2, \quad g(z,\boldsymbol{\xi}) = z(\tau + \xi), \quad h(z,\boldsymbol{\xi}) = z(\tau - \xi),$$

$$f(z,\boldsymbol{\xi}) = \begin{cases} z(\tau^2 + \xi^2), & |z| < R_1, |\tau| < R_2, |\xi| < R_3 \\ 0, & \text{in other points.} \end{cases}$$

Then equation (6.21) assumes the following form

$$-\mathrm{i}(x^3 - p^3)\frac{\partial w}{\partial t} - \mathrm{i}(x^3 + p^3)\frac{\partial w}{\partial y} + 2p\frac{\partial w}{\partial x} - 2x\frac{\partial w}{\partial p} = 0,$$

and function $w(x,p,\mathbf{y})$ can be written as follows

$$w(x,p,\mathbf{y}) = (x^2 + p^2)(\mathrm{e}^{aR_2 + \mathrm{i}\,R_2 t} - \mathrm{e}^{-aR_2 - \mathrm{i}\,R_2 t})(\mathrm{e}^{bR_3 + \mathrm{i}\,R_3 y} - \mathrm{e}^{-bR_3 y - \mathrm{i}\,R_3 y})$$

$$\times \left[\frac{R_3^2 - R_2^2}{(a+\mathrm{i}\,t)(b+\mathrm{i}\,y)} + \frac{2}{(a+\mathrm{i}\,t)(b+\mathrm{i}\,y)^3} + \frac{2}{(a+\mathrm{i}\,t)^3(b+\mathrm{i}\,y)} \right]$$

$$-2(x^2 + p^2)(\mathrm{e}^{aR_2 + \mathrm{i}\,R_2 t} - \mathrm{e}^{-aR_2 - \mathrm{i}\,R_2 t})(\mathrm{e}^{bR_3 + \mathrm{i}\,R_3 y} + \mathrm{e}^{-bR_3 y - \mathrm{i}\,R_3 y}) \frac{R_3}{(a+\mathrm{i}\,t)(b+\mathrm{i}\,y)^2}$$

$$-2(x^2 + p^2)(\mathrm{e}^{aR_2 + \mathrm{i}\,R_2 t} + \mathrm{e}^{-aR_2 - \mathrm{i}\,R_2 t})(\mathrm{e}^{bR_3 + \mathrm{i}\,R_3 y} - \mathrm{e}^{-bR_3 y - \mathrm{i}\,R_3 y}) \frac{R_2}{(a+\mathrm{i}\,t)^2(b+\mathrm{i}\,y)},$$

where

$$a = \frac{2}{3}p^3 + x^2p + \frac{2}{3}p^3 + p^2x - x^2x_0 + \frac{x_0^2}{3} - p^2x_0,$$

$$b = \frac{2}{3}p^3 + x^2p - \frac{2}{3}p^3 - p^2x + x^2x_0 - \frac{x_0^2}{3} + p^2x_0.$$

Example 6.5. Let $n = 1$, $\boldsymbol{\xi} = (\tau, \xi)$, $\mathbf{y} = (t, y)$,

$$R(p, \boldsymbol{\xi}) = R(p), \quad Q(x, \boldsymbol{\xi}) = Q(x),$$

$$g(z, \boldsymbol{\xi}) = -g_{11}(z)\tau^2 - g_{22}(z)\xi^2 + \mathrm{i}\, g_1(z)\tau + \mathrm{i}\, g_2(z)\xi + g_0(z),$$

$$h(z, \boldsymbol{\xi}) = -h_{11}(z)\tau^2 - h_{22}(z)\xi^2 + \mathrm{i}\, h_1(z)\tau + \mathrm{i}\, h_2(z)\xi + h_0(z),$$

$$f(z, \boldsymbol{\xi}) = \mathrm{e}^{-f_{11}(z)\tau^2} - f_{22}(z)\xi^2 + \mathrm{i}\, f_1(z)\tau + \mathrm{i}\, f_2(z)\xi + f_0(z),$$

and let $p > 0$, $x > x_0$, $g_{11} > 0$, $h_{11} > 0$, $f_{11} > 0$. Then equation (6.21) has the following form

$$\Big[g_{11}\big(Q(x)\big)Q'(x) - h_{11}\big(R(p)\big)R'(p)\Big]\frac{\partial^2 w}{\partial t^2} + \Big[g_{22}\big(Q(x)\big)Q'(x) - h_{22}\big(R(p)\big)R'(p)\Big]\frac{\partial^2 w}{\partial y^2}$$
$$+\Big[g_1\big(Q(x)\big)Q'(x) - h_1\big(R(p)\big)R'(p)\Big]\frac{\partial w}{\partial t} + \Big[g_2\big(Q(x)\big)Q'(x) - h_2\big(R(p)\big)R'(p)\Big]\frac{\partial w}{\partial y}$$
$$+\Big[g_0\big(Q(x)\big)Q'(x) - h_0\big(R(p)\big)R'(p)\Big]w + R'(p)\frac{\partial w}{\partial x} - Q'(x)\frac{\partial w}{\partial p} = 0.$$

The solution $w(x, p, \mathbf{y})$ looks like

$$w(x, p, \mathbf{y}) = \exp\left[\int_0^p g_0\big(U(x, p, \eta)\big)\,\mathrm{d}\eta + \int_{x_0}^x h_0\big(V(x, p, \eta)\big)\,\mathrm{d}\eta\right]$$
$$\times \frac{\pi}{\sqrt{f_{11} + \int_0^p g_{11}\big(U(x, p, \eta)\big)\,\mathrm{d}\eta + \int_{x_0}^x h_{11}\big(V(x, p, \eta)\big)\,\mathrm{d}\eta}}$$
$$\times \frac{1}{\sqrt{f_{22} + \int_0^p g_{22}\big(U(x, p, \eta)\big)\,\mathrm{d}\eta + \int_{x_0}^x h_{22}\big(V(x, p, \eta)\big)\,\mathrm{d}\eta}}$$
$$\times \exp\left\{ -\frac{1}{4}\left[\left(f_1 + \int_0^p g_1\big(U(x, p, \eta)\big)\,\mathrm{d}\eta + \int_{x_0}^x h_1\big(V(x, p, \eta)\big)\,\mathrm{d}\eta + t\right)^2\right.\right.$$
$$\times \left(f_{11} + \int_0^p g_{11}\big(U(x, p, \eta)\big)\,\mathrm{d}\eta + \int_{x_0}^x h_{11}\big(V(x, p, \eta)\big)\,\mathrm{d}\eta\right)^{-1}$$

$$+\left(f_{22}+\int_0^p g_{22}\big(U(x,p,\eta)\big)\,\mathrm{d}\eta+\int_{x_0}^x h_{22}\big(V(x,p,\eta)\big)\,\mathrm{d}\eta\right)^{-1}$$

$$\times\left(f_2+\int_0^p g_2\big(U(x,p,\eta)\big)\,\mathrm{d}\eta+\int_{x_0}^x h_2\big(V(x,p,\eta)\big)\,\mathrm{d}\eta\right)^2\Bigg]\Bigg\}.$$

where $U(x,p,\eta)=R(p)-R(\eta)+Q(x)$, $V(x,p,\eta)=Q(x)-Q(\eta)+R(p)$.

Example 6.6. Let $n=1$, $\boldsymbol{\xi}=(\tau,\xi)$, $\mathbf{y}=(t,y)$,

$$Q(x,\boldsymbol{\xi})=x^2,\quad R(p,\boldsymbol{\xi})=p^2,\quad g(z,\boldsymbol{\xi})=z,\quad h(z,\boldsymbol{\xi})=z,$$

$$f(z,\boldsymbol{\xi})=z\,\mathrm{e}^{-4\xi^2-4\tau^2}.$$

Then solution $w(x,p,\mathbf{y})$ has the following representation

$$w(x,p,\mathbf{y})=\pi(p+x)\,\mathrm{e}^{-t^2-y^2}\exp\left(\frac{2}{3}(p^3+x^3)+x^2p+p^2x-x^2x_0+\frac{x_0^3}{3}+p^2x_0\right).$$

Functions $P(x,p,\mathbf{y})$ and $H(x,p,\mathbf{y})$ will be as follows

$$P(x,p,\mathbf{y})=8\pi(x^3-p^3)\delta(t)\delta(y),\quad H(x,p,\mathbf{y})=4\pi(x^2+p^2)\delta(t)\delta(y),$$

and equation (6.21) will be written as

$$2(x^3-p^3)w+2p\frac{\partial w}{\partial x}-2x\frac{\partial w}{\partial p}=0.$$

Example 6.7. Let $n=1$, $\boldsymbol{\xi}=(\tau,\xi)$, $\mathbf{y}=(t,y)$, $p>0$,

$$Q(x,\boldsymbol{\xi})=x,\quad R(p,\boldsymbol{\xi})=p,\quad g(z,\boldsymbol{\xi})=-|\tau|,\quad h(z,\boldsymbol{\xi})=z\xi,$$

$$f(z,\boldsymbol{\xi})=\begin{cases}-\dfrac{\mathrm{i}\tau}{4}z\xi, & |\xi|<R,\\ 0, & \text{in other points.}\end{cases}$$

Then functions $H(x,p,t,y)$, $P(x,p,t,y)$, $w(x,p,t,y)$ can be written as follows

$$H(x,p,t,y)\;=\;p\delta(t)\delta(y)+x\delta(t)\delta(y),$$

$$P(x,p,t,y)\;=\;2t^{-2}\delta(y)-2\mathrm{i}\pi\delta(t)\delta(y),$$

$$w(x,p,t,y)=\frac{tp(x+p)}{(t^2+p^2)^2}\left[\frac{R}{a+\mathrm{i}y}\left(\mathrm{e}^{R(a+\mathrm{i}y)}+\mathrm{e}^{-R(a+\mathrm{i}y)}\right)-\frac{1}{(a+\mathrm{i}y)^2}\left(\mathrm{e}^{R(a+\mathrm{i}y)}-\mathrm{e}^{-R(a+\mathrm{i}y)}\right)\right].$$

Below some correctness classes for formulas (6.22)–(6.24) are presented.

Let θ denote the set of infinitely differentiable functions $\varphi(q)$ satisfying the estimate

$$|D^\alpha\varphi(q)|\le C_\alpha(1+|q|)^{m_\alpha},$$

$$q\in\mathrm{R}^s,\quad s\ge1,\quad C_\alpha\ge0,\quad C_\alpha=\text{const}\,,\quad m_\alpha=\text{const}\,.$$

The set θ is called the set of multiplicators in the space of basic functions (Vladimirov, 1976).

Theorem 6.6. *Let $g(z,\xi)$, $h(z,\xi)$ be infinitely differentiable with respect to z and ξ functions belonging to the set θ in variables z and ξ; let $R(z,\xi)$, $Q(z,\xi)$ be infinitely differentiable with respect to z functions belonging, along with their derivatives $\frac{\partial Q}{\partial x}$ and $\frac{\partial R}{\partial p}$, to the set θ in variable ξ; and let $f(z,\xi)$ be a finite distribution in ξ integrable with respect to z.*

Then formulas (6.22)-(6.24) define functions $w(x,p,\mathrm{y})$, $P(x,p,\mathrm{y})$, $H(x,p,\mathrm{y})$ correctly.

PROOF. It is obvious that, in general, the distribution $H(x,p,\mathrm{y})$ is defined correctly by (6.24).

To ascertain the correctness of the function $P(x,p,\mathrm{y})$, consider the function $g(Q(x,\xi),\xi)\frac{\partial Q}{\partial x}$. We have the following estimate

$$\left|D_{\xi}^{\alpha}\left(g(Q(x,\xi),\xi)\frac{\partial Q}{\partial x}\right)\right| \le \sum_{\beta<\alpha}\binom{\beta}{\alpha}\left|D_{\xi}^{\beta}g(Q(x,\xi),\xi)D_{\xi}^{\alpha-\beta}\frac{\partial Q}{\partial x}\right|.$$

By condition of the theorem, $g(z,\xi)$, $Q(x,\xi)$, $\frac{\partial Q}{\partial x}$ belong to the set θ; therefore, there exist such constants $C \ge 0$, $m \ge 0$ that

$$\left|D_{\xi}^{\alpha}\left(g(Q(x,\xi),\xi)\frac{\partial Q}{\partial x}\right)\right| \le C(1+|\xi|)^m.$$

Having repeated analogous considerations for the function $h\big(R(p,\xi),\xi\big)\frac{\partial R}{\partial p}$, we assure ourselves that function $P(x,p,\mathrm{y})$ is defined correctly by (6.23).

By virtue of the finitenes of the function $f(z,\xi)$ in variable ξ the function

$$f\big(R(p,\xi)+Q(x,\xi),\xi\big)\,\mathrm{e}^{\left(\int\limits_0^p g(R(p,\xi)-R(\eta,\xi)+Q(x,\xi),\xi)\,\mathrm{d}\eta+\int\limits_{x_0}^x h(Q(x,\xi)-Q(\eta,\xi)+R(p,\xi),\xi)\,\mathrm{d}\eta\right)}$$

is finite in ξ, and the function $w(x,p,\mathrm{y})$ is defined correctly by (6.22). The theorem is proved.

Theorem 6.7. *Let function $f(z,\xi)$ belong to the set θ in both variables, let functions $g(z,\xi)$, $h(z,\xi)$, $R(z,\xi)$, $Q(z,\xi)$ be polynomials in z and ξ, and let*

$$\int\limits_0^p g\big(R(p,\xi)-R(\eta,\xi)+Q(x,\xi),\xi\big)\,\mathrm{d}\eta \le -\alpha(x,p)\,|\xi|^{2N}+\mathcal{P}_{2N-1}(\xi),$$

$$\int\limits_{x_0}^x h\big(Q(x,\xi)-Q(\eta,\xi)+R(p,\xi),\xi\big)\,\mathrm{d}\eta \le -\beta(x,p)\,|\xi|^{2M}+\mathcal{P}_{2M-1}(\xi),$$

where $\alpha(x,p)>0$, $\beta(x,p)>0$ for all x, p, and $\mathcal{P}_{2N-1}$, $\mathcal{P}_{2M-1}$ are polynomials of powers no greater than $2N-1$, $2M-1$ respectively.

Then formulas of Theorem 6.5 are defined correctly.

Proof. It is obvious that, in general, the distribution $H(x,p,\mathbf{y})$ is defined correctly by (6.24).

To ascertain the correctness of the function $P(x,p,\mathbf{y})$, consider the function $g(Q(x,\boldsymbol{\xi}),\boldsymbol{\xi})\dfrac{\partial Q}{\partial x}$. By the above condition, functions $g(z,\boldsymbol{\xi})$, $Q(x,\boldsymbol{\xi})$ are polynomials in z and $\boldsymbol{\xi}$. Therefore, function $g(Q(x,\boldsymbol{\xi}),\boldsymbol{\xi})\dfrac{\partial Q}{\partial x}$ is also a polynomial in x and $\boldsymbol{\xi}$. Similarly, function $h(R(p,\boldsymbol{\xi}),\boldsymbol{\xi})\dfrac{\partial R}{\partial p}$ is also a polynomial in p and $\boldsymbol{\xi}$. We can conclude that, generally speaking, the distribution $P(x,p,\mathbf{y})$ is defined correctly by (6.23).

To prove the correctness of the function $w(x,p,\mathbf{y})$, consider the function $e^{\left(\int_0^p g(R(p,\boldsymbol{\xi})-R(\eta,\boldsymbol{\xi})+Q(x,\boldsymbol{\xi}),\boldsymbol{\xi})\,d\eta\right)}$.

By condition of the theorem

$$\int_0^p g\big(R(p,\boldsymbol{\xi})-R(\eta,\boldsymbol{\xi})+Q(x,\boldsymbol{\xi}),\boldsymbol{\xi}\big)\,d\eta \le -\alpha(x,p)\,|\boldsymbol{\xi}|^{2N}+\mathcal{P}_{2N-1}(\boldsymbol{\xi}),$$

where $\alpha(x,p)>0$ for all x, p.

Therefore, the following estimate holds

$$\exp\left(\int_0^p g\big(R(p,\boldsymbol{\xi})-R(\eta,\boldsymbol{\xi})+Q(x,\boldsymbol{\xi}),\boldsymbol{\xi}\big)\,d\eta\right) \le \exp\left(-\alpha(x,p)\,|\boldsymbol{\xi}|^{2N}+\mathcal{P}_{2N-1}(\boldsymbol{\xi})\right).$$

Since the power of the polynomial $\mathcal{P}_{2N-1}(\boldsymbol{\xi})$ is no greater than $2N-1$, the function $\exp\left(\int_0^p g\big(R(p,\boldsymbol{\xi})-R(\eta,\boldsymbol{\xi})+Q(x,\boldsymbol{\xi}),\boldsymbol{\xi}\big)\,d\eta\right)$ decreases faster than any power of $|\boldsymbol{\xi}|^{-1}$ when $\boldsymbol{\xi}$ tends to infinity. Following the same lines for $\exp\left(\int_{x_0}^x h\big(Q(x,\boldsymbol{\xi})-Q(\eta,\boldsymbol{\xi})+R(p,\boldsymbol{\xi}),\boldsymbol{\xi}\big)\,d\eta\right)$ and taking into account that $f(z,\boldsymbol{\xi})$ belongs to the set θ in $\boldsymbol{\xi}$, we conclude that $w(x,p,\mathbf{y})$ is defined correctly by (6.22). The theorem is proved.

Now we present the following separation theorem.

Theorem 6.8. *Let $\hat{w}(x,p,\boldsymbol{\xi})$ be the inverse Fourier transform of the function $w(x,p,\mathbf{y})$ in the variable $\mathbf{y}$.*

Let function $R(p,\boldsymbol{\xi})$ be even in p. Then the following formulas hold true

$$\begin{aligned}\ln\hat{w}(x,p,\boldsymbol{\xi})^{+} &= \ln\frac{1}{(2\pi)^{n+1}}f\big(R(p,\boldsymbol{\xi})+Q(x,\boldsymbol{\xi}),\boldsymbol{\xi}\big)\\ &\quad+\int_{x_0}^{x} h\big(Q(x,\boldsymbol{\xi})-Q(\eta,\boldsymbol{\xi})+R(p,\boldsymbol{\xi}),\boldsymbol{\xi}\big)\,d\eta, \qquad (6.26)\\ \ln\hat{w}(x,p,\boldsymbol{\xi})^{-} &= \int_0^p g\big(R(p,\boldsymbol{\xi})-R(\eta,\boldsymbol{\xi})+Q(x,\boldsymbol{\xi}),\boldsymbol{\xi}\big)\,d\eta,\end{aligned}$$

where + and − denote the even and odd parts of $\ln \hat{w}(x,p,\boldsymbol{\xi})$.

If, in addition to the above conditions, $x_0 = 0$, *and* $Q(x,\boldsymbol{\xi})$ *is even in* x, *then*

$$\begin{aligned} \ln \hat{w}(x,p,\boldsymbol{\xi})^{+}_{+} &= \ln \frac{1}{(2\pi)^{n+1}} f\big(R(p,\boldsymbol{\xi}) + Q(x,\boldsymbol{\xi}),\boldsymbol{\xi}\big), \\ \ln \hat{w}(x,p,\boldsymbol{\xi})^{\pm} &= \int_{x_0}^{x} h\big(Q(x,\boldsymbol{\xi}) - Q(\eta,\boldsymbol{\xi}) + R(p,\boldsymbol{\xi}),\boldsymbol{\xi}\big)\,\mathrm{d}\eta, \end{aligned} \tag{6.27}$$

where $^{+}_{+}$ *denotes evenness in* p *and* x, *and* $\pm$ *denotes evenness in* p *and oddness in* x.

PROOF.

$$\hat{w}(x,p,\boldsymbol{\xi}) = (2\pi)^{-(n+1)} f\big(R(p,\boldsymbol{\xi}) + Q(x,\boldsymbol{\xi}),\boldsymbol{\xi}\big) \times \mathrm{e}^{\int_0^p g(R(p,\boldsymbol{\xi})-R(\eta,\boldsymbol{\xi})+Q(x,\boldsymbol{\xi}),\boldsymbol{\xi})\,\mathrm{d}\eta + \int_{x_0}^{x} h(Q(x,\boldsymbol{\xi})-Q(\eta,\boldsymbol{\xi})+R(p,\boldsymbol{\xi}),\boldsymbol{\xi})\,\mathrm{d}\eta}.$$

Since $R(p,\boldsymbol{\xi})$ is even in p, we obtain

$$\hat{w}(x,-p,\boldsymbol{\xi}) = (2\pi)^{-(n+1)} f\big(R(p,\boldsymbol{\xi}) + Q(x,\boldsymbol{\xi}),\boldsymbol{\xi}\big) \times \mathrm{e}^{-\int_0^p g(R(p,\boldsymbol{\xi})-R(\eta,\boldsymbol{\xi})+Q(x,\boldsymbol{\xi}),\boldsymbol{\xi})\,\mathrm{d}\eta + \int_{x_0}^{x} h(Q(x,\boldsymbol{\xi})-Q(\eta,\boldsymbol{\xi})+R(p,\boldsymbol{\xi}),\boldsymbol{\xi})\,\mathrm{d}\eta},$$

whence formulas (6.26) are obtained.

Now let $Q(x,\boldsymbol{\xi})$ be even in x, and let $x_0 = 0$. Then

$$\ln \hat{w}(-x,p,\boldsymbol{\xi})^{+} = \ln \frac{1}{(2\pi)^{n+1}} f\big(R(p,\boldsymbol{\xi})+Q(x,\boldsymbol{\xi}),\boldsymbol{\xi}\big) + \int_{x_0}^{x} h\big(Q(x,\boldsymbol{\xi})-Q(\eta,\boldsymbol{\xi})+R(p,\boldsymbol{\xi}),\boldsymbol{\xi}\big)\,\mathrm{d}\eta,$$

whence formulas (6.27) are obtained.

COROLLARY 6.1. *If the conditions of the theorem are satisfied, and* $h(0,\boldsymbol{\xi}) = 0$, $Q(0,\boldsymbol{\xi}) = 0$, *then the following equalities hold true*

$$\begin{aligned} \ln \hat{w}(0,p,\boldsymbol{\xi})^{-} &= \int_0^p g\big(R(p,\boldsymbol{\xi}) - R(\eta,\boldsymbol{\xi}),\boldsymbol{\xi}\big)\,\mathrm{d}\eta, \\ \ln \hat{w}(x,0,\boldsymbol{\xi})^{\pm} &= \int_0^x h\big(Q(x,\boldsymbol{\xi}) - Q(\eta,\boldsymbol{\xi}),\boldsymbol{\xi}\big)\,\mathrm{d}\eta, \\ R(p,\boldsymbol{\xi}) + Q(x,\boldsymbol{\xi}) &= f^{-1}\Big((2\pi)^{n+1}\exp\big(\ln \hat{w}(x,0,\boldsymbol{\xi})^{+}_{+}\big),\boldsymbol{\xi}\Big). \end{aligned}$$

PROOF. These formulas are ascertained by direct substitution of values $R(0,\boldsymbol{\xi}) = 0$, $Q(0,\boldsymbol{\xi}) = 0$ into formulas (6.26) and (6.27).

Theorem 6.9. *Let* $x_0 = 0$, $H = \frac{1}{2}\tilde{\alpha}(\mathbf{y})p^2 + \frac{1}{2}\tilde{\beta}(\mathbf{y})x^2$, $w_0(p,\mathbf{y}) = w|_{x=0}$, $w_1(x,\mathbf{y}) = w|_{p=0}$; *let* $\hat{w}_0(p,\boldsymbol{\xi})$, $\hat{w}_1(x,\boldsymbol{\xi})$ *be the inverse Fourier transforms of functions* w_0. w_1 *in variable* $\mathbf{y}$;

and let $\alpha(\boldsymbol{\xi})$, $\beta(\boldsymbol{\xi})$ be the inverse Fourier transforms of functions $\tilde{\alpha}(\mathbf{y})$, $\tilde{\beta}(\mathbf{y})$. Then the following formulas hold

$$
\begin{aligned}
f(z,\boldsymbol{\xi}) &= (2\pi)^{n+1}\sqrt{\hat{w}_0\left(\sqrt{\frac{2z}{\alpha(\boldsymbol{\xi})}},\boldsymbol{\xi}\right)\hat{w}_0\left(-\sqrt{\frac{2z}{\alpha(\boldsymbol{\xi})}},\boldsymbol{\xi}\right)}, \quad z\ge 0,\\
g(z,\boldsymbol{\xi}) &= \frac{\sqrt{2\alpha(\boldsymbol{\xi})}}{\pi}\frac{\mathrm{d}}{\mathrm{d}z}\int_0^z \ln\left[\frac{\hat{w}_0\left(\sqrt{2\eta/\alpha(\boldsymbol{\xi})},\boldsymbol{\xi}\right)}{\hat{w}_0\left(-\sqrt{2\eta/\alpha(\boldsymbol{\xi})},\boldsymbol{\xi}\right)}\right]\frac{\mathrm{d}\eta}{\sqrt{z-\eta}},\\
h(z,\boldsymbol{\xi}) &= \frac{\sqrt{2\beta(\boldsymbol{\xi})}}{\pi}\frac{\mathrm{d}}{\mathrm{d}z}\int_0^z \ln\left[\frac{\hat{w}_1\left(\sqrt{2\eta/\beta(\boldsymbol{\xi})},\boldsymbol{\xi}\right)}{\hat{w}_1\left(-\sqrt{2\eta/\beta(\boldsymbol{\xi})},\boldsymbol{\xi}\right)}\right]\frac{\mathrm{d}\eta}{\sqrt{z-\eta}}.
\end{aligned}
$$

PROOF. Since $H = \frac{1}{2}\tilde{\alpha}(\mathbf{y})p^2 + \frac{1}{2}\tilde{\beta}(\mathbf{y})x^2$, by virtue of (6.24) we obtain

$$\frac{1}{2}\tilde{\alpha}(\mathbf{y})p^2 = R(p,\boldsymbol{\xi}), \quad \frac{1}{2}\tilde{\beta}(\mathbf{y})x^2 = Q(x,\boldsymbol{\xi}).$$

Then

$$\hat{w}_0(p,\boldsymbol{\xi}) = \frac{1}{(2\pi)^{n+1}} f\left(\frac{1}{2}\tilde{\alpha}(\mathbf{y})p^2,\boldsymbol{\xi}\right)\exp\left(\int_0^p g\left(\frac{1}{2}\tilde{\alpha}(\mathbf{y})p^2 - \frac{1}{2}\tilde{\alpha}(\mathbf{y})\eta^2\right)\mathrm{d}\eta\right),$$

whence we obtain

$$f\left(\frac{1}{2}\tilde{\alpha}(\mathbf{y})p^2,\boldsymbol{\xi}\right) = (2\pi)^{n+1}\sqrt{\hat{w}_0(p,\boldsymbol{\xi})\hat{w}_0(-p,\boldsymbol{\xi})}.$$

Denoting $\frac{1}{2}\tilde{\alpha}(\mathbf{y})p^2 = z$ we get

$$f(z,\boldsymbol{\xi}) = (2\pi)^{n}\sqrt{\hat{w}_0\left(\sqrt{\frac{2z}{\alpha(\boldsymbol{\xi})}},\boldsymbol{\xi}\right)\hat{w}_0\left(-\sqrt{\frac{2z}{\alpha(\boldsymbol{\xi})}},\boldsymbol{\xi}\right)}.$$

We have the following relationship for g

$$\int_0^p g\left(\frac{1}{2}\tilde{\alpha}(\mathbf{y})p^2 - \frac{1}{2}\tilde{\alpha}(\mathbf{y})\eta^2\right)\mathrm{d}\eta = \ln\frac{\hat{w}_0(p,\boldsymbol{\xi})}{\hat{w}_0(-p,\boldsymbol{\xi})}.$$

Introducing a new variable $\zeta = \frac{1}{2}\tilde{\alpha}(\mathbf{y})p^2 - \frac{1}{2}\tilde{\alpha}(\mathbf{y})\eta^2$ and denoting $z = \frac{1}{2}\tilde{\alpha}(\mathbf{y})p^2$, we get

$$\sqrt{\frac{1}{2\alpha(\boldsymbol{\xi})}}\int_0^z \frac{g(\zeta)}{\sqrt{z-\zeta}}\,\mathrm{d}\zeta = \ln\left[\frac{\hat{w}_0\left(\sqrt{2z/\alpha(\boldsymbol{\xi})},\boldsymbol{\xi}\right)}{\hat{w}_0\left(\sqrt{-2z/\alpha(\boldsymbol{\xi})},\boldsymbol{\xi}\right)}\right].$$

Having solved this equation we arrive at

$$g(z,\xi)=\frac{\sqrt{2\alpha(\xi)}}{\pi}\frac{\mathrm{d}}{\mathrm{d}z}\int\limits_0^z \ln\left[\frac{\hat{w}_0\left(\sqrt{2\eta/\alpha(\xi)},\xi\right)}{\hat{w}_0\left(-\sqrt{2\eta/\alpha(\xi)},\xi\right)}\right]\frac{\mathrm{d}\eta}{\sqrt{z-\eta}}.$$

Formula for $h(z,\xi)$ is obtained similarly. The theorem is proved.

6.5 FORMULAS FOR SOLUTIONS AND COEFFICIENTS IN NONLINEAR EQUATIONS

6.5.1 Introduction

In this section we consider nonlinear multidimensional evolutionary systems of the first and second orders with respect to the time variable. Such systems are very important for description of real natural and societal processes. In this case, the nonlinearity is the essence of the problem in question and characterizes the interaction of the object.

Well-known examples are the nonlinear Schrödinger equations; the Turing equation system; the equation of the diffusion-reaction type; the nonlinear system describing the vibrational processes of connected oscillators; the populations theory equations; in particular, the equations of Kolmogorov, Petrovskii, Piskunov and so on (Cole, 1948; Zaslavskii *et al.*, 1982; Kolmogorov *et al.*, 1985; Haken, 1987; Svirizhev, 1987; Akhromeeva *et al.*, 1988). In differential equation theory, nonlinear equations are again widely met with. There exist general results in boundary-value problem solvability, a lot of particular solutions, group analysis, inverse problems and so on.

Here, we seek for not only solutions but also nonlinear terms of equations under consideration. In this case, we assume the solution to have a certain analytic form. In particular, we require some symmetry of solutions. It turns out in this case that for the desired vector-functions we obtain explicit formulas. These formulas have an essential generality, which may be used for investigation of both direct and inverse multidimensional problems for nonlinear equations. Some of these results were announced in Anikonov (1993a, 1994b).

Let R^n be the real Euclidean space, $x=(x_1,x_2,\dots,x_n)$, $n\geq 1$, $D\subset\mathrm{R}^n$ be a domain in R^n and

$$\mathbf{L}=\sum_{i,j=1}^{n} a_{ij}(x)\frac{\partial^2}{\partial x_i\partial x_j}+\sum_{i=1}^{n} a_i(x)\frac{\partial}{\partial x_i}$$

be a linear differential operator of the second order with continuous complex valued coefficients $a_{ij}(x)$, $a_i(x)$, $x\in D$.

We shall consider the following systems

$$\rho(x)\frac{\partial w}{\partial t}=\mathbf{L}w+\lambda(x,w), \tag{6.28}$$

$$\rho(x)\frac{\partial^2 w}{\partial t^2}=\mathbf{L}w+\lambda(x,w), \tag{6.29}$$

where $w(x,t)=(w_1,w_2,\dots,w_m)$, $m\geq 1$, $\lambda(x,z)=(\lambda_1,\lambda_2,\dots,\lambda_m)$ are the sought for complex-valued functions, $x\in D$, $\alpha<t<\beta$, $z\in\mathrm{R}^m$; $\rho(x)$ is a given continuous function; α, β are constants.

Note that our main restriction is that the solutions $w(x,t)$ of equations (6.28), (6.29) have one of the following representations

$$w(x,t) = u(x)F\left(\varphi(t) + v(x)\right), \tag{6.30}$$

$$w(x,t) = u(x)F\big(\varphi_1(t)v_1(x), \dots, \varphi_m(t)v_m(x)\big), \tag{6.31}$$

where $u(x)$ is sufficiently big smooth function, $x \in D$; F, φ, v are given sufficiently smooth mappings. The representation of $w(x,t)$ in the form (6.30) and (6.31) denotes certain a *priori* assumed symmetry, though the vector-functions entering these formulas are not known. Representations (6.30), (6.31) allow us to write down the formulas also for the vector-functions $\lambda(x,z)$ entering into (6.28), (6.29), which also may be determined by the vector-functions F, φ, v, u.

6.5.2 Formulations and proofs of the theorems

Let F, b, v, u be either twice differentiable or analytical mappings

$$F : \mathrm{R}^m \to \mathrm{R}^m\,, \quad b : \mathrm{R}^m \to \mathrm{R}^m,$$

$$v : D \to \mathrm{R}^m\,, \quad u : D \to \mathrm{R}^1,$$

and a mapping $z = F(y)$, $y \in \mathrm{R}^m$, $z \in \mathrm{R}^m$ have the inverse $y = F^{-1}(z)$, $u(x) \neq 0$, $x \in D$.

Let $\varphi(t)$, $\alpha < t < \beta$ solve the ordinary differential equations system

$$\frac{d\varphi}{dt} = b(\varphi). \tag{6.32}$$

The following theorems hold.

Theorem 6.10. *The vector-functions $w(x,t)$, $\lambda(x,z)$, $x \in D$, $\alpha < t < \beta$, $z \in \mathrm{R}^m$ defined as*

$$w(x,t) = u(x)F\big(\varphi(t) + v(x)\big)$$

$$\lambda(x,z) = \sum_{k=1}^{m} \frac{\partial F\big(F^{-1}(z/u)\big)}{\partial y_k}\left[\rho u b_k\left(F^{-1}\left(\frac{z}{u}\right) - v(x)\right) - 2\sum_{i,j=1}^{n} a_{ij}(x)\frac{\partial u}{\partial x_i}\frac{\partial v_k}{\partial x_j} - u\mathbf{L}v_k\right]$$

$$-u\sum_{k,l=1}^{m} \frac{\partial^2 F\big(F^{-1}(z/u)\big)}{\partial y_k \partial y_l}\sum_{i,j=1}^{n} a_{ij}(x)\frac{\partial v_k}{\partial x_i}\frac{\partial v_l}{\partial x_j} - z\frac{\mathbf{L}u}{u}$$

satisfy the nonlinear system of equations

$$\rho(x)\frac{\partial w}{\partial t} = \mathbf{L}w + \lambda(x,w).$$

Theorem 6.11. *The vector - functions $w(x,t)$, $\lambda(x,z)$, $x \in D$, $\alpha < t < \beta$, $z \in \mathrm{R}^m$ defined as*

$$w(x,t) = u(x)F\big(\varphi(t) + v(x)\big)$$

$$\lambda(x,z) = \sum_{k=1}^{m} \frac{\partial F(F^{-1}(z/u))}{\partial y_k}\left[\rho u \sum_{l=1}^{m} \frac{\partial b_k(F^{-1}(z/u)-v(x))}{\partial \varphi_l} b_l\left(F^{-1}\left(\frac{z}{u}\right)-v(x)\right) - 2\sum_{i,j=1}^{n} a_{ij}(x)\frac{\partial u}{\partial x_i}\frac{\partial v_k}{\partial x_j} - u\mathbf{L}v_k\right]$$
$$+\sum_{k,l=1}^{m} \frac{\partial^2 F(F^{-1}(z/u))}{\partial y_k \partial y_l}\left[\rho b_k\left(F^{-1}\left(\frac{z}{u}\right)-v(x)\right) b_l\left(F^{-1}\left(\frac{z}{u}\right)-v(x)\right) - \sum_{i,j=1}^{n} a_{ij}(x)\frac{\partial v_k}{\partial x_i}\frac{\partial v_l}{\partial x_j}\right] - z\frac{\mathbf{L}u}{u}$$

satisfy the nonlinear system of equations

$$\rho(x)\frac{\partial^2 w}{\partial t^2} = \mathbf{L}w + \lambda(x,w).$$

Proof of Theorems 6.10 and 6.11. Since $w(x,t)$ is represented in the form

$$w(x,t) = u(x)F(\varphi(t)+v(x))$$

we have

$$\frac{\partial w}{\partial t} = u\sum_{k=1}^{m} \frac{\partial F(\varphi+v)}{\partial y_k}\frac{\mathrm{d}\varphi_k}{\mathrm{d}t},$$
$$\frac{\partial^2 w}{\partial t^2} = u\sum_{k,l=1}^{m} \frac{\partial^2 F(\varphi+v)}{\partial y_k \partial y_l}\frac{\mathrm{d}\varphi_k}{\mathrm{d}t}\frac{\mathrm{d}\varphi_l}{\mathrm{d}t} + u\sum_{k=1}^{m}\frac{\partial F(\varphi+v)}{\partial y_k}\frac{\mathrm{d}^2\varphi_k}{\mathrm{d}t^2},$$
$$\frac{\partial w}{\partial x_i} = \frac{\partial u}{\partial x_i}F + u\sum_{k=1}^{m}\frac{\partial F(\varphi+v)}{\partial y_k}\frac{\partial v_k}{\partial x_i},$$
$$\frac{\partial^2 w}{\partial x_i \partial x_j} = \frac{\partial^2 u}{\partial x_i \partial x_j}F(\varphi+v) + u\sum_{k,l=1}^{m}\frac{\partial^2 F(\varphi+v)}{\partial y_k\partial y_l}\frac{\partial v_k}{\partial x_i}\frac{\partial v_l}{\partial x_j} + \sum_{k=1}^{m}\frac{\partial F(\varphi+v)}{\partial y_k}\left[\frac{\partial u}{\partial x_i}\frac{\partial v_k}{\partial x_j} + \frac{\partial u}{\partial x_j}\frac{\partial v_k}{\partial x_i} + u\frac{\partial^2 v_k}{\partial x_i \partial x_j}\right].$$

Since $\frac{\mathrm{d}\varphi_k}{\mathrm{d}t} = b_k(\varphi_1,\varphi_2,\ldots,\varphi_m)$, $k=1,2,\ldots,m$, then

$$\frac{\mathrm{d}^2\varphi_k}{\mathrm{d}t^2} = \sum_{l=1}^{m}\frac{\partial b_k}{\partial \varphi_l}\frac{\mathrm{d}\varphi_l}{\mathrm{d}t}.$$

Therefore,

$$\frac{\partial w}{\partial t} = u\sum_{k=1}^{m}\frac{\partial F(\varphi+v)}{\partial y_k}b_k(\varphi),$$
$$\frac{\partial^2 w}{\partial t^2} = u\sum_{k,l=1}^{m}\frac{\partial^2 F(\varphi+v)}{\partial y_k\partial y_l}b_k(\varphi)b_l(\varphi) + u\sum_{k=1}^{m}\frac{\partial F(\varphi+v)}{\partial y_k}\sum_{l=1}^{m}\frac{\partial b_k(\varphi)}{\partial \varphi_l}b_l(\varphi).$$

Hence, we obtain the following relations

$$\rho(x)\frac{\partial w}{\partial t} - \mathbf{L}w = \rho u \sum_{k=1}^{m} \frac{\partial F(\varphi+v)}{\partial y_k} b_k(\varphi)$$
$$- \left\{ F(\varphi+v)\mathbf{L}u + \sum_{k=1}^{m} \frac{\partial F(\varphi+v)}{\partial y_k}\left[2\sum_{i,j=1}^{n} a_{ij}(x)\frac{\partial v_k}{\partial x_i}\frac{\partial u}{\partial x_j} + u\mathbf{L}v_k\right]\right. \quad (6.33)$$
$$\left. +u\sum_{k,l=1}^{m}\frac{\partial^2 F(\varphi+v)}{\partial y_k \partial y_l}\sum_{i,j=1}^{n} a_{ij}(x)\frac{\partial v_k}{\partial x_i}\frac{\partial v_l}{\partial x_j}\right\},$$

$$\rho(x)\frac{\partial^2 w}{\partial t^2} - \mathbf{L}w = \rho u \left[\sum_{k,l=1}^{m}\frac{\partial^2 F(\varphi+v)}{\partial y_k \partial y_l} b_k(\varphi) b_l(\varphi) + \sum_{l,k=1}^{m}\frac{\partial F(\varphi+v)}{\partial y_k}\frac{\partial b_k(\varphi)}{\partial \varphi_l} b_l(\varphi)\right]$$
$$- \left\{ F(\varphi+v)\mathbf{L}u + \sum_{k=1}^{m} \frac{\partial F(\varphi+v)}{\partial y_k}\left[2\sum a_{ij}(x)\frac{\partial v_k}{\partial x_i}\frac{\partial u}{\partial x_j} + u\mathbf{L}v_k\right]\right. \quad (6.34)$$
$$\left. +u\sum_{k,l=1}^{m}\frac{\partial^2 F(\varphi+v)}{\partial y_k \partial y_l}\sum_{i,j=1}^{n} a_{ij}(x)\frac{\partial v_k}{\partial x_i}\frac{\partial v_l}{\partial x_j}\right\}.$$

Our aim is to prove that the right sides of (6.33) and (6.34) are equal to $\lambda(x,z)$ for $z = w$ correspondingly. From Theorem 6.10, 6.11 conditions it follows that the mapping $z = F(y)$ has the inverse $y = F^{-1}(z)$. Therefore, from the formula $w(x,t) = uF\big(\varphi(t) + v(x)\big)$, we obtain the equality

$$\varphi = F^{-1}\left(\frac{w}{u}\right) - v.$$

Substituting in formulas for $\lambda(x,z)$ from Theorem 6.10, 6.11 the solution $w(x,t)$ instead of z and taking into account the relation $F^{-1}(w/u) = \varphi + v$, we obtain (6.33) and (6.34) correspondingly. Theorems 6.10, 6.11 are proved.

REMARK 6.1. For $m = 1$, $u(x) = 1$, $\rho(x) = 1$ the formulas of Theorems 6.10, 6.11 become as follows

$$w(x,t) = F\big(\varphi(t) + v(x)\big),$$

$$\lambda(x,z) = F'\big(F^{-1}(z)\big)\left[b\big(F^{-1}(z) - v(x)\big) - \mathbf{L}v\right] - F''\big(F^{-1}(z)\big)\sum_{i,j=1}^{n} a_{ij}(x)\frac{\partial v}{\partial x_i}\frac{\partial v}{\partial x_j}, \quad (6.35)$$

$$F'(y) = \frac{\mathrm{d}F}{\mathrm{d}y}, \quad F''(y) = \frac{\mathrm{d}^2F}{\mathrm{d}y^2}, \quad y \in \mathrm{R}^1,$$

$$w(x,t) = F\big(\varphi(t) + v(x)\big),$$

$$\lambda(x,z) = F'\big(F^{-1}(z)\big)\left[b'\big(F^{-1}(z) - v(x)\big)b\big(F^{-1}(z) - v(x)\big) - \mathbf{L}v\right]$$
$$+F''\big(F^{-1}(z)\big)\left[b^2\big(F^{-1}(z) - v(x)\big) - \sum_{i,j=1}^{n} a_{ij}(x)\frac{\partial v}{\partial x_i}\frac{\partial v}{\partial x_j}\right]. \quad (6.36)$$

In particular, for $\mathbf{L} = \Delta = \sum_{j=1}^{n} \partial^2/\partial x_j^2$, $n \geq 1$, $\varphi(t) = \omega t$, $v(x) = \sum_{i=1}^{n} q_i x_i = (q, x)$, we may set $b = \omega$ and formulas (6.35), (6.36) will take the form (6.37), (6.38) with the functions $\lambda(z)$ independent of x

$$\begin{cases} w = F\big(\omega t + (q, x)\big), \\ \lambda(z) = F'\big(F^{-1}(z)\big)\omega - F''\big(F^{-1}(q)\big)\,|q|^2, \end{cases} \tag{6.37}$$

$$\begin{cases} w = F\big(\omega t + (q, x)\big), \\ \lambda(z) = F''\big(F^{-1}(z)\big)\big[\omega^2 - |q|^2\big]. \end{cases} \tag{6.38}$$

For $\lambda(z) = 0$, $F''(y) \neq 0$ we obtain the variance ratio $\omega^2 = |q|^2$. Note, also, that in the case $m = 1$ we have $\varphi(t) = B^{-1}(t)$, where $B^{-1}(t)$ is the inverse function to

$$B(y) = \int \frac{dy}{b(y)}.$$

In particular, if $b(y) = y^l$, $l < 1$ then

$$B(y) = \frac{1}{1-l} y^{1-l}, \quad \varphi(t) = \big[t(1-l)\big]^{\frac{1}{1-l}}.$$

Now, consider the case when the solution $w(x, t)$ has the form

$$w(x, t) = u(x) F\big(\varphi_1(t) v_1(x), \ldots, \varphi_m(t) v_m(x)\big).$$

We use the following notation

$$w(x, t) = u(x) F\big(\varphi(t) v(x)\big).$$

Assume that all restrictions on F, v, b, u hold. Suppose that all components $v_k(x)$ of the vector $v = (v_1, \ldots, v_m)$ do not attain zero in the domain D, i. e., $v_k(x) \neq 0$, $x \in D$. Finally, denote by $F_k^{-1}(z)$, $k = 1, 2, \ldots, m$ the components of the inverse mapping $F^{-1}(z)$, $z \in \mathrm{R}^m$ to the mapping $z = F(y)$.

Theorem 6.12. *The vector-functions* $w(x, t)$, $\lambda(x, z)$ *defined as follows*

$$w(x, t) = u(x) F\big(\varphi_1(t) v_1(x), \ldots, \varphi_m(t) v_m(x)\big),$$

$$\begin{aligned} \lambda(x, z) \; = \; & \sum_{k=1}^{m} \frac{\partial F\big(F^{-1}(z/u)\big)}{\partial y_k} \left[\rho u v_k b_k \left(\frac{F_1^{-1}(z/u)}{v_1}, \ldots, \frac{F_m^{-1}(z/u)}{v_m} \right) \right. \\ & \left. - 2F_k^{-1}\left(\frac{z}{u}\right) \left(\sum_{i,j=1}^{n} a_{ij}(x) \frac{\partial v}{\partial x_i} \frac{\partial u}{\partial x_j} - u \mathbf{L} v_k \right) \right] \\ & -u \sum_{k,l=1}^{m} \frac{\partial^2 F\big(F^{-1}(z/u)\big)}{\partial y_k \partial y_l} \frac{F_k^{-1}(z/u)}{v_k} \frac{F_l^{-1}(z/u)}{v_l} \sum_{i,j=1}^{n} a_{ij} \frac{\partial v_k}{\partial x_i} \frac{\partial v_l}{\partial x_j} - z \frac{\mathbf{L} u}{u} \end{aligned} \tag{6.39}$$

satisfy the nonlinear system of equations

$$\rho(x)\frac{\partial w}{\partial t} = \mathbf{L}w + \lambda(x,w).$$

Theorem 6.13. *The vector-functions $w(x,t)$, $\lambda(x,z)$ defined as follows*

$$w(x,t) = u(x)F\big(\varphi_1(t)v_1(x),\dots,\varphi_m(t)v_m(x)\big),$$

$$\lambda(x,z) = \sum_{k=1}^{m}\frac{\partial F\big(F^{-1}(z/u)\big)}{\partial y_k}\Bigg[\rho u v_k\sum_{l=1}^{m}\frac{\partial b_k}{\partial\varphi_l}\left(\frac{F_1^{-1}(z/u)}{v_1},\dots,\frac{F_m^{-1}(z/u)}{v_m}\right)$$
$$\times b_l\left(\frac{F_1^{-1}(z/u)}{v_1},\dots,\frac{F_m^{-1}(z/u)}{v_m}\right) - \frac{F_k^{-1}(z/u)}{v_k}\left(2\sum_{i,j=1}^{n}a_{ij}(x)\frac{\partial v_k}{\partial x_i}\frac{\partial u}{\partial x_j} + u\mathbf{L}v_k\right)\Bigg] \qquad (6.40)$$
$$+u\sum_{k,l=1}^{m}\frac{\partial^2 F\big(F^{-1}(z/u)\big)}{\partial y_k\partial y_l}\Bigg[v_k v_l b_k\left(\frac{F_1^{-1}(z/u)}{v_1},\dots,\frac{F_m^{-1}(z/u)}{v_m}\right)$$
$$\times b_l\left(\frac{F_1^{-1}(z/u)}{v_1},\dots,\frac{F_m^{-1}(z/u)}{v_m}\right) - \frac{F_k^{-1}(z/u)}{v_k}\frac{F_l^{-1}(z/u)}{v_l}\sum_{i,j=1}^{n}a_{ij}(x)\frac{\partial v_k}{\partial x_i}\frac{\partial v_l}{\partial x_j}\Bigg] - z\frac{\mathbf{L}u}{u}$$

satisfy the nonlinear system

$$\rho(x)\frac{\partial^2 w}{\partial t^2} = \mathbf{L}w + \lambda(x,w).$$

Proof of the theorems 6.12, 6.13. Taking into account the relations $\mathrm{d}\varphi_k/\mathrm{d}t = b_k(\varphi_1,\dots,\varphi_m)$, $k = 1,2,\dots,m$, and

$$w(x,t) = u(x)F\big(\varphi_1(t)v_1(x),\dots,\varphi_m(t)v_m(x)\big)$$

we obtain

$$\frac{\partial w}{\partial t} = u\sum_{k,l=1}^{m}\frac{\partial F(\varphi v)}{\partial y_k}v_k b_k(\varphi),$$

$$\frac{\partial^2 w}{\partial t^2} = u\sum_{k,l=1}^{m}\left(\frac{\partial^2 F(\varphi v)}{\partial y_k\partial y_l}v_k v_l b_k(\varphi)b_l(\varphi) + \frac{\partial F(\varphi v)}{\partial y_k}v_k\frac{\partial b_k(\varphi)}{\partial\varphi_l}b_l(\varphi)\right),$$

$$\frac{\partial w}{\partial x_i} = \frac{\partial u}{\partial x_i}F(\varphi v) + u\sum_{k=1}^{m}\frac{\partial F(\varphi v)}{\partial y_k}\varphi_k\frac{\partial v_k}{\partial x_i},$$

$$\frac{\partial^2 w}{\partial x_i\partial x_j} = \frac{\partial^2 u}{\partial x_i\partial x_j}F(\varphi v) + \sum_{k=1}^{m}\frac{\partial F(\varphi v)}{\partial y_k}\left[\varphi_k\left(\frac{\partial u}{\partial x_i}\frac{\partial v_k}{\partial x_j} + \frac{\partial u}{\partial x_j}\frac{\partial v_k}{\partial x_i}\right) + u\varphi_k\frac{\partial^2 v}{\partial x_i\partial x_j}\right]$$
$$+u\sum_{k,l=1}^{m}\frac{\partial^2 F(\varphi v)}{\partial y_k\partial y_l}\left[\varphi_k\varphi_l\frac{\partial v_k}{\partial x_i}\frac{\partial v_l}{\partial x_j}\right].$$

Hence it follows that

$$\rho\frac{\partial w}{\partial t} - \mathbf{L}w = \rho u \sum_{k=1}^{m} \frac{\partial F(\varphi v)}{\partial y_k} v_k b_k(\varphi)$$
$$-\mathbf{L}uF - \sum_{k=1}^{m} \frac{\partial F(\varphi v)}{\partial y_k}\left(u\varphi_k \mathbf{L}v_k + 2\varphi_k \sum_{i,j=1}^{n} a_{ij}(x)\frac{\partial u}{\partial x_i}\frac{\partial v_k}{\partial x_j}\right) \tag{6.41}$$
$$-u\sum_{k,l=1}^{m}\frac{\partial^2 F(\varphi v)}{\partial y_k \partial y_l}\varphi_k\varphi_l \sum_{i,j=1}^{n} a_{ij}(x)\frac{\partial v_k}{\partial x_i}\frac{\partial v_l}{\partial x_j},$$

$$\rho\frac{\partial^2 w}{\partial t^2} - \mathbf{L}w = \rho u \left[\sum_{k,l=1}^{m} \frac{\partial^2 F(\varphi v)}{\partial y_k \partial y_l} v_k v_l b_k(\varphi) b_l(\varphi) + \sum_{k=1}^{m}\frac{\partial F(\varphi v)}{\partial y_k} v_k \sum_{l=1}^{m} \frac{\partial b_k(\varphi)}{\partial \varphi_l} b_l(\varphi)\right]$$
$$-\mathbf{L}uF - \sum_{k=1}^{m} \frac{\partial F(\varphi v)}{\partial y_k}\left(u\varphi_k \mathbf{L}v_k + 2\varphi_k \sum_{i,j=1}^{n} a_{ij}(x)\frac{\partial u}{\partial x_i}\frac{\partial v_k}{\partial x_j}\right) \tag{6.42}$$
$$-u\sum_{k,l=1}^{m}\frac{\partial^2 F(\varphi v)}{\partial y_l \partial y_k}\varphi_k\varphi_l \sum_{i,j=1}^{n} a_{ij}(x)\frac{\partial v_k}{\partial x_i}\frac{\partial v_l}{\partial x_j}.$$

The relation $w = uF(\varphi_1 v_1, \ldots, \varphi_m v_m)$ gives

$$\varphi_k = \frac{F_k^{-1}(w/u)}{v_k}, \quad k = 1, 2, \ldots, m. \tag{6.43}$$

Substituting in the formulas for $\lambda(x, z)$ from Theorems 6.12, 6.13 instead of z the solution $w = F(\varphi v)$ and taking into account (6.43), we obtain (6.41) and (6.42)

$$\rho(x)\frac{\partial w}{\partial t} = \mathbf{L}w + \lambda(x, w), \quad \rho(x)\frac{\partial^2 w}{\partial t^2} = \mathbf{L}w + \lambda(x, w)$$

which was to be proved.

REMARK 6.2. For $m = 1$, $u(x) = 1$, $\rho(x) = 1$ formulas (6.39), (6.40) take the following form

$$w(x, t) = F\big(\varphi(t)v(x)\big),$$

$$\lambda(x, z) = F'(F^{-1}(z))\left[vb\left(\frac{F^{-1}(z)}{v}\right) - \mathbf{L}v\right] \tag{6.44}$$
$$-F''(F^{-1}(z))\left(\frac{F^{-1}(z)}{v}\right)\sum_{i,j=1}^{n} a_{ij}(x)\frac{\partial v}{\partial x_i}\frac{\partial v}{\partial x_j},$$

$$w(x, t) = F\big(\varphi(t)v(x)\big),$$

$$\lambda(x, z) = F'(F^{-1}(z))\left[b'\left(\frac{F^{-1}(z)}{v}\right) b\left(\frac{F^{-1}(z)}{v}\right) - \frac{F^{-1}(z)}{v}\mathbf{L}v\right] \tag{6.45}$$
$$+F''(F^{-1}(z))\left[v^2 b^2\left(\frac{F^{-1}(z)}{v}\right) - \left(\frac{F^{-1}(z)}{v}\right)^2 \sum_{i,j=1}^{n} a_{ij}(x)\frac{\partial v}{\partial x_i}\frac{\partial v}{\partial x_j}\right].$$

6.5.3 Nonlinear equations and differential inequalities

REMARK 6.3. If we consider the nonlinear evolutionary equation

$$\rho(x)\frac{\partial w}{\partial t} = \mathbf{L}w + \mu(x)w + \lambda(x,w), \tag{6.46}$$

where $\lambda(x,y) \leq 0$ or $\lambda(x,y) \geq 0$ for all $x \in D$, $y \in \mathrm{R}^1$, then for each real solution $w(x,t)$ of (6.46) we have

$$\rho\frac{\partial w}{\partial t} \leq \mathbf{L}w + \mu(x)w \tag{6.47}$$

or

$$\rho\frac{\partial w}{\partial t} \geq \mathbf{L}w + \mu(x)w. \tag{6.48}$$

Therefore, it is appropriate to investigate the differential inequalities (6.47) and (6.48). In this paragraph, we shall obtain the formulas both for solution $w(x,t)$ of differentional inequality (6.47) and for coefficients ρ, μ of this inequality by means of three functions $F(t,y)$, $\varphi(x,t)$, $\tau(x,t)$. The restrictions imposed on these functions will be given below.

Consider the differential operator $\mathbf{L}$

$$\mathbf{L} = \sum_{i,j=1}^{n} a_{ij}(x,t)\frac{\partial^2}{\partial x_i \partial x_j} + \sum_{i=1}^{n} a_i(x,t)\frac{\partial}{\partial x_i}, \quad x \in D, \quad t \in (\alpha,\beta)$$

where $a_{ij}(x,t)$, $a_j(x,t)$ are continuous and

$$\sum_{i,j=1}^{n} a_{ij}\xi_i\xi_j > 0, \quad \forall \xi \in \mathrm{R}^n, \quad \xi \neq 0.$$

Theorem 6.14. *Let twice differentiable functions $\varphi(x,t) > 0$, $\tau(x,t)$ satisfy the differential equation*

$$\left[\left(\sum_{i,j=1}^{n} a_{ij}\frac{\partial \tau}{\partial x_i}\frac{\partial \tau}{\partial x_j}\right)\frac{\partial \tau}{\partial t} - \mathbf{L}\tau\right]\varphi = 2\sum_{i,j=1}^{n} a_{ij}\frac{\partial \varphi}{\partial x_i}\frac{\partial \tau}{\partial x_j} \tag{6.49}$$

and a function $F(t,y)$, $\alpha < t < \beta$, $y \in \mathrm{R}^1$ satisfy the differential inequality

$$\frac{\partial F}{\partial t} \leq \frac{\partial^2 F}{\partial y^2}. \tag{6.50}$$

Then the function

$$w(x,t) = \varphi(x,t)F\big(t,\tau(x,t)\big)$$

satisfies the following differential inequality

$$\left(\sum_{i,j=1}^{n} a_{ij}\frac{\partial \tau}{\partial x_i}\frac{\partial \tau}{\partial x_j}\right)\frac{\partial w}{\partial t} \leq \mathbf{L}w + \frac{\left[\left(\sum_{i,j=1}^{n} a_{ij}\frac{\partial \tau}{\partial x_i}\frac{\partial \tau}{\partial x_j}\right)\frac{\partial \varphi}{\partial t} - \mathbf{L}\varphi\right]}{\varphi}w. \tag{6.51}$$

Moreover, if in (6.50) the equality holds, then in (6.51) the equality holds also.

PROOF. From the following evident equations

$$\frac{\partial w}{\partial t} = \varphi_t F + \varphi F_t + \varphi \tau_t F_y,$$

$$\frac{\partial w}{\partial x_i} = \frac{\partial \varphi}{\partial x_i} F + \varphi \frac{\partial \tau}{\partial x_i} F_y,$$

$$\frac{\partial^2 w}{\partial x_i \partial x_j} = \frac{\partial^2 \varphi}{\partial x_i \partial x_j} F + \left(\frac{\partial \varphi}{\partial x_i} \frac{\partial \tau}{\partial x_j} + \frac{\partial \varphi}{\partial x_j} \frac{\partial \tau}{\partial x_i} \right) F_y + \varphi \frac{\partial^2 \tau}{\partial x_i \partial x_j} F_y + \varphi \frac{\partial^2 \tau}{\partial x_i \partial x_j} F_{yy}$$

we obtain the relations

$$\mathbf{L} w = F \mathbf{L} \varphi + \varphi F_y \left[2 \sum_{i,j=1}^{n} a_{ij} \frac{\partial \varphi}{\partial x_i} \frac{\partial \tau}{\partial x_j} + \mathbf{L} \tau \right] + F_{yy} \varphi \sum_{i,j=1}^{n} a_{ij} \frac{\partial \tau}{\partial x_i} \frac{\partial \tau}{\partial x_j}.$$

From these two relations, taking into account (6.49), we obtain the relation

$$\begin{aligned} &\left(\sum_{i,j=1}^{n} a_{ij} \frac{\partial \tau}{\partial x_i} \frac{\partial \tau}{\partial x_j} \right) \frac{\partial w}{\partial t} - \mathbf{L} w \\ &= \left[\left(\sum_{i,j=1}^{n} a_{ij} \frac{\partial \tau}{\partial x_i} \frac{\partial \tau}{\partial x_j} \right) \frac{\partial \varphi}{\partial t} - \mathbf{L} \varphi \right] F + \varphi \sum_{i,j=1}^{n} a_{ij} \frac{\partial \tau}{\partial x_i} \frac{\partial \tau}{\partial x_j} [F_t - F_{yy}]. \end{aligned} \tag{6.52}$$

Since $F = w/\varphi$, $\varphi > 0$, $\sum_{i,j=1}^{n} a_{ij} \frac{\partial \tau}{\partial x_i} \frac{\partial \tau}{\partial x_j} \geq 0$, $F_t - F_{yy} < 0$, from (6.52) we obtain (6.51).

REMARK 6.4. If in (6.51) we know the coefficient $\rho(x,t)$ or $\mu(x,t)$, then, by (6.49), we have correspondingly the closed system of two equations. In the case when ρ is given we get

$$\begin{aligned} \left[\left(\sum_{i,j=1}^{n} a_{ij} \frac{\partial \tau}{\partial x_i} \frac{\partial \tau}{\partial x_j} \right) \frac{\partial \tau}{\partial t} - \mathbf{L} \tau \right] \varphi &= 2 \sum_{i,j=1}^{n} a_{ij} \frac{\partial \varphi}{\partial x_i} \frac{\partial \tau}{\partial x_j}, \\ \sum_{i,j=1}^{n} a_{ij}(x,t) \frac{\partial \tau}{\partial x_i} \frac{\partial \tau}{\partial x_j} &= \rho(x,t). \end{aligned}$$

When $\mu(x,t)$ is given we get

$$\begin{aligned} \left[\left(\sum_{i,j=1}^{n} a_{ij} \frac{\partial \tau}{\partial x_i} \frac{\partial \tau}{\partial x_j} \right) \frac{\partial \tau}{\partial t} - \mathbf{L} \tau \right] \varphi &= 2 \sum_{i,j=1}^{n} a_{ij} \frac{\partial \varphi}{\partial x_i} \frac{\partial \tau}{\partial x_j}, \\ \left(\sum_{i,j=1}^{n} a_{ij} \frac{\partial \tau}{\partial x_i} \frac{\partial \tau}{\partial x_j} \right) \frac{\partial \varphi}{\partial t} - \mathbf{L} \varphi &= \varphi \mu. \end{aligned}$$

The function $F(t, y)$ remains arbitrary, satisfying the inequality (6.50).

6.5.4 The formulas for others nonlinearities

We consider multidimensional nonlinear second order equations of the Bürgers type.

Theorem 6.15. *Let a twice differentiable function $\varphi(x,t)$, $x \in D \subset \mathrm{R}^n$, $\alpha < t < \beta$ solve the nonlinear equation*

$$|\mathrm{grad}_x \varphi|^2 \frac{\partial \varphi}{\partial t} = \Delta \varphi$$

and a three times differentiable function $F(t,y) \neq 0$, $\alpha < t < \beta$, $y \in \mathrm{R}^1$ solve the parabolic equation $\partial F/\partial t = \partial^2 F/\partial y^2$.

Then the function

$$w(x,t) = -2\frac{\partial F(t,\varphi(x,t))/\partial y}{F(t,\varphi(x,t))}$$

satisfies the equation

$$|\mathrm{grad}_x \varphi|^2 \frac{\partial w}{\partial t} = \Delta w - w\sum_{j=1}^{n} \frac{\partial w}{\partial x_j}\frac{\partial \varphi}{\partial x_j}.$$

PROOF. Let the conditions of Theorem 6.15 hold. Then

$$\frac{\partial w}{\partial t} = -2\frac{F'_y(t,\varphi)}{F(t,\varphi)} = -2\frac{(F''_{yy}F - F'^2_y)\varphi_t + F''_{yt}F - F'_yF_t}{F^2},$$

$$\frac{\partial^2 w}{\partial x_j^2} = -2\frac{F''_{yy}F - F'_y}{F^2}\frac{\partial^2 \varphi}{\partial x_j^2} - \frac{2(F'''_{yyy}F - F''_{yy}F'_y)}{F^2}\left(\frac{\partial \varphi}{\partial x_j}\right)^2 + 4\frac{F_y(F''_{yy}F - F'^2_y)}{F^3}\left(\frac{\partial \varphi}{\partial x_j}\right)^2.$$

Hence we obtain the equalities

$$\begin{aligned}
|\mathrm{grad}_x \varphi|^2 \frac{\partial w}{\partial t} &= -2\,|\mathrm{grad}_x \varphi|^2 \left[\frac{F''_{yy}F - F'^2_y}{F^2}\frac{\partial \varphi}{\partial t} - \frac{F'_yF'_t}{F^2} + \frac{F''_{yt}}{F}\right], \\
-w\sum_{j=1}^{n} \frac{\partial w}{\partial x_j}\frac{\partial \varphi}{\partial x_j} &= -4\frac{F'_y(F''_{yy}F - F'^2_y)}{F^3}\,|\mathrm{grad}_x \varphi|^2,
\end{aligned}$$

$$\Delta w = -2\frac{F''_{yy}F - F'^2_y}{F^2}\Delta\varphi - \left[2\frac{F'''_{yyy}F - F''_{yy}F'_y}{F^2}\,|\mathrm{grad}_x \varphi|^2 - 4\frac{F'_y(F''_{yy}F - F'^2_y)}{F^3}\,|\mathrm{grad}_x \varphi|^2\right].$$

Then taking into account $|\mathrm{grad}_x \varphi|^2\,\partial\varphi/\partial t = \Delta\varphi$, $\partial F/\partial t = \partial^2 F/\partial y^2$, we obtain

$$|\mathrm{grad}_x \varphi|^2 \frac{\partial w}{\partial t} = \Delta w - w\sum_{j=1}^{n} \frac{\partial w}{\partial x_j}\frac{\partial \varphi}{\partial x_j}.$$

REMARK 6.5. It is clear that each function $\varphi(x)$ harmonic in $D \subset \mathrm{R}^n$ solves the equation $|\mathrm{grad}_x \varphi|^2\,\partial\varphi/\partial t = \Delta\varphi$. Therefore, from Theorem 6.15, it follows that

$$w(x,t) = -2\frac{F'_y(t,\varphi)}{F(t,\varphi)},$$

where $\Delta\varphi = 0$, $x \in D$, $\partial F/\partial t = \partial^2 F/\partial y^2$ and $w(x,t)$ solves the equation

$$|\text{grad}_x \varphi|^2 \frac{\partial w}{\partial t} = \Delta w - w \sum_{j=1}^{n} \frac{\partial w}{\partial x_j} \frac{\partial \varphi}{\partial x_j}.$$

REMARK 6.6. For $n = 1$, $\varphi(x) = x$ the function

$$w(x,t) = -2 \frac{F'_y(t,x)}{F(t,x)}$$

solves the Bürgers equation

$$\frac{\partial w}{\partial t} + w \frac{\partial w}{\partial x} = \frac{\partial^2 w}{\partial x^2}.$$

This fact was obtained earlier by Cole (1948).

Now consider the nonlinear equations of the Schrödinger type

$$\rho(x) \frac{\partial w}{\partial t} = f(w) \Delta w.$$

Theorem 6.16. *Let $F(t,y)$ solve the equation*

$$\frac{\partial F}{\partial t} = f(F) \frac{\partial^2 F}{\partial y^2}$$

where $f(p)$, $p \in \mathrm{R}^1$ is a given continuous or analytical complex-valued function and $\varphi(x)$, $x \in D$ solve the equation

$$\mathbf{L}\varphi \equiv \sum_{i,j=1}^{n} a_{ij}(x) \frac{\partial^2 \varphi}{\partial x_i \partial x_j} + \sum_{j=1}^{n} a_j(x) \frac{\partial \varphi}{\partial x_j} = 0.$$

Then the function

$$w(x,t) = F\big(t, \varphi(x)\big)$$

solves the nonlinear equation

$$\left(\sum_{i,j=1}^{n} a_{ij}(x) \frac{\partial \varphi}{\partial x_i} \frac{\partial \varphi}{\partial x_j} \right) \frac{\partial w}{\partial t} = f(w) \mathbf{L} w.$$

Example 6.8. Let $\mathbf{L} = \Delta = \sum_{j=1}^{n} \partial^2/\partial x_j^2$. Then the function

$$w(x,t) = \sqrt{\frac{2\alpha}{\gamma}} \frac{l^{\left[\mathrm{i} \frac{a}{2} \varphi(x) - \left(\frac{a^2}{4} - \alpha\right) t\right]}}{\cosh\left[\sqrt{\alpha}\left(\varphi(x) - u_0 - at\right)\right]},$$

where a, u_0, α, γ are constants and $\Delta\varphi = 0$, satisfies the nonlinear equation of the Schrödinger type

$$\mathrm{i}\, |\text{grad}\, \varphi|^2 \frac{\partial w}{\partial t} + \left(1 + \gamma\, |w|^2\right) \Delta w = 0.$$

In conclusion, we consider the equation

$$\rho(x)\frac{\partial^2 w}{\partial t^2} = \mathbf{L}w + \mu(x)\frac{\partial w}{\partial t} + \lambda(x)w,$$

where

$$\mathbf{L} = \sum_{i,j=1}^{n} a_{ij}(x)\frac{\partial^2}{\partial x_i \partial x_j}, \quad x \in D, \quad \alpha < t < \beta.$$

Theorem 6.17. *Let twice differentiable functions $\varphi(x) \neq 0$, $\tau(x)$ satisfy the system*

$$\sum_{i,j=1}^{n} a_{ij}\frac{\partial \tau}{\partial x_i}\frac{\partial \tau}{\partial x_j} = \rho(x),$$

$$\sum_{i,j=1}^{n} 2a_{ij}\frac{\partial \tau}{\partial x_i}\frac{\partial \varphi}{\partial x_j} + \varphi \mathbf{L}\tau + \mu\varphi = 0.$$

Then the functions

$$w(x,t) = \varphi(x)F\big(\tau(x) + t\big),$$

$$\lambda(x) = -\frac{\mathbf{L}\varphi}{\varphi},$$

where $F(y) \in C^2(\mathrm{R}^1)$ satisfy the equation

$$\rho(x)\frac{\partial^2 w}{\partial t^2} = \mathbf{L}w + \mu(x)\frac{\partial w}{\partial t} + \lambda(x)w.$$

6.5.5 Some applications

The formulas of Theorems 6.10 – 6.13 contain arbitrary mappings and solutions of autonomous systems of ordinary differential equations. Therefore, generally speaking, the applications of these formulas may be connected with the problems of geometry, differential topology, qualitative theory of ordinary differential equations. Other applications may be connected with *a priori* information about $\lambda(x,z)$, in particular, on its zeros.

Example 6.9. In physics and biology, often, for the function $\lambda(x,z)$ we have the relations (Kolmogorov *et al.*, 1985; Svirizhev, 1987)

$$\lambda(x,0) = 0, \quad \lambda(x,1) = 0.$$

These relations at the same time lead to the differential equations for the functions $u(x)$, $v(x)$ from the formulas for $w(x,t)$ and $\lambda(x,z)$ in Theorems 6.10 – 6.13. We confine ourself to the case $m = 1$ and the representation of functions $w(x,t)$ and $\lambda(x,z)$ from Theorem 6.10.

In this case, if $\lambda(x,0) = \lambda(x,1) = 0$, we have, generally speaking, the system of nonlinear equations relative to the functions $u(x)$, $v(x)$, $x \in D$

$$F'(F^{-1}(0))\left[\rho u b(F^{-1}(0) - v(x)) - 2\sum_{i,j=1}^{n} a_{ij}\frac{\partial u}{\partial x_i}\frac{\partial v}{\partial x_j} - u\mathbf{L}v\right]$$

$$-uF''(F^{-1}(0))\sum_{i,j=1}^{n} a_{ij}\frac{\partial v}{\partial x_i}\frac{\partial v}{\partial x_j} = 0,$$

$$F'\left(F^{-1}\left(\frac{1}{u}\right)\right)\left[\rho ub\left(F^{-1}\left(\frac{1}{u}\right)-v(x)\right)-2\sum_{i,j=1}^{n}a_{ij}\frac{\partial u}{\partial x_i}\frac{\partial v}{\partial x_j}-u\mathbf{L}v\right]$$

$$-uF''\left(F^{-1}\left(\frac{1}{u}\right)\right)\sum_{i,j=1}^{n}a_{ij}\frac{\partial v}{\partial x_i}\frac{\partial v}{\partial x_j}-\frac{\mathbf{L}u}{u}=0.$$

Another application is connected with the experimental data. Consider the equation of Kolmogorov, Petrovskii, Piskunov

$$\frac{\partial w}{\partial t}=k\frac{\partial^2 w}{\partial x^2}+\lambda(w),\ \ k>0,\ \ x\in\mathrm{R}^1,\ t>0.$$

The profile $F(x)$ of the wave front set may be taken from the experiments of the population investigation (Kolmogorov *et al.*, 1985; Svirizhev, 1987). In accordance with the theory and practice, the density $w(x,t)$ may be determined for large t by the formula $w(x,t)=F(\omega t+x)$. We choose the function $F(x)$ depending on parameter α:

$$F_\alpha(x)=\frac{1}{\pi}\mathrm{arctg}\,\alpha x+\frac{1}{2}.$$

This function satisfies the estimate $0<F_\alpha(x)<1$ and

$$\lim_{\alpha\to\infty}F_\alpha(x)=\begin{cases}1, & \text{if}\ \ x>0\\ 0, & \text{if}\ \ x<0.\end{cases}$$

From Theorem 6.10 for $m=1$, $n=1$, $\mathbf{L}=k\partial^2/\partial x^2$, $v(x)=x$, $b=\omega$, we obtain

$$w(x,t)=\frac{1}{\pi}\mathrm{arctg}\,\alpha(\omega t+x)+\frac{1}{2},$$

$$\lambda(z)=\frac{\alpha\omega}{\pi}\sin^2\pi z\left(1+\frac{\alpha^2k^2}{\omega}\sin 2\pi z\right),\ \ 0\le z\le 1.$$

Hence we obtain the conditions $\lambda(0)=\lambda(1)=0$ and for $\lambda(z)\ge 0$ the estimate $\omega\ge\alpha^2k^2$, which corresponds to the practically justified conditions (Kolmogorov *et al.*, 1985; Svirizhev, 1987).

References

Akhromeeva T.S, Kurdyumov S.P, Malinetskii G.G., and Samarskii A.A. (1988). The paradoxes of the nonstationary structure universe. In: *Computers and Nonlinear Phenomena*. Samarskii A.A., Ed., Nauka, Moscow, 192p., pp. 44–123 (in Russian).

Alekseev B.V. (1983). *Mathematical Kinetics of Reacting Gases.* Nauka, Moscow (in Russian).

Anikonov D.S. (1985). Uniqueness of the determination of a coefficient of the transport equation for a special type of source. *Sov. Math. Dokl.* **32**, 511–515.

Anikonov Yu.E. (1973). Some partial solution for one inverse kinematic problem. In: *Mathematical Problems of Geophysics*. Lavrent'ev M.M. and Alekseev A.S., Eds, Computer Center, SO AN SSSR, Novosibirsk, pp. 30–60 (in Russian).

Anikonov Yu.E. (1978). *Some Methods of Studying Multidimensional Inverse Problems for Differential Equations.* Nauka, Novosibirsk (in Russian).

Anikonov Yu.E. (1979). Formulas and inequalities in an inverse kinematic problem. *Sov. Math. Dokl.* **20**, 311–313.

Anikonov Yu.E. (1980). An inverse kinematic problem of seismology and some questions of stellar dynamics. *Sov. Math. Dokl.* **21**, 649–652.

Anikonov Yu.E. (1983). On the uniqueness classes of the solution for the problems of integral geometry. In: *Methods of Investigation of Incorrect Problems of Mathematical Physics.* Computer Center, SO AN SSSR, Novosibirsk, 16–17 (in Russian).

Anikonov Yu.E. (1986). *Pseudodifferential Operators and Inverse Problems.* Preprint №86–671. Computer Center, SO AN SSSR, Novosibirsk (in Russian).

Anikonov Yu.E. (1990a). Formulas in multidimensional inverse problems for evolution equations. *Sov. Math. Dokl.* **41**, 385–388.

Anikonov Yu.E. (1990b). Inverse problems for nonlinear equation and synergetics. *Sov. Phys. Dokl.* **35**, 930–932.

Anikonov Yu.E. (1991a). Representation of the solutions of equations with variable coefficients and inverse problems. *Sov. Phys. Dokl.* **36**, 431–433.

Anikonov Yu.E. (1991b). Inverse problems in mathematical physics and biology. *Sov. Phys. Dokl.* **36**, 436–438.

Anikonov Yu.E. (1991c). Formulas for solution of some inverse problems for evolution equations. *Sov. Phys. Dokl.* **36**, 569–570.

Anikonov Yu.E. (1991d). Inverse problem formulas for second order equations. *Sov. Phys. Dokl.* **36**, 679.

Anikonov Yu.E. (1992a). Multidimensional inverse kinematic problem. *The Arabian Journal for Science and Engineering.* **17**, 635–638.

Anikonov Yu.E. (1992b). New methods and results in multidimensional inverse problems for kinetic equations. In: *Ill-posed problems in Nature Sciences, Proceedings of the international conference held in Moscow (Russia) August 19 – 25, 1991.* Tikhonov A.N., Ed. VSP, Utrecht, pp. 244–253.

Anikonov Yu.E. (1992c). Generating functions, evolution equations, and inverse problems. *Sov. Phys. Dokl.* **37**, 172–174.

Anikonov Yu.E. (1992d). Inverse problem in mathematical biology. *Sib. Math. J.* **31**, 385–388.

Anikonov Yu.E. (1992e). Multidimensional inverse problems for evolution equations with discrete time. *Sov. Phys. Dokl.* **37**, 174–176.

Anikonov Yu.E. (1993a). Methods for studying multidimensional inverse problems for evolution equations. *Phys. Dokl.* **38**, 329–331.

Anikonov Yu.E. (1993b). Some results of theory of inverse problem for kinetic equations. *J. Inv. Ill-posed Problems* **1**, 1–16.

Anikonov Yu.E. (1993c). An estimate in a problem of integral geometry. *Russian Acad. Sci. Dokl. Math.* **47**, 483–484.

Anikonov Yu.E. (1994a). Formulas in multidimensional inverse problem for evolution equation. *Dokl. Acad. Nauk* **334**, 263–265 (in Russian).

Anikonov Yu.E. (1994b). The formulas in the inverse problems for differential equations and inequalities. *Dokl. Akad. Nauk. Russia.* **377**, 23–24 (in Russian).

Anikonov Yu.E. (1995a). *Multidimensional Inverse and Ill-posed Problems for Differential Equation.* VSP, Utrecht, The Netherland, Tokyo, Japan, 138p.

Anikonov Yu.E. (1995b). Uniqueness theorem for inverse problems of kinematics and integral geometry. *J. Inv. and Ill-posed Problems* **3**, 103-120.

Anikonov Yu.E. (1995c). Integral geometry. In: *Computerized Tomography. Proceedings of the Fourth International Symposium.* Novosibirsk, Russia, pp. 31–48.

Anikonov Yu.E. (1995d). Constructive methods of studying multidimensional inverse kinematic problems. *J. Inv. and Ill-posed Problems* **3**, 173–211.

Anikonov Yu.E. (1995e). Analytical representation of solutions to multidimensional inverse problems for evolution equations. *J. Inv. and Ill-posed Problems* **3**, 259–267.

Anikonov Yu.E. (1995f). Formulas in inverse problem for kinetic equation. *Dokl. Russian Acad. Nauk* **340**, 5–6.

Anikonov Yu.E. (1995g). Relations connected in tomography problems. *Dokl. Russian Acad. Nauk* **342**, 167–168.

Anikonov Yu.E. (1995h). On mathematical modelling of ethnical processes. *Dokl. Russian Acad. Nauk* **345**, 7–9.

Anikonov Yu.E. and Ayupova N.B. (1995). On the mathematical models in the problems of ethnogeny and evolution of populations. *J. Inv. and Ill-posed Problems* **3**, 359–381.

Anikonov Yu.E. and Lorenzi A. (1994). Existence results for parabolic identification problems. *J. Inv. and Ill-posed Problems* **3**, 181–210.

Anikonov Yu.E. and Pestov L.N. (1993). Integral geometry and the structure of Riemannian spaces. *J. Inv. and Ill-Posed Problems* **1**, 177–192.

Anikonov Yu.E. and Shneiberg I.Ya. (1991). Radon transformation with variable absorption. *Sov. Phys. Dokl.* **36**, 34–35.

Anikonov Yu. E. and Smirnov O.N. (1992). Algebraic questions of the theory of multidimensional inverse problems for nonlinear evolution equations. *Sov. Phys. Dokl.* **37**, 226–228.

Anikonov Yu. E. and Stepanov V.N. (1991). An inversion formula in a problem of integral geometry. *Sov. Math. Dokl.* **43**, 678–679.

Anikonov Yu. E. and Stepanov V.N. (1994). The geometry of convex surfaces and inverse problems of scattering theory. *Sib. Math. J.* **35**, 845–862.

Anikonov Yu.E. and Vishnevskii M.P. (1994). Reduction of multidimensional inverse problems to initial-boundary value problem in Hilbert spaces. *Sib. Math. J.* **35**, 439–458.

Anikonov D.S., Prohorov I.V., and Kovtanjuk A.E. (1993). Diagnostics of scattering and absorbing medium by methods of X-ray tomography. *J. Inv. and Ill-posed Problems* **1**, 259–281.

Annina M.B. (1987). An inverse problem for evolution equation. In: *Methods for Studying Inverse and Ill-Posed Problems.* Computer Center, SO AN SSSR, Novosibirsk, pp. 41–48 (in Russian).

Ayupova N.B. (1986). Formulas in problems of determination of sources. In: *Questions of correctness and methods of investigations of inverse problems* Lavrent'ev M.M., Ed., Computer Center SO AN SSSR, Novosibirsk, pp. 47-57 (in Russian).

Bellini S., Piacentini M., Cafforio C., and Rocco F. (1979). Compensation of tissue absorption in emission tomography. *IEEE Trans. Acoust. Speech Signal Processes* **27**, 213–218.

Bers L., John F., and Schechter M. (1974). *Partial Differential Equation* Amer.Math. Soc., Providence, RI.

Blaschke W. (1930). *Vorlesungen über Differentialgeometrie und geometrische Grundlagen Einsteins Relativitatstheorie. VI. Elementare Differentialgeometrie.* Springer Verlag, Berlin.

Blaschke W. (1956). *Kreis und Kugel.* Veit, Leipzig.

Bogolyubov N.N. (1981). *Lectures on Quantum Statistics.* Nauka, Moscow (in Russian).

Bubnov B.A.(1985). Necessary and sufficient conditions for solvability of initial boundary value problems for ultrahyperbolic and elliptic equations in classes of finite smoothness. In: *Nonclassical Equations of Mathematical Physics.* Vragov V.N., Ed., Institute of Mathematics, Novosibirsk, pp. 38–53 (in Russian).

Bubnov B.A. (1986). Existence and uniqueness of solution of inverse problems for parabolic and elliptic equations. In: *Nonclassical Equations of Mathematical Physics.* Vragov V.N., Ed., Institute of Mathematics, Novosibirsk, pp. 25–29 (in Russian).

Butkovskii A.G. (1985). *Phase Portraits of Controlled Dynamical Systems.* Nauka, Moscow (in Russian).

Case K.M. and Zweifel P.F. (1967). *Linear Transport Theory.* Addison Wesley, London.

Chin Long Chiang (1970). *Introduction to Stochastic Processes in Biostatistics*, NY.

Cole R.H. (1948). *Underwater Explosions.* Princeton, New Jersey.

Davydov A.S. (1986). Nonlinear biophysics. In: *Mathematical Modelling. Processes in Nonlinear Media.* Moscow, Nauka, pp. 223–260 (in Russian).

Fantappié L. (1933). Intégration par quadrature de l'équation parabolique général a coefficients constants sur les caractéristiques. *C.R.Acad.Sci.* **197**, 969–970.

Ferziger J.H. and Kaper H.G. (1972). *Mathematical Theory of Transport Processes in Gases.* North-Holland, Amsterdam.

Finch D.V. (1985). Cone-beam reconstruction with sources on curve. *SIAM J. Appl. Math.* **45**, 665–673.

Finch D.V. and Hertle A. (1987). The exponential Radon transform. In: *Integral Geometry, Proc. AMS-ISM-SIAM Joint Summer Res. Conf., Brunswick/Maine 1984, Contmp. Math. 63*, pp. 67–73.

Forrester J.W. (1974). *World Dynamics.* Cambridge (Mass.)

Gel'fand I.M. and Shilov G.E. (1959). *Generalized Functions and their Operations.* Gos. Izd. Fiz-Mat. Lit., Moscow (in Russian).

Gel'fand I.M., Graev M.I., and Vilenkin N.Ya. (1966). *Generalized Functions, Vol 5: Integral Geometry and Representations Theory.* Academic Press, New York-London.

Gel'fand I.M., Gindikin S.G., and Graev M.I. (1980). Integral geometry in affine and projective spaces, In: *Current problems in mathematics*, 16, Vsesojuz Inst. Nauch i Tehnc. Informacii, Moscow, pp. 53–226 (in Russian).

Gumilev L.N. (1989). *Ethnogeny and the Biosphere of the Earth.* Leningrad, LGU (in Russian).

Haken G. (1987). *Advanced Synergetics, Ser. Synergetics*, Vol.20, Springer, Berlin.

Helgason S. (1980). *The Radon Transform.* Birkhaser Verlag, Boston.

Ianelli M., Miller F., and Pugliese A. (1992). Analytical and numerical results for the age-structured S-I-S epidemic model with mixed inter-intracohort transmission. *SIAM J. Math. Anal.* **23**, 662–688.

Isakov V.M. (1991). Inverse parabolic problems with the final overdetermination. *Comm. Pure Appl. Math.* **44**, 185–209.

Izrael' Yu.A. (1976). Tolerable anthropogenic burden on the environment. In: *Comprehensive Analysis of the Environment.* Proceedings of the Soviet-American symposium. Leningrad, Gidrometeoizdat, pp. 12–25 (in Russian).

Kato T. (1966) *Perturbation Theory for Linear Operators. Die Grundlehren der mathematische Wissenschaften*, 132, Springer-Verlag, Berlin, Heidelberg, New York.

Kirillov A.A. (1961). On a problem of I.M.Gelfand. *Sov. Math. Dokl.* **2**, 268–269.

Klimontivich Yu.D. (1982). *Statistical Physics.* Nauka, Moscow (in Russian).

Kolmogorov A.N., Petrovskii I.G., and Piskunov N.S. (1985). The analysis of diffusion equations, combined with the growth of the quantity of material, and its application to a biological problem. In: *Kolmogorov A.N. Mathematics and Mechanics, Selected Works.* Nikol'skii S.M. Ed., Nauka, Moscow, pp. 221–246 (in Russian).

Kompaneets A.S. (1977). *Physico-Chemical and Relativistic Gas Dynamics.* Nauka, Moscow (in Russian).

Kuchment P.A. and L'vin S.Ya. (1990). Paley-Wiener theorem for exponential Radon transform. *Acta Appl.Math.* **18**, 251–260.

Kugaenko A.A. (1991). *Synthesis of Dynamical Models of Domestic Economy and Methods of Forecasting Socio–Economic Processes.* Prometej, Moscow (in Russian).

Ladyzhenskaya O.A., Solonnikov V.A., and Ural'tzeva N.N. (1968). *Linear and Quasilinear Equations of Parabolic Type. Translations of Mathematical Monographs*, 23, American Mathematical Society, Providence RI.

Lavrent'ev M.M., Romanov V.G., and Shishatskii S.P. (1986). *Ill-Posed Problems in Mathematical Physics and Analysis. Translations of Mathematical Monographs*, 64, American Mathematical Society, Providence RI.

Levin J.J. (1959). On the matrix Riccati equation. *Proc. Amer.Math Soc.* **10**, 519–524.

Lifshitz and Pitaevskii L.P. (1979). *Physical Kinetics.* Nauka, Moscow (in Russian).

Lions J.-L. (1968). *Contrôle Optimal de Systémes Gouvernés par des Équations aux Dérivées Partielles*, Dunod, Gautier-Villars.

Lions J.-L. (1973). Sur le contrôle optimal de systéms distribues. Enseignement math., II Ser.19, 125–166.

Lions J.-L. and Magenes E. (1972). *Non Homogeneous Boundary Value Problems and Aplications. V1* Springer-Verlag, Berlin, Heidelberg, New York.

Marchuk G.I. (1989). Mathematical modelling in the problem of environment preservation. In: *Mathematical Modelling.* Moscow, Nauka, pp. 238–253 (in Russian).

Murray J.D. (1977). *Lectures on Non-Linear Differential Equation Models in Biology.* Oxford, England, Clarendon Press.

Nakagiry S. and Yamamoto M. (1991). *Determination of Constant Parameters in Diffusion Equation* [Preprint, No. 107], Univ. of Tokyo, Tokyo.

Natterer F. (1983). Computerized tomography with unknown sources. *SIAM J.Appl. Math.* **43**, 1201–1212.

Oberhettinger F. (1973). *Fourier Transforms of Distributions and their Inverses.* New York, London.

Odum E. (1983). *Basic Ecology.* CBS College Publishing, New York, London.

Poluektov R.A., Pinakh Yu.A., and Shvitov I.A. (1980). *Dynamical Models of Ecological Systems.* Gidrometeoizdat, Leningrad (in Russian).

Prigogine I. (1980). *From Being to Becoming: Time and Complexity in the Physical Sciences.* W.H.Freeman, San Francisco.

Prilepko A.I. and Kostin A.B. (1992). On certain inverse problems for parabolic equations with the final and integral observation, *Mat. Sb.* **183**, 46–68 .

Prilepko A.I. and Vasin I.A. (1989). Inverse initial-boundary-value problem for linearized nonstationary Navier-Stokes equations, *Differ. Equations* **25**, 85–92.

Prüss J. (1981). Equilibrium solutions of age-specific population dynamics of several species, *J. Math. Biol.* **11**, 65–84.

Radon J. (1917). Über die Bestimmung von Funktionen durch ihre Integral wärte längs Gewisser Mannigfaltigkeiten. *Ber. Verh. Sachs. Akad. Wiss. Leipzig., Math.-Nat. Kl.* **68**, 262–277.

Riordan J. (1980). *An Introduction to Combinatorial Analysis.* Priceton Univ. Press

Röhrl H. and Wischnewsky M.B. (1976). Subalgebras that are cyclic as submodules. *Manuscripta Math* **19**, 195–209.

Semyanistyi V.I. (1960). On some integral transformations in Euclidean space. *Dokl. Akad. Nauk SSSR* **134**, 536–539 (in Russian).

Semyanistyi V.I. (1961). Homogeneous functions and some problems of integral geometry in space of constant curvature. *Sov. Math. Dokl.* **2**, 59–62.

Semyanistyi V.I. (1966). Certain problems of integral geometry in pseudo-Euclidean and non-Euclidean spaces. *Trudy Sem. Vektor. Tensor. Anal.* **13**, 244–302 (in Russian).

Sevostyanov B.A. (1971). *Branching Processes* Nauka, Moscow.

Shishko N.P. (1975). On uniqueness of the solution of two inverse problems. *Sib. Mat. J.* **16**, 309–313.

Solmon D.C. and Hazou I.A. (1988). Inversion of the exponential X-ray transform. *Analysis Math. Methods Appl. Sci.* **10**, 561–574.

Solmon D.C. and Hazou I.A. (1989). Filtered-backprojection and the exponential Radon transform. *J. Math. Anal. Appl.* **141**, 109–119.

Svirezhev Yu.M. (1967). *Nonlinear Waves, Dissipative Structures, and Catastrophes in Ecology.* Nauka, Moscow, 361p, (in Russian).

Svirezhev Yu.M., Lagofet D.O. (1978). *Stability of Biological Communities.* Moscow, Nauka (in Russian).

Tikhonov A.N. and Arsenin V.Y. (1977). *Solutions of Ill-posed Problems. Scripta series in Mathematics*, John F., Ed., Willey & Sons, New York.

Tikhonov I.V. and Ehjdel'man Yu.S. (1994). Questions of correctness of direct and inverse problems for evolution equations of the special type. *Mat. Zametki* **56**, 99–113 (in Russian).

Tikhonov A.N., Arsenin V.Ya., and Timonov A.A. (1987). *Mathematical Problems of Computer Tomography.* Moscow, Nauka (in Russian).

Tuy H.K. (1983). An inverse formula for cone-beam reconstruction. *SIAM J. Appl. Math.* **43**, 546–552.

Vishnevskii M.P. (1992). On solvability of one inverse boundary value problem for a parabolic equation with convergence. *Sib. Math. J.* **33**, 42–49.

Vladimirov V.S. (1976) *Distributions in Mathematical Physics.* Nauka, Moscow (in Russian).

Volterra V. (1931). *Leçons sur les Théorie de la Lutte pour la Vie.* Gauthier- Villar, Paris.

Watt K. (1968) *Ecology and Resource Management: A Quantitative Approach.* McGraw Hill, New York.

Zaslavskii G.M., Mejtlis V.P., and Filonenko N.N. (1982). *The wave interaction in inhomogeneous media.* Sibirsk. Otdel. Akad. Nauk SSSR., Nauka, Novosibirsk (in Russian).

Bei Fragen zur Produktsicherheit wenden Sie sich bitte an:
If you have any questions regarding product safety,
please contact:

Walter de Gruyter GmbH
Genthiner Straße 13
10785 Berlin
productsafety@degruyterbrill.com